地理信息系统二次开发

刘亚静 姚纪明 郭力娜 王政 张永彬 编著

武汉大学出版社

图书在版编目(CIP)数据

地理信息系统二次开发/刘亚静等编著.—武汉:武汉大学出版社,2014.8
ISBN 978-7-307-13887-2

Ⅰ.地… Ⅱ.刘… Ⅲ.地理信息系统—系统开发 Ⅳ.P208

中国版本图书馆 CIP 数据核字(2014)第 167648 号

责任编辑:鲍　玲　　责任校对:汪欣怡　　版式设计:韩闻锦

出版发行:武汉大学出版社　　(430072　武昌　珞珈山)
　　　　　(电子邮件:cbs22@whu.edu.cn　网址:www.wdp.com.cn)
印刷:黄石市华光彩色印务有限公司
开本:787×1092　1/16　　印张:24.5　　字数:593 千字
版次:2014 年 8 月第 1 版　　2014 年 8 月第 1 次印刷
ISBN 978-7-307-13887-2　　定价:49.00 元

版权所有,不得翻印;凡购买我社的图书,如有质量问题,请与当地图书销售部门联系调换。

前　言

　　GIS 及相关技术应用于交通规划管理、地质调查与矿产资源的开发和利用、林业、农业、景观生态等人们生活和生产等各个领域，为相关研究及决策提供可靠的信息收集和评价分析，得到了社会各个应用领域的认可。现有的工具地理信息系统软件功能较强，但是不能满足不同专业业务流程的所有问题，因此进行集成式的二次开发可以针对不同应用领域对 GIS 的功能需求进行软件功能的拓展，弥补基础地理信息系统软件功能的不足，也可以弥补独立开发带来的开发周期长、难度大，且能力、财力、实践各个方面都受限制的不足，还可以弥补宏语言二次开发功能较弱的缺点。

　　SuperMap GIS 7C 是超图软件全新架构的新一代云端一体化 GIS 平台软件，协助客户打造强云富端、互联互享、安全稳定、灵活可靠的 GIS 系统，同时二、三维一体化技术贯穿所有产品，有助于构建更加绚丽和实用的真三维应用。SuperMap iObjects .NET 是 SuperMap 公司推出的专题地理信息系统开发组件，通过开发组件提供的模块以及对应的类的方法和属性，用户可以根据各种需求分析来进行组件式专题地理信息系统的开发。另外，SuperMap GIS 7C 还提供了 SuperMap iClient 7C for JavaScript，弥补了传统 WebGIS 开发策略的不足，进行 WebGIS 功能的开发。以上两部分都实现了地理信息系统的灵活运用和拓展，满足了个性需求，提高了用户工作效率。

　　本书主要介绍了如何利用 .NET 平台，进行地理信息系统的二次开发，主要包括地图显示与制图，地理数据的复杂模拟与分析、空间分析、水文分析以及三维浏览与分析功能的实现。

　　本书共两大部分，第一部分包括第 1 章至第 10 章，第 1 章主要回顾了 GIS 的相关概念以及进行 GIS 二次设计的方法、必要性以及 GIS 开发的模式。第 2 章主要介绍了 SuperMap Objects 产品的架构、主要特点、主要的模块以及主要的功能介绍。第 3 章重点阐述了地图基本操作功能的实现。第 4 章主要介绍了如何用代码实现空间数据的基本操作。第 5 章主要描述了地图制图及各类专题图制作的实现方法。第 6 章主要介绍了如何实现空间数据图查属性和属性查图功能以及拓扑查询功能。第 7 章重点讲述了各种网络分析功能的实现以及拓扑分析功能的实现。第 8 章主要讲述了三维空间数据浏览、网络分析、缓冲区分析等功能的实现方法。第 9 章主要介绍了填充伪洼地、计算流向、计算累计汇水量以及量算等功能的水文分析实现方法。第 10 章主要介绍了等值线、等值面等表面分析以及动态分析等功能实现方法。第二部分主要介绍了 WebGIS 定义、特点、数据模型以及传统 WebGIS 开发的策略和 SuperMap iClient 7C for JavaScript 进行 WebGIS 开发的实现方法，并用实例描述了 WebGIS 开发的应用。

　　本书由河北联合大学矿业工程学院刘亚静编著，参与编写的人员还有姚纪明、郭力娜、王政、张永彬老师。全书完成后由刘亚静负责统稿。参与资料收集和整理工作的人员还有贾雪珊、赵兰、时静。

  本书在编写的过程中，得到了河北联合大学领导、老师和同事们的大力支持，得到了研究生们的大力帮助，对他们的帮助、支持和劳动深表感谢，同时引用了其他书籍以及超图软件公司的各种资料，在此表示感谢。

  由于作者的水平、经验有限，书中难免会有一些缺点和错误，希望得到广大同行专家、读者的批评和指正。

作 者

2014 年 6 月于唐山

目 录

第1章 地理信息系统开发概述 1
  1.1 地理信息系统的概念 1
  1.2 地理信息系统的主要类型及应用领域 1
  1.3 地理信息系统设计特点 3
  1.4 地理信息系统的设计方法 3
  1.5 地理信息系统设计的内容 9
  1.6 应用型地理信息系统的设计过程 9
  1.7 地理信息系统开发的方式 11

第2章 C#语言环境下的 SuperMap Objects 组件式开发 13
  2.1 语言基础 13
  2.2 开发环境 16
  2.3 SuperMap Objects 简介 17

第3章 地图的基本操作 34
  3.1 打开地图 34
  3.2 基本操作 37

第4章 空间数据的基本操作 40
  4.1 工作空间管理 40
  4.2 工作树状结构 57
  4.3 数据源管理 68
  4.4 数据集管理 81
  4.5 投影转换 93

第5章 地图制图及专题图制作 102
  5.1 地图制图 102
  5.2 专题图制作 123

第6章 查询功能 173
  6.1 图查属性与属性查图 173
  6.2 空间查询 179

第 7 章 空间分析 ········ 186
7.1 网络分析 ········ 186
7.2 拓扑分析 ········ 257

第 8 章 三维浏览 ········ 265
8.1 气泡 ········ 265
8.2 飞行管理 ········ 289

第 9 章 水文分析 ········ 306

第 10 章 表面分析与动态分段 ········ 316
10.1 表面分析 ········ 316
10.2 动态分段 ········ 322

第 11 章 WebGIS 设计与开发 ········ 336
11.1 WebGIS 的定义及其组成 ········ 336
11.2 WebGIS 的特点 ········ 337
11.3 传统 WebGIS 的分类 ········ 338
11.4 WebGIS 的数据模型 ········ 339
11.5 传统 WebGIS 的实现方法 ········ 340
11.6 SuperMap iClient 7C for JavaScript ········ 346
11.7 矿用对象采销信息管理系统的设计与实现 ········ 368

参考文献 ········ 382

第1章 地理信息系统开发概述

1.1 地理信息系统的概念

从1963年Rofer F. Tomlinson提出地理信息系统（Geographic Information System，GIS）这一术语以来，已经有半个多世纪的时间，地理信息系统作为获取、处理、管理和分析地理空间数据的重要工具、技术和学科，得到了专家和学者的广泛关注，发展迅速，但是一直没有一个统一的概念。大家从不同的角度对GIS进行理解：从技术和应用的角度，GIS是解决空间问题的工具、方法和技术；从学科的角度，GIS是在地理学、地图学、测量学和计算机科学等学科基础上发展起来的一门学科，具有独立的学科体系；从功能上，GIS具有空间数据的获取、存储、显示、编辑、处理、分析、输出和应用等功能；从系统学的角度，GIS具有一定的结构和功能，是一个完整的系统。虽然角度不同，但是所有的定义都是从三个方面来考虑的：① GIS的理论和方法基础：地理信息科学、系统工程、信息科学、计算机科学等；② GIS使用的工具：计算机软硬件系统；③ GIS处理的对象：具有空间内涵的地理数据；④ GIS的功能：具有采集、存储、显示、处理、分析和输出等功能。因此，地理信息系统可以这样定义：作为信息处理技术的一种，是以计算机技术为依托，以具有空间内涵的地理数据为处理对象，运用系统工程和信息科学的理论和方法，集采集、存储、显示、处理、分析、输出地理信息于一体的计算机系统。

1.2 地理信息系统的主要类型及应用领域

1.2.1 地理信息系统的主要类型

1. 工具型地理信息系统

工具型地理信息系统也称地理信息系统开发平台或外壳，是可供其他系统调用或用户进行二次开发的操作平台。它具有地理信息系统的基本功能，如对各种地理空间数据进行输入、处理、管理、查询、分析和输出，可供其他系统调用或允许用户进行二次开发，以建立应用型的地理信息系统的操作平台。它具有以下优点：① 对计算机硬件适用性强；② 数据管理和操作效率高，功能强；③ 具有普遍性和易于扩展性；④ 操作简便且容易掌握。目前，国内外已有的商品化的工具型地理信息系统软件包括：ArcGIS、MapInfo、GeoMedia、MapGIS、SuperMap、GeoStar、CityStar等。

2. 应用型地理信息系统

由于工具型地理信息系统软件通常只具有一些通用的功能，不能满足不同行业具体的应用需求，因此应用型地理信息系统的开发工作显得非常重要。应用型地理信息系统是根据用户的需求和应用目的而设计的一种解决一类或多类实际应用问题的某一应用领域或者是某一特定区域的地理信息系统，除了具有地理信息系统基本功能外，还有解决地理空间实休及空间信息的分布规律、分布特性及相互依赖关系的应用模型和方法。它可以在比较成熟的工具型地理信息系统的基础上进行二次开发，因此工具型地理信息系统是进行应用型地理信息系统开发的捷径。应用型地理信息系统一般具有明确的应用目的和使用对象。应用型地理信息系统按研究对象的性质和内容又可分为专题地理信息系统和区域地理信息系统。

（1）专题地理信息系统（Thematic GIS）

专题地理信息系统是具有有限目标和专业特点的地理信息系统，为特定专门目的服务，应用范围和用户对象一般比较明确。例如：医疗救助地理信息系统、交通地理信息系统、房产地理信息系统、水资源管理地理信息系统、矿产资源地理信息系统、森林动态监测地理信息系统、移动用户管理地理信息系统、环境地理信息系统、物业管理地理信息系统等，都属于专题地理信息系统。

（2）区域地理信息系统（Regional GIS）

区域地理信息系统主要以区域综合研究和全面信息服务为目标，可以有不同的规模，如国家级的、地区或省级的、市级和县级等为各不同级别行政区服务的区域信息系统，也可以按自然分区或流域为单位的区域信息系统。例如：加拿大地理信息系统、北京市交通地理信息系统、上海市环境管理信息系统、河北省土地分等定级信息系统等都属于区域地理信息系统。

实用型地理信息系统在我国最早是由陈俊、宫鹏提出。他们在《实用地理信息系统》一书中作了详细论述，认为实用型地理信息系统就是从实用的角度来探讨地理信息系统的理论和技术。实用，英文为 Practice，《韦伯大词典》中解释 Practice 的意思为：To perform and work repeatedly so as to become proficient，译成中文的意思是"不断地实践来达到娴熟和精湛"。实用的目的是使实践过程优化，使这个实践过程在不断提高中得到完善。因此，可以推断定义，实用地理信息系统应该是在使用地理信息系统的过程中，不断地提高和完善，使得地理信息系统的应用趋向成熟。

1.2.2 地理信息系统的相关应用领域

① 政府企业管理和决策：GIS 可用于企业生产经营管理、税收、地籍管理、宏观规划、开发评价管理、交通工程、公共设施使用、道路维护、市区设计、经济发展等一系列政府企业管理和决策中。

② 公共安全与卫生：GIS 主要用于流行病数据的可视化、流行病数据的空间分析、流行病模型的建立、医疗设施的规划、可达性、可用性及结果，地震、火灾、旱涝等事件的信息采集与上报，灾情分析与救援对策等公共安全与卫生领域。

③ 环境评价和监测系统：GIS 主要用于环境影响评价、污染评价、灌溉适宜性评价、灾害监测（森林火灾、洪水灾情、救灾抢险等）、生态系统的研究、生物圈遗迹管理、自然资源管理等有关环境评价和监测领域。

④ 土地和资源评价管理：GIS 广泛应用于土地规划管理、土地分等定级、水资源清

查、矿产资源评价(矿产预测、矿产评价、工程地质、地质灾害)等土地和资源评价管理方面。

⑤ GIS 用于公共供应网络(电、气、水、废水)、电信网络、交通领域、区域和城市规划、道路工程。

⑥ 地图数字制图与生产：由于 GIS 强有力的数据管理、处理、显示和制图功能，测绘制图部门可利用 GIS 技术实现制图计算机设计、数据存储、编辑加工及自动化生产，采用 GIS 技术便于地图修改，地图更新，缩短成图周期，大大提高了劳动部门生产率。

此外，GIS 应用于商业中，例如消费者分析、商圈分析、选址分析以及企业内部的物流；金融领域，主要进行客户服务、客户挖掘、网店/ATM 定位；保险行业，主要进行市场分析、客户挖掘、网点配置；房地产领域，主要用于房地产评估、房地产管理；物流领域，主要进行交通路线的选择、机构设施地理位置的选择、车辆运输的动态管理，等等。总之，地理信息系统已经深入到人们的生活和生产各个领域。

1.3　地理信息系统设计特点

地理信息系统源于多个学科，涉及多门技术，且在众多领域得到应用，使得它对计算机软硬件以及数据有较高的要求。因此，地理信息系统的设计有自身的特点：

① 地理信息系统的处理对象为空间数据，具有数据量庞大、实体类型繁多，实体间的关系复杂等特点。与一般的信息系统相比，地理信息系统处理的是与地理实体位置、属性、拓扑关系等相关的多源空间数据，数据结构复杂，需采用相关的地理数据模型进行表达。与传统的层次模型、网状模型以及关系模型相比，面向对象模型对地理数据表达具有很大的优势。因此，地理信息系统不仅需要对系统业务流程进行分析，更重要的是必须对系统所涉及的地理实体类型以及实体之间的各种关系进行分析和描述，并采用合理的数据模型进行描述。

② 地理信息系统是以空间数据为驱动。地理信息系统中数据是血液，因此空间数据的管理占据非常重要的位置，也就是说地理信息系统在某种意义上可以说是一种空间数据库，地理信息系统的功能是为空间数据库提供服务，其主要任务是对空间数据进行分析统计处理并提供辅助决策。因此，与一般的信息系统以业务为导向的建设系统的思想不同，地理信息系统设计是以空间数据为导向进行系统的建设。

③ 地理信息系统工程投资大、周期长、风险大、涉及部门繁多。在地理信息系统的设计中，项目计划管理是一个十分重要的部分，在项目计划管理中，需要完成以下工作：估计系统建设的投资效益，评估系统建设的风险性和必要性；制定系统的建设进度安排，保证系统建设的高效性，建立系统建设的组织机构和进行人员协调等工作。

1.4　地理信息系统的设计方法

1.4.1　结构化生命周期方法

结构化生命周期方法是在早期国内外比较流行的信息系统开发方法，在系统开发中得到了广泛的应用和推广，尤其在开发复杂的大系统时，显示了无比的优越性。它也是迄今为止开发方法中应用最普遍最成熟的一种。

1. 基本思想

将软件工程学与系统工程的理论和方法引入计算机系统的研制开发中，按照用户至上的原则，采用结构化、模块化和自顶向下的思想对系统进行分析和设计。具体来说，它将整个信息系统开发过程划分为独立的六个阶段，包括系统开发准备阶段、调查研究及可行性研究阶段、系统分析阶段、系统设计阶段、系统实施阶段、维护和评价阶段，这六个阶段构成信息系统的生命周期。

2. 开发过程

(1) 系统开发准备阶段

当现行系统不能适应新形势的要求时，用户将提出开发新系统的要求。有关人员进行初步调查，然后组成专门的新系统开发领导小组，制订新系统开发的进度和计划，负责新系统开发中的一切工作。系统开发准备阶段虽不属系统分析与设计的正式工作阶段，它却是不可缺少的。如果新系统开发采取外包方式，本阶段还包括招投标过程。

(2) 调查研究及可行性研究阶段

本阶段要回答的关键问题是"到底要解决什么问题"。在成本和时间的限制条件下能否解决问题？是否值得做？系统分析员采用各种方式进行调查研究，了解现行系统界限、组织分工、业务流程、资源及薄弱环节等，绘制现行系统的相关图表。在此基础上，与用户协商方案，提出初步的新系统目标，并进行系统开发的可行性研究，提交可行性报告。

(3) 系统分析阶段

系统分析阶段是新系统的逻辑设计阶段，本阶段要回答的关键问题是"如何做出目标系统"。系统分析阶段是新系统的逻辑设计阶段。

系统分析旨在对现行系统进行调查研究的基础上，使用一系列的图表工具进行系统的目标分析，划分子系统以及功能模块，构造出新系统的逻辑模型，确定其逻辑功能需求，交付新系统的逻辑功能说明书。

系统分析也是新系统方案的优化过程，数据流程图是新系统逻辑模型的主要组成部分，它在逻辑上描述了新系统的功能、输入、输出和数据存储等，从而摆脱了所有的物理内容。

(4) 系统设计阶段

本阶段要回答的关键问题是"如何在计算机中实现目标系统"。系统设计阶段又称新系统的物理设计阶段，系统分析员根据新系统的逻辑模型进行物理模型的设计，并具体选择一个物理的计算机信息处理系统。系统设计的关键是模块化。这个阶段分两个步骤：总体设计和详细设计，其中总体设计是解决软件系统的模块划分、模块层次结构及数据库设计。详细设计是解决每个模块的控制流程、内部算法和数据结构的设计。

(5) 系统实施阶段

本阶段要解决的问题是"正确地实现已做的设计"，即"如何编写正确的、可维护的程序(代码)"。系统实施是新系统付诸实现的实践阶段，主要是实现系统设计阶段所完成的新系统物理模型。首先，要进行计算机系统设备的安装和调试工作。然后，程序员根据程序模块进行程序的设计、代码编写和调试工作。此外，系统分析人员还要对用户及操作人员进行培训，编制操作、使用手册和有关文档，帮助用户熟悉、使用新系统。

(6) 维护和评价阶段

系统的维护和评价是系统生命周期的最后一个阶段，也是很重要的阶段，新系统是否有持久的生命力取决于此阶段的工作。GIS 是复杂的大系统，要能适应系统内、外部环境，各种人为因素和机器因素的影响，这就需要进行系统维护。

3. 优缺点

生命周期法的突出优点是强调系统开发过程的整体性和全局性，强调在整体优化的前提下考虑具体的分析设计问题，即自顶向下的观点。它从时间角度把软件开发和维护分解为若干阶段，每个阶段有各自相对独立的任务和目标。降低了系统开发的复杂性，提高了可操作性。另外，每个阶段都对该阶段的成果进行严格的审批，发现问题及时反馈和纠正，保证了软件质量，特别是提高了软件的可维护性。实践证明，生命周期法大大提高了软件开发的成功率。

但是，生命周期法开发周期较长，因为开发顺序是线性的，各个阶段的工作不能同时进行，前阶段所犯的错误必然带入后一阶段，而且是越是前面犯的错误对后面工作的影响越大，更正错误所花的工作量就越大。而且，在功能经常要变化的情况下，难以适应变化要求，不支持反复开发。

1.4.2 原型法

原型法(Prototyping)是 20 世纪 80 年代随着计算机软件技术的发展，提出的一种从设计思想、工具、手段都全新的系统开发方法。它摒弃了那种一步步周密细致地调查分析，然后逐步整理出文字档案，最后才能让用户看到结果的繁琐做法。其核心是，用交互的方法，快速建立起来的原型取代了形式的、僵硬的(不允许更改的)大部分的规格说明，用户通过在计算机上实际运行和试用原型系统而向开发者提供真实的、具体的反馈意见。

1. 基本思想

在投入大量的人力、物力之前，在限定的时间内，用最经济的方法开发出一个可实际运行的系统模型，用户在运行使用整个原型的基础上，通过对其评价，提出改进意见，对原型进行修改，统一使用，评价过程反复进行，使原型逐步完善，直到完全满足用户的需求为止。

2. 开发过程

(1) 确定用户的基本需求

由用户提出对新系统的基本要求，如功能、界面的基本形式，所需要的数据，应用范围，运行环境等，开发者根据这些信息估算开发该系统所需的费用，并建立简明的系统模型。

(2) 构造初始原型

系统开发人员在明确了对系统基本要求和功能的基础上，依据计算机模型，以尽可能快的速度和尽可能多的开发工具来建造一个结构仿真模型，即快速原型构架。之所以称为原型构架，是因为这样的模型是系统总体结构，子系统以上部分的高层模型。由于要求快速，这一步骤要尽可能使用一些软件工具和原型制造工具，以辅助进行系统开发。

(3) 运行、评价、修改原型

快速原型框架建造成后，就要交给用户立即投入试运行，各类人员对其进行试用、检查分析效果。由于构造原型中强调的是快速，省略了许多细节，一定存在许多不合理的部分。所以，在试用中要充分进行开发人员和用户之间的沟通，尤其是要对用户提出的不满意的地方进行认真细致的反复修改、完善，直到用户满意为止。

(4) 形成最终的系统

如果用户和开发者对原型比较满意，则将其作为正式原型。经过双方继续细致地工作，把开发原型过程中的许多细节问题逐个补充、完善、求精，最后形成一个适用的地理信息系统。

3. 优缺点

原型法符合人们认识事物的规律，系统开发循序渐进，反复修改，确保较好的用户满意度；开发周期短，费用相对少；由于有用户的直接参与，系统更加贴近实际；易学易用，减少用户的培训时间；应变能力强。但是，原型法不适合大规模系统的开发；开发过程管理要求高，整个开发过程要经过"修改—评价—再修改"的多次反复；用户过早看到系统原型，误认为系统就是这个模样，易使用户失去信心；开发人员易将原型取代系统分析；缺乏规范化的文档资料。适合处理过程明确、简单系统、涉及面窄的小型系统，不适合设计大型、复杂系统；难以模拟存在大量运算、逻辑性强的处理系统；管理基础工作不完善、处理过程不规范；大量批处理系统。

1.4.3　面向对象设计方法

面向对象编程(Object Oriented Programming，OOP)是一种计算机编程架构。OOP 的一条基本原则是计算机程序是由单个能够起到子程序作用的单元或对象组合而成。OOP 达到了软件工程的三个主要目标：重用性、灵活性和扩展性。为了实现整体运算，每个对象都能够接收信息、处理数据和向其他对象发送信息。

1. 面向对象的基本概念

① 对象：对象是要研究的任何事物。从一本书到一家图书馆，单独整数到整数列庞大的数据库、极其复杂的自动化工厂、航天飞机都可看作对象，它不仅能表示有形的实体，也能表示无形的(抽象的)规则、计划或事件。对象由数据(描述事物的属性)和作用于数据的操作(体现事物的行为)构成一独立整体。从程序设计者来看，对象是一个程序模块，从用户来看，对象为他们提供了所希望的行为，对象之间的操作通常称为方法。

② 类：类是对象的模板，即类是对一组有相同数据和相同操作的对象的定义，一个类所包含的方法和数据描述一组对象的共同属性和行为。类是在对象之上的抽象，对象则是类的具体化，是类的实例。类可有其子类，也可有其他类，形成类层次结构。

③ 消息：消息是对象之间进行通信的一种规格说明。一般它由三部分组成：接收消息的对象、消息名及实际变元。

2. 面向对象主要特征

(1) 封装性

封装是一种信息隐蔽技术，它体现于类的说明，是对象的重要特性。封装使数据和加工该数据的方法(函数)封装为一个整体,以实现独立性很强的模块,使得用户只能见到对象的外特性(对象能接收哪些消息,具有哪些处理能力),而对象的内特性(保存内部状态的私有数据和实现加工能力的算法)对用户是隐蔽的。封装的目的在于把对象的设计者和对象的使用者分开,使用者不必知晓行为实现的细节,只需用设计者提供的消息来访问该对象。

(2)继承性

继承性是子类自动共享父类之间数据和方法的机制。它由类的派生功能体现。一个类直接继承其他类的全部描述,同时可修改和扩充。继承具有传递性。继承分为单继承(一个子类只有一个父类)和多重继承(一个类有多个父类)。类的对象是各自封闭的,如果没有继承性机制,则类对象中数据、方法就会出现大量重复。继承不仅支持系统的可重用性,而且还促进系统的可扩充性。

(3)多态性

对象根据所接收的消息而做出动作。同一消息为不同的对象接收时可产生完全不同的行动,这种现象称为多态性。利用多态性用户可发送一个通用的信息,而将所有的实现细节都留给接收消息的对象自行决定,如是,同一消息即可调用不同的方法。例如:Print 消息被发送给一图或表时调用的打印方法与将同样的 Print 消息发送给一正文文件而调用的打印方法会完全不同。多态性的实现受到继承性的支持,利用类继承的层次关系,把具有通用功能的协议存放在类层次中尽可能高的地方,而将实现这一功能的不同方法置于较低层次,这样,在这些低层次上生成的对象就能给通用消息以不同的响应。在 OOPL 中可通过在派生类中重定义基类函数(定义为重载函数或虚函数)来实现多态性。

综上可知,在面对对象方法中,对象和传递消息分别表现事物及事物间相互联系的概念。类和继承是适应人们一般思维方式的描述范式。方法是允许作用于该类对象上的各种操作。这种对象、类、消息和方法的程序设计范式的基本点在于对象的封装性和类的继承性。通过封装能将对象的定义和对象的实现分开,通过继承能体现类与类之间的关系,以及由此带来实体的多态性,从而构成了面向对象的基本特征。

面向对象设计是一种把面向对象的思想应用于软件开发过程中,指导开发活动的系统方法,是建立在"对象"概念基础上的方法学。对象是由数据和容许的操作组成的封装体,与客观实体有直接对应关系,一个对象类定义了具有相似性质的一组对象。而继承性是对具有层次关系的类的属性和操作进行共享的一种方式。所谓面向对象就是基于对象概念,以对象为中心,以类和继承为构造机制,来认识、理解、刻画客观世界和设计、构建相应的软件系统。面向对象程序设计方法的步骤如下:

① 识别系统里的类;
② 给每个类提供属性和完整的一组操作;
③ 明确地使用继承来表现共同点。

3. 面向对象程序设计方法的优点

由这个定义,我们可以看出:面向对象设计就是"根据需求决定所需的类、类的操作以及类之间关联的过程"。

面向对象出现以前,结构化程序设计是程序设计的主流,结构化程序设计又称为面向过程的程序设计。在面向过程程序设计中，问题被看作一系列需要完成的任务，函数（在此泛指例程、函数、过程）用于完成这些任务，解决问题的焦点集中于函数。其中函数是面向过程的，即它关注如何根据规定的条件完成指定的任务。在多函数程序中，许多重要的数据被放置在全局数据区，这样它们可以被所有的函数访问。每个函数都可以具有它们自己的局部数据。这种结构很容易造成全局数据在无意中被其他函数改动，因而程序的正确性不易保证。面向对象程序设计的出发点之一就是弥补面向过程程序设计中的一些缺点：对象是程序的基本元素，它将数据和操作紧密地连接在一起，并保护数据不会被外界的函数意外地改变。

比较面向对象程序设计和面向过程程序设计，还可以得到面向对象程序设计的其他优点：

① 数据抽象的概念可以在保持外部接口不变的情况下改变内部实现，从而减少甚至避免对外界的干扰；

② 通过继承大幅减少冗余的代码，并可以方便地扩展现有代码，提高编码效率，也降低了出错概率，降低软件维护的难度；

③ 结合面向对象分析、面向对象设计，允许将问题域中的对象直接映射到程序中，减少软件开发过程中中间环节的转换过程；

④ 通过对对象的辨别、划分可以将软件系统分割为若干相对独立的部分，在一定程度上更便于控制软件复杂度；

⑤ 以对象为中心的设计可以帮助开发人员从静态（属性）和动态（方法）两个方面把握问题，从而更好地实现系统；

⑥ 通过对象的聚合、联合可以在保证封装与抽象的原则下实现对象对内在结构以及外在功能上的扩充，从而实现对象由低到高的升级。

综上所述，结构化生命周期法是传统的地理信息系统分析设计方法，具有较为成熟和完善、整体性好等特点，但是缺乏灵活性、开发周期长，且对系统需求要求较高，而在实际的 GIS 设计过程中，系统的需求是在系统设计过程中逐步明确的，因此，进行 GIS 设计开发往往会出现重复性劳动、开发周期长、用户接受度低等问题。原型法能解决系统需求不确定性的问题，采用原型法进行 GIS 设计也有其不足，就是系统的整体性差，重复劳动多。在实际系统开发过程中，原型法常用于小型 GIS 软件设计，而在大型 GIS 软件设计中，采用原型法与其他软件设计方法相结合来进行软件设计。其中，原型法主要用来确定系统的需求。

面向对象技术有利于现实世界实体的表达和系统开发，与人类思维方法一致，便于描述客观世界。开发的软件性能稳定、易于重用和维护。因此，面向对象设计方法在 GIS 设计中具有很大的优势，但是面向对象设计方法目前仍不成熟和完善，尤其是在 GIS 领域中的应用。考虑到 GIS 应用的特点以及 GIS 应用的多样化，想要找到一种适用于所有 GIS 软件开发的设计方法几乎是不可能的。进行 GIS 设计方法的选择需要考虑多方面的因素，例如：系统规模的大小、系统应用类型、系统需求明确程度等。

通常，小型 GIS 软件设计常采用原型法进行开发。大型 GIS 软件设计多采用结构化生命周期法或是面向对象方法进行开发。考虑到 GIS 设计需求不确定性的特点，通常也在需求分析阶段应用原型法来确认用户需求。

1.5　地理信息系统设计的内容

1. 系统总体设计

系统总体设计主要采用结构化程序设计工具(层次图、HIPO 图以及结构图)或者面向对象程序设计方法(类图和用例图)来确定系统总体架构与软、硬件配置,根据系统分析成果进行系统功能模块的划分,建立模块的层次结构及调用关系,确定模块间的接口及人机界面,设计数据库总体结构,以及 GIS 接口设计和 GIS 用户界面的设计。

2. 系统详细设计

系统详细设计就是根据总体设计的系统结构图,采用程序流程图、N-S 盒式图、问题分析图、类程序设计语言等结构化系统详细设计的方法或者是序列图、活动图等面向对象的系统详细设计工具进行系统的详细设计,具体包括系统功能模块的划分,模块的数据设计(输入、输出数据)、模块详细的算法、数据表示和数据结构以及实现的功能和使用的数据之间的关系。确定模块的接口细节及模块间的调度关系,描述每个模块的流程逻辑。

3. 空间数据库设计

空间数据库设计是对数据进行空间特征描述、逻辑预处理,确定空间数据库的数据模型以及数据结构,提出空间数据库相关功能的实现方案。将设计的空间数据库系统的结构体系进行编码实现。将收集来的空间数据入库,建立空间数据库管理信息系统。主要进行空间数据库概念设计、逻辑设计和功能设计。其中,功能设计包括空间数据输入设计、空间数据检索设计、空间数据输出设计、空间数据更新设计以及空间数据共享设计。

4. 系统功能设计

系统功能设计包括总体模块功能设计、属性数据库管理系统结构与功能设计、图形数据库管理系统结构与功能设计。

5. 应用模型和方法设计

主要是采用主成分分析法、系统聚类分析法、层次分析法、模糊综合评价法、地统计分析方法、人工神经网络方法、元胞自动机模型等方法,根据系统功能和性能的需要进行统计分析模型、预测模型、决策模型、模拟模型、GIS 空间分析模型等应用模型的设计。

6. 输入、输出设计

在充分考虑空间数据输入设计原则的基础上,进行图形数据输入、属性数据输入方式的设计,GIS 产品输出形式和输出介质设计等。

1.6　应用型地理信息系统的设计过程

应用型地理信息系统的建立过程是一项庞大的系统工程,为了使系统开发达到预期最优化目标,必须合理地提出任务,最优化地设计任务以及最有效地运行。根据地理信息系统工程学思想,采用科学的开发步骤和技术,对系统建立的全过程进行控制和协调,应用型地理信息系统开发和实现的过程见表 1-1:

表 1-1　　　　　　　　　　　应用型地理信息系统设计过程

阶段	内容	用户	领导	开发人员
系统分析	需求分析	① 提出所要解决的问题 ② 指出所需要的信息 ③ 详细介绍现行系统 ④ 提供各种资料和数据	① 批准开始研究 ② 组织开发队伍 ③ 进行必要的培训	① 吸取用户要求 ② 回答用户的问题 ③ 详细调查现行系统 ④ 搜集资料和数据 ⑤ 总结和分析
系统分析	可行性研究	① 评价现行系统 ② 协助提出各种方案 ③ 选择最适宜的方案	① 审查可行性报告 ② 决定是否开发	① 提出多种备选方案 ② 与用户一起讨论各方案的优劣 ③ 开发的费用估计和时间估计
系统设计	总体设计	① 讨论子系统模块的合理性并提出看法 ② 对设备选择发表看法	① 鼓励用户参加系统设计 ② 要求开发人员多听用户意见	① 说明系统目标和功能 ② 子系统和模块划分 ③ 计算机系统选择
系统设计	详细设计	① 讨论设计和用户界面的合理性 ② 提出修改意见	① 听取用户有关系统界面的反映 ② 批准转入系统实施	① 软件设计 ② 代码设计 ③ 功能设计 ④ 数据库设计 ⑤ 用户界面设计 ⑥ 输入、输出设计
系统实施	编程	随时准备回答一些具体的业务问题	监督编程进度	分头进行编程和调试
系统实施	调试	① 评价系统的总调试 ② 检查用户界面的良好性	① 监督调试的进度 ② 协调用户与开发人员的不同意见	① 模块调试 ② 分调（子系统调试） ③ 总调（系统调试）
系统实施	培训	接受培训	① 组织培训 ② 批准系统转换	① 编写用户手册 ② 进行培训
运行维护	运行和维护	① 按系统的要求定期输入数据 ② 使用系统的输出 ③ 提出修改和扩充意见	① 监督用户严格执行操作规程 ② 批准适应性和完善性维护 ③ 准备对系统全面评价	① 按系统要求进行数据处理工作 ② 积极稳妥地进行维护
运行维护	系统评价	参加系统评价	组织系统评价	① 参加系统评价 ② 总结经验教训

1.7 地理信息系统开发的方式

1. 独立开发

独立开发是指不依赖于任何 GIS 工具软件,从空间数据的采集、编辑到数据的处理分析及结果输出,所有的算法都由开发者独立设计,然后选用某种程序设计语言,如 Visual C++、Delphi 等,在一定的操作系统平台上编程实现。这种方式的好处在于无须依赖任何商业 GIS 工具软件,减少了开发成本,但一方面对于大多数开发者来说,能力、时间、财力方面的限制使其开发出来的产品很难在功能上与商业化 GIS 工具软件相比,而且在购买 GIS 工具软件上省下的钱可能还抵不上开发者在开发过程中绞尽脑汁所花的代价。

2. 宿主型二次开发

宿主型二次开发是指基于 GIS 平台软件上进行应用系统开发。大多数 GIS 平台软件都提供了可供用户进行二次开发的脚本语言,如 ESRI 的 ArcView 提供了 Avenue 语言,MapInfo 公司的 MapInfo Professional 提供了 MapBasic 语言,等等。用户可以利用这些脚本语言,以原 GIS 软件为开发平台,开发出自己的针对不同应用对象的应用程序。这种方式省时省心,但进行二次开发的脚本语言作为编程语言,功能极弱,用它们来开发应用程序仍然不尽如人意,并且所开发的系统不能脱离 GIS 平台软件,是解释执行的,效率不高。

3. 基于 GIS 组件的二次开发

大多数 GIS 软件生产商都提供商业化的 GIS 组件,如 ESRI 公司的 MapObjects、MapInfo 公司的 MapX 等,这些组件都具备 GIS 的基本功能,开发人员可以基于通用软件开发工具尤其是可视化开发工具,如 Delphi、Visual C++、Visual Basic、Power Builder 等为开发平台,进行二次开发,实现地理信息系统的各种功能。

4. 组件式 GIS 系统的特点

①小巧灵活、价格便宜。

由于传统 GIS 结构的封闭性,往往使得软件本身变得越来越庞大,不同系统的交互性差,系统的开发难度大。在组件模型下,各组件都集中地实现与自己最紧密相关的系统功能,用户可以根据实际需要选择所需控件,最大限度地降低了用户的经济负担。组件化的 GIS 平台集中提供空间数据管理能力,并且能以灵活的方式与数据库系统连接。在保证功能的前提下,系统表现得小巧灵活,而其价格仅是传统 GIS 开发工具的十分之一,甚至更少。这样,用户便能以较好的性能价格比获得或开发 GIS 应用系统。

②无需专门 GIS 开发语言,直接嵌入 MIS 开发工具。

传统 GIS 往往具有独立的二次开发语言,对用户和应用开发者而言存在学习上的负担,而且使用系统所提供的二次开发语言,开发往往受到限制,难以处理复杂问题。而组件式 GIS 建立在严格的标准之上,不需要额外的 GIS 二次开发语言,只需实现 GIS 的基本功能函数,按照 Microsoft 的 ActiveX 控件标准开发接口。这有利于减轻 GIS 软件开发者的负担,而且增强了 GIS 软件的可扩展性。GIS 应用开发者,不必掌握额外的 GIS 开发语言,只需熟悉基于 Windows 平台的通用集成开发环境,以及 GIS 各个控件的属性、方法和事件,就可以完成应用系统的开发和集成。目前,可供选择的开发环境有很多,如 Visual

C++、Visual Basic、Visual FoxPro、Borland C++、Delphi、C++Builder 以及 Power Builder 等都可直接成为 GIS 或 GMIS 的优秀开发工具，它们各自的优点都能够得到充分发挥。这与传统 GIS 专门性开发环境相比，是一种质的飞跃。

③强大的 GIS 功能。

新的 GIS 组件都是基于 32 位系统平台的，采用 InProc 直接调用的形式，所以无论是管理大数据的能力还是处理速度方面均不比传统 GIS 软件逊色。小小的 GIS 组件完全能提供拼接、裁剪、叠合、缓冲区等空间处理能力和丰富的空间查询与分析能力。

④开发便捷。

由于 GIS 组件可以直接嵌入 MIS 开发工具中，对于广大的开发人员来讲，就可以自由选用他们熟悉的开发工具。而且，GIS 组件提供的 API 形式非常接近 MIS 工具的模式，开发人员可以像管理数据库表一样熟练地管理地图等空间数据，无须对开发人员进行特殊的培训。在 GIS 或 MIS 的开发过程中，开发人员的素质与熟练程度是十分重要的因素。这将使大量 MIS 开发人员能够较快地过渡到 GIS 的开发工作中，从而大大加速了 GIS 的发展。

⑤更加大众化。

组件式技术已经成为业界标准，用户可以像使用其他 ActiveX 控件一样使用 GIS 控件，使非专业的普通用户也能够开发和集成 GIS 应用系统，推动了 GIS 大众化进程。组件式 GIS 的出现使 GIS 不仅是专家们的专业分析工具，同时也成为普通用户对地理相关数据进行管理的可视化工具。

第 2 章　C#语言环境下的 SuperMap Objects 组件式开发

2.1　语言基础

2.1.1　C#概述

C#(C sharp)是微软公司在 2000 年发布的一种新的程序设计语言，是微软基于 .NET Framework 平台设计的，它集合了 C、C++和 Visual Basic 的特点，具有强大的功能而且使用简便，本节对其进行简单的介绍。

1. 什么是 C#语言

C#语言是在 C、C++和 Java 语言的基础上重新构造的，其语法与 C++和 Java 语言都比较相似，是一种基于 .NET Framework、完全面向对象的、类型安全的编程语言，也是 .NET 的首选编程语言。从开发效率上讲，C#为应用程序开发人员提供了快速开发手段，并且保有 C++语言的特点和优点。从继承性上讲，C#语言几乎综合了目前流行的所有高级语言的优点，提供了一种语法简洁、功能完善而又容易使用的外在表现形式。

C#语言的设计目的是简化网络应用，使用 C#语言能够快速构建基于 Windows 和 Internet 的应用程序和组件，开发人员可以使用 C#开发多种类型的应用程序。

2. .NET Framework

说到 C#，就不得不提 .NET Framework，.NET Framework 简称 .NET，是微软为开发应用程序创建的一个富有革命性的新平台。在这个平台中，可以开发出运行在 Windows 上的几乎所有的应用程序，而微软也将推出运行在其他操作系统上的版本。简单地说，.NET Framework 就是一个创建、部署和运行应用程序的多语言平台环境，包含了一个庞大的代码库，各种 .NET 语言都可以共用这些代码。

2.1.2　C#程序设计基础

C#程序是由一系列的语句组成，其目的是为了完成一项任务。程序设计就是规划并创建程序的过程。在编写较复杂的 C#程序前，必须了解它的一些基本要素，包括标识符、数据类型、变量、常量、运算符和表达式等，此外，还需了解 C#中提供的一些常用类和结构的使用方法。

1. 标识符

标识符是程序中用户定义的一些有意义的名称，例如变量和函数的名称。C#的标识符有如下规则：

① 一个合法的 C#标识符，是以字母或者下划线开头、其后可以跟任意一个字母、数字或者下划线。

② C#的标识符严格区分大小写，即使两个标识符的字母的大小写不同，也被认为是完全不同的标识符，如 xyz 和 xYz 是两个不同的标识符。

③ 关键字也可以作为标识符，只要关键字前加上@前缀。

④ 标识符不能包含空格、标点符号、运算符等其他符号。

⑤ 标识符不能与 C#关键字名和类库名相同。

2. C#中数据类型

数据类型是用来区分不同的数据。由于数据在存储时所需要的容量各不相同，不同的数据就必须要分配不同的内存空间来存储，因此要将数据划分成不同的数据类型。C#数据类型主要分为值类型和引用类型两大类。

（1）值类型

从用户角度来看，变量是存储信息的基本单元；从系统角度来看，变量是计算机内存中的一个存储空间。用户定义变量时必须指出其数据类型，系统会按照该数据类型为变量分配相应大小的内存存储空间，用于存放变量的值。

值类型的变量内含变量值本身，C#的值类型可以分为简单类型、结构类型和枚举类型。其中，简单类型包括整数类型、浮点类型、小数类型、字符类型和布尔类型等。

（2）引用类型

引用类型是 C#中和值类型并列的类型，它的引入主要是因为值类型比较简单，不能描述结构复杂、抽象能力比较强的数据。引用类型所存储的实际数据是当前引用值的地址，因此引用类型数据的值会随所指向的值的不同而变化，同一个数据也可以有多个引用。这与简单类型数据是不同的，简单类型数据存储的是自身的值，而引用类型存储的是将自身的值直接指向到某个对象的值。它就像一面镜子，虽然我们从镜子中看到了物体，但物体并不在镜子中，只不过是将物体反射过来罢了。

C#中引用类型数据有四种：类类型（class-type）、数组类型（array-type）、接口类型（interface-type）和委托类型（delegate-type）。例如，C#中经常用到的两个类类型：对象类（object）和字符串类（string）。object 是 C#中所有类型的基类，包括所有的值类型和引用类型，C#中的所有类型都直接或间接地从 object 类中继承而来。因此，对一个 object 的变量可以赋予任何类型的值。对 object 类型的变量声明采用 object 关键字，这个关键字是在.NET Framework 的命名空间 System 中定义的，是类 System.Object 的别名。C#还定义一个 string 类，表示一个 Unicode 字符系列，专门用于对字符串的操作。同样，这个类也是在.NET Framework 的命名空间 System 中定义的，是类 System.String 的别名。

（3）类型转换

数据类型在一定条件下是可以相互转换的，C#允许使用两种转换方式：隐式转换和显式转换。隐式转换是系统默认的、不需要加以声明就可以进行的转换。在隐式转换过程中，编译器不需要对转换进行详细的检查就能安全地执行转换，例如数据从 int 类型到 long 类型的转换。显式类型又叫强制类型转换，与隐式转换相反，显式转换需要用户明确地指定转换类型；一般在不存在该类型的隐式转换时才使用。

（4）装箱和拆箱

装箱和拆箱是 C#类型系统中重要的概念。它们允许将任何类型的数据转换为对象，同时也允许任何类型的对象转换到与之兼容的数据类型。其实拆箱是装箱的逆过程。装箱转换是指将一个值类型的数据隐式地转换成一个对象类型的数据。把一个值类型装箱，就是创建一个 object 类型的实例，并把该值类型的值复制给这个 object 实例。拆箱转换是指将一个对象类型的数据显式地转换成一个值类型数据。拆箱操作分为两步：首先检查对象实例，确保它是给定值类型的一个装箱值，然后把实例的值复制到值类型数据中。

3. C#中的变量和常量

在程序执行过程中，其值不发生改变的量称为常量，其值可变的量称为变量。它们可与数据类型集合起来进行分类。在程序中，变量是可以不经说明而直接引用的，而常量则必须先定义后使用。

变量是在程序的运行过程中使其值可以发生变化的量，可以在程序中使用变量来存储各种各样的数据，并对它们进行读、写、运算等操作。从用户角度看，变量是用来描述一条信息的名称，在变量中可以存储各种类型的信息，例如某人的姓名、年龄等。而从系统的角度看，变量就是程序中的基本存储单元，它既表示这块内存空间的地址，也表示这块内存空间存储的数据。

常量是在程序执行中其值保持固定不变的量。常量一般分为直接常量和符号常量，常量的类型可以是任何一种值类型或引用类型。直接常量是指把程序中不变的量直接硬编码为数值和字符串值。符号常量是通过关键字 const 声明的常量，包括常量的名称和它的值。

4. C#循环型语句

在 C#中，循环结构有 while、do/while、for 和 foreach 语句。除 foreach 语句之外，在其他各种情况中，当一个布尔表达式为 true 时，执行一个特定的简单语句或复合语句。先判断循环型结构 while 语句，该语句的执行过程是，先计算布尔表达式之值，若该值为 true，则执行循环体中的嵌入语句；否则，退出该循环体，执行 while 语句后面的第一条语句。之后判断结构 do/while 语句，do/while 和 while 语句非常类似，两者都是循环到布尔表达式或计算值为 false 为止，do 循环体（嵌入语句）至少要执行一次，而 while 循环有可能执行零次。因此，如果需要确保嵌入的循环语句至少要执行一次，则使用 do/while 语句。for 语句是最常用的循环语句，for 语句由三部分组成。首先初始化，由一个循环初始值的表达式表示，用于循环开始时执行，且只执行一次，这个表达式对变量进行说明并初始化，例如 int i=0，它说明 i 变量是局部于循环本身的变量，在循环执行完毕后，它即终止存在。然后是布尔表达式，即测试条件，在初始循环之前和每次执行循环体与循环变量增量之后进行测试，其决定是否能再次循环。最后是步长，即增量，可为任何表达式，用于控制循环继续的增量计数器，改变循环变量的值，以使循环条件为假，从而使循环终止，退出循环。for 循环的嵌入语句是每次循环所执行的语句，它既可包含一个语句，又可包含一个语句块，当然也可包含另一个循环语句，使之构成嵌套的循环。

foreach 语句是 C#特别为数组和集合的迭代而设计的专门语句，它能让迭代通过数组和集合的元素，而且类似于 VB 的 for each 结构。foreach 循环的执行过程：每次循环时，从集合中取出一个新的元素值，放到只读变量中去，括号中的整个表达式返回值为 true 就执行 foreach 块中的语句。一旦集合中的元素均已被访问，整个表达式的值为 false，就转到 foreach 循环后面的第一条可执行语句。

5. 转移语句

在循环体中,只要使用 break 语句,就能跳出循环,即控制转移到本循环的后继语句,如果在循环体内有嵌套的循环就转到外层循环,否则程序就继续执行该循环之后的下一个语句。通常,break 语句与 if 语句有关,即在一定条件下使用 brcak。break 语句的语法简单,break 语句没有圆括号或参数。

像 break 语句一样,continue 语句允许改变一个循环的执行。在循环体内任何一个地方,只要使用 continue 语句,就可以继续执行本循环体中的其他语句,从而转到本循环的首部,继续执行循环。同样,continue 语句与 if 语句有关,即要在一定条件下使用 continue 语句。

2.2 开发环境

2.2.1 硬件要求

1. 最低硬件要求

处理器:1.00GHz

内存:512MB

硬盘:40GB

如需体验三维效果,可参考以下配置:

CPU:酷睿 i3 或同级别处理器,主频 2.00GHz 以上

内存:2GB(64 位系统建议 4GB)

硬盘空间:40GB 或以上

显卡:512MB 或以上显存,需要安装显卡驱动

2. 推荐硬件要求

处理器:2.00GHz 以上主频

内存:2GB

硬盘容量:100GB

如需体验三维效果,可参考以下配置:

CPU:酷睿 i7 或同级别处理器

内存:4GB 或以上(64 位系统建议 16GB 以上)

硬盘空间:100GB 或以上

2.2.2 软件要求

1. 操作系统要求

操作系统可以是 Microsoft Windows XP SP3、Microsoft Windows Server 2003 SP、Microsoft Windows、Server 2008 系列、Microsoft Windows 7 系列、Microsoft Windows 8 系列中的任一种。

2. 基础软件要求

基础软件为 Microsoft .NET Framework 4.0。

3. 开发环境要求

开发环境可以是 Microsoft Visual Studio 2010 或 Microsoft Visual Studio 2012。

4. 对数据库的支持

支持如下数据库：SQL Server 2000/2005/2008/2012、Oracle 9i/10g/11g/12c、PostgreSQL 8.3 及以上版本、DB2 9.7/DB2 10.5。

2.3 SuperMap Objects 简介

2.3.1 SuperMap Objects 概述

1. SuperMap iObjects .NET 7C 的定义

SuperMap iObjects .NET 是基于 Microsoft 的 .NET 技术开发的一款产品，它是 SuperMap Objects 家族中的一员，基于 SuperMap 共相式 GIS 内核开发的组件式 GIS 开发平台。共相式 GIS 内核采用标准 C++编写，实现基础的 GIS 功能，在此基础上，SuperMap iObjects .NET 组件采用 C++/CLI 进行封装，是纯 .NET 的组件，而不是通过 COM 封装或者中间件运行的组件，比通过中间件调用 COM 的方式在性能上将有极大的提高；SuperMap iObjects .NET 支持所有 .NET 开发语言，如 C#、VB .NET、C++/CLI 等，并且，在实际应用中，相比 COM 组件，.NET 组件更适合 .NET 开发人员的编码习惯。

SuperMap iObjects .NET 7C 具有丰富、强大的 GIS 功能，可以用来构建处理地图和地理信息的系统，用户通过该产品可以完成以下工作：

① 创建和使用地图；
② 编辑和处理地理数据；
③ 管理数据库中的地理信息；
④ 分析和研究地理信息；
⑤ 共享和显示地理信息；
⑥ 地理信息的仿真与虚拟现实；
⑦ 在一系列应用程序中使用地理信息。

2. SuperMap Universal 系列产品架构

GIS 是信息技术的重要组成部分，必须紧跟信息技术快速发展的步伐，但商品化的 GIS 平台软件总是落后于主流信息技术，从 DOS 到 Windows、从文件到数据库以及组件技术的应用，莫不如此。主流信息技术每踏一步，GIS 开发商都要耗费巨大的人力、物力来追随，非常被动。

SuperMap 一直在思索一个问题：如何从根本上解决 GIS 开发商在主流信息技术面前的被动局面。通过一系列的探索和思考，并借鉴柏拉图哲学中"共相"的概念，SuperMap 提出了"共相式 GIS"体系结构，即对 GIS 平台软件进行共相式的概括和抽象，提炼出独立于具体技术环境之外的 GIS 技术框架和核心功能，这部分可以和信息技术的发展变化相对分离，只关注于 GIS 核心技术的发展和功能扩展；而外围部分则可以以较小的代价紧随信息技术的发展潮流，以避免因为技术变革导致整个软件的重新构造。超图在"共相式 GIS" 理念的指导下，确定了新 GIS 结构——SuperMap Universal 产品体系结构，如图 2-1 所示。

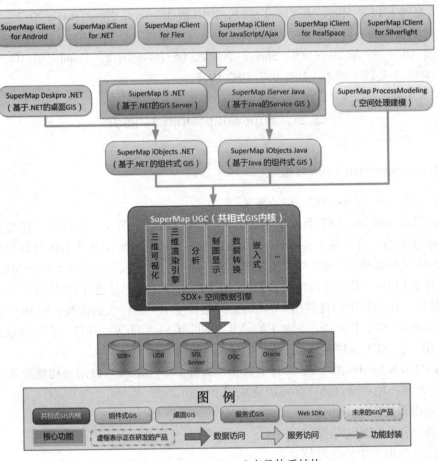

图 2-1　SuperMap Universal 产品体系结构

2.3.2　SuperMap iObjects.NET 7C 的主要特点

① 灵活的安装与便捷的开发。

SuperMap iObjects .NET 7C 产品提供了多种安装与部署方式，不仅提供了安装包方式进行快速安装，包括定制安装和完全安装；还提供了 zip 包，通过解压和简单的部署方式，完成 SuperMap iObjects.NET 7C 使用和运行环境的配置，同样也可以实现自定义部署。因此，多种安装方式可以满足不同用户的需求，体现一切以人为本的服务理念。

SuperMap iObjects .NET 7C 是专门面向二次开发者设计的组件式开发平台，基于该平台开发的应用软件易于分发和再部署，用户只需要将 SuperMap iObjects .NET 7C 运行库文件与 VC++2008 重分发包连同所开发的应用系统一起打包分发即可，只要使用该应用系统的目标机器上安装了 .NET Framework 2.0 及以上版本，即可保证系统的正常运行。这种分发与再部署，有效地降低了成本，为用户最大限度地创造价值。

② 适度的封装"粒度"，易于开发。

封装"粒度"是 GIS 组件接口很重要的指标，它和组件提供的接口是否易用有重大关系。如果"粒度"封装过粗，则在开发时很难做到功能的扩展和灵活开发；如果"粒度"过细，则会导致对象数量过于庞大，就有可能导致系统初始化速度慢，而且也将导致理解和掌握该组件群非常困难，即使进行基本功能开发也会耗费大量成本编写代码。SuperMap iObjects .NET 7C 组件对象封装"粒度"适中，使用灵活且易于掌握，是大型全组件式 GIS 软件开发平台，各个组件对象既可以协同工作，也可以任意裁剪，具有高度的伸缩性和灵活性。

③ 高度的可伸缩性。

SuperMap iObjects .NET 7C 是全组件式 GIS 开发平台，各个 GIS 组件可以像搭积木一样灵活地拆分和组合，用户可以使用全部组件来开发大型 GIS 项目，SuperMap iObjects .NET 7C 的高度可伸缩性，可以让开发者充分考虑项目的 GIS 需求和项目经费等多个因素，灵活地选购并组合开发，获得高软件性价比，降低开发的成本和风险。SuperMap iObjects .NET 7C 组件同样适合开发中小型的 GIS 系统，其高度的可伸缩性，可以帮助开发者灵活地控制系统的规模，随需求进行扩展和压缩，从而减小系统维护的成本，保证中小型 GIS 系统的稳定性。

④ 功能强大，并内嵌大型空间数据库引擎。

SuperMap iObjects .NET 7C 内置了 SuperMap GIS 6R 最新的空间数据库引擎技术——SuperMap SDX+，它为 SuperMap GIS 6R 中的所有产品提供了访问不同引擎存储的空间数据的能力，采用先进的空间数据存储技术、空间索引技术和数据查询技术，实现了具有"空间-属性数据一体化"、"矢量-栅格数据一体化"和"空间信息-业务信息一体化"的集成式空间数据引擎技术。经过多年的研发和应用完善，SuperMap SDX+已成为一个运行稳定、功能成熟、性能卓越的空间数据库引擎，可支持目前流行的多种商用数据库平台，如 Oracle、SQL Server、PostgreSQL、DB2 等，这些数据库可以运行在多种操作系统平台上，既可以搭建同类型数据库之间的多节点集群，也可以搭建异构数据库和异构操作系统的分布式集群。因此，SuperMap iObjects .NET 7C 完全能够胜任各种大型 GIS 系统建设，是 GIS 系统建设的理想选择。

⑤ 二、三维一体化技术。

三维 GIS 技术的快速发展无疑引领了新一代 GIS 技术的巨大变革，但是，相对于三维 GIS，二维 GIS 数据模型更加简单、更抽象、更综合，在分析和建模等方面相对成熟，在各行业中已经被广泛应用。为了充分利用二维 GIS 的优越性以及兼顾行业已有的海量数据基础，二、三维一体化的 GIS 才是 GIS 软件未来的发展方向。SuperMap iObjects .NET 7C 突破了二维 GIS 与三维 GIS 割裂的局面，构建了二维与三维一体化的 GIS 平台，实现了数据管理一体化、应用开发一体化、功能模块一体化、表达一体化、符号系统一体化、分析功能一体化。

2.3.3 SuperMap iObjects .NET 7C 模块架构

SuperMap iObjects .NET 7C 具有更加合理的组件划分，它是由一系列模块构成的，其中包括数据模块、数据转换模块、数据处理模块、拓扑模块、地图模块、排版打印模块、三维模块、三维空间分析模块、三维网络分析模块、地址匹配模块、空间分析模块、网络分析模块、公交分析模块、地形分析模块、控件模块、海图模块。其中，数据模块为核心模块，主要专注于对空间数据的处理，其他模块依赖于数据模块的同时又相对独立，具体如图 2-2 所示。

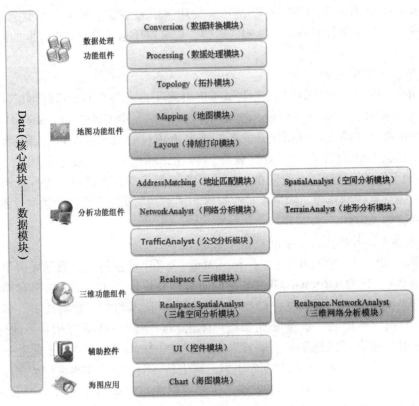

图 2-2 SuperMap 数据模块图

每个模块对应的程序集以及相应的功能介绍见表 2-1。

表 2-1　　　　　　　　　　　　SuperMap 功能模块表

功能模块	对应的程序集	功能概要
数据模块	SuperMap.Data.dll	核心模块，提供对空间数据及其属性的全面的操作和处理，包括创建、管理、访问和查询等功能，同时还提供数据版本管理功能。此外，数据模块还包含与拓扑和布局排版打印相关的数据操作功能
数据转换模块	SuperMap.Data.Conversion.dll	提供多种栅格数据、矢量数据的转换功能
数据处理模块	SuperMap.Data.Processing.dll	提供数据处理，包括三维影像、地形和模型数据的缓存生成功能
拓扑模块	SuperMap.Data.Topology.dll	提供对矢量数据的拓扑预处理、拓扑检查、拓扑错误自动修复和拓扑处理等功能

续表

功能模块	对应的程序集	功能概要
地图模块	SuperMap.Mapping.dll	提供综合的地图显示、渲染、编辑以及强大的出图等功能；提供制作各种专题图的功能，包括标签专题图（包括分段标签专题图和高级标签专题图）、统计专题图、分段专题图、点密度专题图等；同时，地图模块还提供制图表达的功能
排版打印模块	SuperMap.Layout.dll	提供布局排版打印等功能，SuperMap iObjects .NET 的布局排版与二维地图使用同一套对象模型，同时，支持 CMYK 颜色模型，支持海量数据打印。另外，还提供标准图幅图框，使布局排版更为专业化，方便特定领域制图的需要
三维模块	SuperMap.Realspace.dll	提供数据显示、分析，二、三维一体化的三维场景展示等功能，同时，全球尺度的地形数据以及全球尺度的高分辨率影像数据都可以加载到三维模型中进行显示；支持海底三维；支持自定义几何体 Mesh 功能。另外，可以在三维窗口中进行各种方式的漫游、浏览，并且可以进行选择、查询和定位等操作
三维空间分析模块	SuperMap.Realspace.SpatialAnalyst.dll	提供在场景中进行空间分析的功能，目前该模块提供了三维通视性分析功能
三维网络分析模块	SuperMap.Realspace.NetworkAnalyst.dll	提供三维网络数据集的构建和创建流向；提供查找源和汇、上下游追踪、上游最近设施查找等三维设施网络分析功能；提供最佳路径分析等三维交通网络分析功能
地址匹配模块	SuperMap.Analyst.AddressMatching.dll	提供中文地址模糊匹配搜索的功能，该功能基于一个地址词典，可以对地图中的多个图层进行地址匹配
空间分析模块	SuperMap.Analyst.SpatialAnalyst.dll	提供基于矢量数据的空间分析，如叠加分析、缓冲区分析；提供完备的基于栅格数据的空间分析功能，包含栅格代数运算、距离栅格、栅格统计分析、插值分析、地形构建和计算、可视性分析等；提供矢栅转换、聚合、重采样、重分级、镶嵌和裁剪等功能
网络分析模块	SuperMap.Analyst.NetworkAnalyst.dll	提供全面的网络分析功能，涵盖交通网络分析（包括选址分区分析、旅行商分析、物流配送分析、最佳路径分析、最近设施查找分析等）、设施网络分析（包括检查环路、查找共同上下游、查找连通弧段、上下游路径分析、查找源和汇、上下游追踪等）

续表

功能模块	对应的程序集	功能概要
公交分析模块	SuperMap.Analyst.TrafficAnalyst.dll	提供公交换乘分析、查找经过站点的线路、查找线路上的站点等主要功能。不仅支持丰富的线路和站点信息设置，如公交票价信息、发车时间和间隔等，还提供避开线路或站点、优先线路或站点、站点归并容限、站点捕捉容限、最大步行距离、换乘策略和偏好的设置，以及对换乘时步行线路的支持，结合高效、准确和灵活的查找算法，为使用者提供最优的公交换乘方案
地形分析模块	SuperMap.Analyst.TerrainAnalyst.dll	提供填充洼地、计算流向、计算流长、计算累积汇水量、流域划分及提取矢量水系等水文分析功能及网格剖分功能
控件模块	SuperMap.UI.Controls.dll	提供粗粒度的基础控件，方便用户快捷开发，控件模块提供了符号编辑器控件、符号管理控件、工作空间管理器控件、图层管理控件等多种控件
海图模块	SuperMap.Chart.dll	提供基于 S-57 数字海道测量数据传输标准的海图数据（*.000）的导入与导出；提供基于 S-52 显示标准的电子海图的显示；提供海图环境配置、创建海图、标准海图显示风格设置、海图物标属性的查看与编辑等功能；提供海图数据与 GIS 数据的统一管理，支持海图数据和陆地数据的整合，实现海陆一体化存储、显示与发布

2.3.4 SuperMap Objects 7C 的安装

SuperMap iObjects .NET 7C 产品以 zip 包的形式提供，方便用户快速部署产品运行环境与开发环境，需用户进行传统的安装过程。

下面介绍如何通过 SuperMap iObjects .NET 7C zip 包完成 SuperMap iObjects .NET 7C 产品的部署。

1. 产品部署

（1）快速部署

① SuperMap iObjects .NET 7C zip 包的提供形式是一个压缩包。首先，需要将压缩包解压到计算机的某个路径下，该路径将作为产品应用程序所在的路径。

② 双击产品根目录下的 Install.bat 文件，即可进行 SuperMap iObjects .NET 7C 运行环境与开发环境的部署。

（2）产品部署的具体内容

① 在系统环境变量 Path 中添加产品应用程序 Bin 目录的路径。

② 修改产品提供的 SampleCode 源代码工程对产品程序集的引用，保证工程编译通过。

③ 安装运行环境 Microsoft .NET Framework 4.0。

④ 安装许可驱动和为期 90 天的试用许可。

⑤ 注册 Microsoft Visual Studio 工具箱，将产品提供的控件添加到工具箱中。

上述的操作可以实现通过 SuperMap iObjects .NET 7C zip 包进行 SuperMap iObjects .NET 7C 部署，可以说是完全部署，推荐用户使用，此时，用户可以开始使用 SuperMap iObjects .NET 7C。如果用户不需要部署上述所列的全部内容，而需要有选择性地部署需要的内容，请参见下面"自定义部署"的说明；如果用户不需要有选择性地部署，而是完全满意完全部署的操作，那么可以跳过"自定义部署"部分。

2. 自定义部署

产品根目录下的 Install.bat 文件执行，其实质是调用了 Tools 和 Support 目录下的一系列文件的执行，从而实现完全部署，其中，环境变量 Path 的配置是 SuperMap iObjects .NET 7C 产品运行和使用的必须操作，其他内容的部署是可选的，因此，对于高级用户，可以不必使用产品根目录下的 Install.bat 文件实现完全部署，而是通过 Tools 和 Support 目录下提供的可执行文件，选择性地进行必要内容的部署。下面分别详细介绍 Tools 和 Support 目录下的文件内容及各个文件的作用。

（1）Tools 目录

① SetEnv.bat：双击执行该文件，实现将产品应用程序 Bin 目录的路径添加到系统环境变量 Path 中。

② RunReference.bat：双击执行该文件，实现修改产品提供的 SampleCode 源代码工程对产品程序集的引用，保证工程编译通过 RunReference.bat 调用了 SampleCode Reference.exe 作为处理工具。

③ Register_VisualStudio.vbs：双击执行该文件，实现与 Microsoft Visual Studio 开发环境集成，包括注册开发模板、添加程序集引用以及注册工具箱。

④ Redistribute.bat：批处理文件，可直接运行，它将执行安装目录下的 Support 文件中相应的安装程序，安装 Microsoft .NET Framework 4.0。

（2）Support 目录

① dotNetFx40_Full_x86_x64.exe：双击执行该文件，安装运行环境 Microsoft .NET Framework 4.0。

② SuperMap Trial License.exe：在 cmd 中以 SuperMap Trial License.exe-i-nosmg 形式执行本程序安装许可驱动并获得为期 90 天的试用许可。SuperMap Trial License.v2c 文件是备用的试用许可。当使用 SuperMap Trial License.exe 安装试用许可失败时，可以通过 SuperMap 许可中心的更新许可功能加载该文件，使试用许可生效。

3. 产品卸载

如果用户想清除对 SuperMap iObjects .NET 7C 运行环境和开发环境的部署，即卸载 SuperMap iObjects .NET 7C，可以通过双击产品根目录下的 UnInstall.bat 文件，可以将部署的内容进行清除。

产品根目录下的 UnInstall.bat 文件执行，其实质也是调用了 Tools 目录下的一系列文件的执行，从而实现产品部署内容的完全清除。如果采用的是自定义部署，用户也可以根据需要，通过 Tools 目录下提供的可执行文件，选择性地清除所部署的内容。这些可执行文件包括：

① unsetEnv.bat：双击执行该文件，可以清除写入系统环境变量 Path 中的产品应用程

序 Bin 目录的路径。

② UnRegister_VisualStudio.vbs：清除与 Microsoft Visual Studio 开发环境集成，包括开发模板、工具箱的反注册和移除程序集引用。

4．许可配置说明

SuperMap GIS 7C 全系列产品采用统一的 SuperMap 许可中心来管理软件所有与许可相关的工作，包括配置许可信息、许可信息的内外管理以及查看目标机器上详细的许可状态和使用情况等。

（1）获取许可

使用 SuperMap iObjects .NET 7C 需要拥有该产品许可，SuperMap GIS 7C 产品许可分为试用许可和正式许可。

① 试用许可：试用许可不需要用户单独获取，SuperMap GIS 7C 系列产品默认提供了 90 天的试用许可。用户只要通过产品根目录下的 Install.bat 文件完成 SuperMap iObjects .NET 7C 产品的部署，或者部署 SuperMap 许可中心，即可获得试用许可。

② 正式许可：正式许可的提供形式有两种：软许可和硬件许可。硬件许可又分为单机加密锁和网络加密锁。用户通过配置许可，即可使正式许可生效。

a．软许可：软许可是以离线或在线方式获得合法的软件运行许可，激活到本机，即可生效。软许可分为单机软许可和网络软许可。如果激活单机软许可，则只能为本机提供许可服务；如果激活网络软许可，则可以为当前网络中的计算机提供许可服务。

b．硬件许可：硬件许可是以硬件加密锁（简称"硬件锁"）的形式获得合法的软件运行许可。目前 SuperMap 硬件锁生产厂家为 SafeNet。硬件锁分为以下两种：单机锁只提供一个授权许可，需与 SuperMap GIS 产品安装在同一台计算机上。单机锁外观为绿色磨砂。网络锁可安装在网络中任意一台计算机上，可以提供多个授权许可。安装有网络锁的计算机称为许可服务器，网络中许可范围内的客户端无论是否安装许可驱动都能使用该网络锁。网络锁的外观为红色磨砂。

（2）配置许可信息

① 配置软许可：SuperMap 许可中心提供以软件激活方式配置软许可。用户通过 SuperMap 许可中心获取本机信息，并将信息提交给超图软件来获取正式许可，再将正式许可更新到本机，从而完成许可的配置。软件激活方式的具体步骤如下。

a．获取本机信息：在 SuperMap 许可中心，打开"导出信息"页，如图 2-3 所示，选择"软件激活"，在"生成位置"处指定生成的本机信息文件所存放的路径，然后，单击"生成"按钮，在指定的路径下生成 *.c2v 文件。

b．将本机信息提交给超图软件：将上面步骤生成的本机信息文件（*.c2v）提交给超图软件，超图软件将根据用户的申请生成 *.v2c 正式许可件并返回给用户，用户通过该文件配置正式许可。

c．许可生效：在 SuperMap 许可中心，打开"许可更新"页，将用户获得的 *.v2c 正式许可文件指定到"文件位置"处，然后点击"更新"按钮，即可使许可生效。

注意：许可服务器上激活网络软许可后，无法再转移该网络软许可。

② 配置硬件许可：在 Windows 操作系统下，单机锁和网络锁的客户端，都不需要安装驱动程序即可运行 SuperMap GIS 7C 系列产品；网络锁的许可服务器端需要安装许可驱

2.3 SuperMap Objects 简介

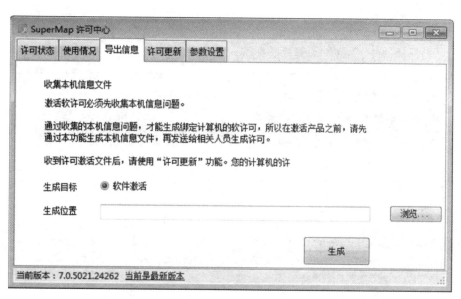

图 2-3 SuperMap 许可中心导出信息图

动。如果当前网络或当前网段中没有找到可用的硬件锁许可，请按照以下步骤进行许可配置。

　　a. 打开 SuperMap 许可中心，在"SuperMap 许可中心"对话框上选择"参数设置"选项卡；

　　b. 在列表中填入许可服务器 IP 或服务器机器名称，默认为空；

　　c. 点击"应用"按钮，即可生效。

注意：a. 硬件锁插入计算机后，锁上的信号指示灯点亮说明硬件锁有效。

　　b. 在 Windows 操作系统下，硬件锁插入后会被识别为 USB 设备，可直接运行。

　　c. 对于两种硬件锁，在同一台计算机，会优先使用单机锁。

　　d. 如果在虚拟机上使用硬件锁，需要通过虚拟机软件的相关设置将硬件锁设备连接到虚拟机上。

（3）管理许可信息

通过 SuperMap 许可中心，用户可以完成查看许可状态、查看许可使用情况、更新许可等许可管理操作。

① 查看许可状态：SuperMap 许可中心的"许可状态"页面，显示了目标机器上所具有的 SuperMap GIS 7C 系列产品的许可信息，信息的组织方式按照产品进行分类，每一类下面的每条记录对应该产品的一个许可模块，如图 2-4 所示。

每条许可模块记录展示了详细的许可信息，每个字段的具体含义请参见以下内容。

　　a. 名称：许可模块名称。

　　b. 类型：显示许可类型是试用许可还是正式许可。

　　c. 起始时间：显示该许可模块的使用期限。

　　d. 连接状态：显示该许可模块的可用的资源，以及当前已经使用的资源。

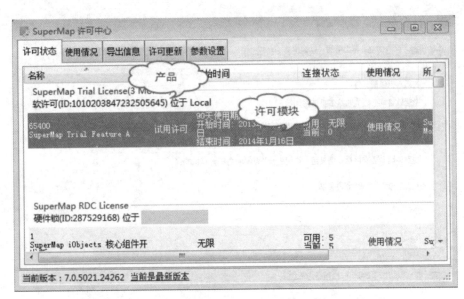

图 2-4　SuperMap 许可中心许可状态图

　　e. 使用情况：显示该许可模块的当前状态。
　　f. 所属产品：产品名称及许可类型信息。
　② 查看许可使用情况：SuperMap 许可中心的"使用情况"页面显示了目标机器上所有 SuperMap GIS 7C 系列产品许可模块在目标机器上当前被使用的详细信息，其中每条记录对应一个许可模块，如图 2-5 所示。

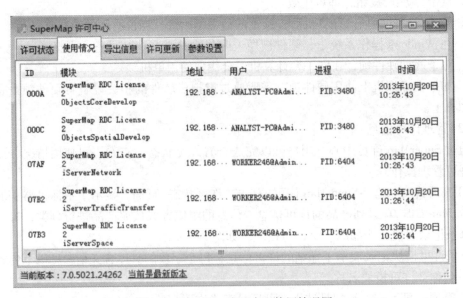

图 2-5　SuperMap 许可中心使用情况图

其中，记录的每个字段表达的内容如下。

a. ID：许可模块对应的 ID 值。

b. 模块：许可模块的名称。

c. 地址：许可模块的许可位置，如果是来源于本地，则显示 Local；否则显示对应机器的 IP 地址。

d. 用户：许可模块的许可所在的计算机的用户名和计算机名。

e. 进程：显示当前许可模块被使用的进程。

f. 时间：当前许可模块被连接的时间。

③ 更新许可：SuperMap 许可中心的"许可更新"页用来进行更新许可的工作。当用户获得了软件激活的正式许可文件（*.v2c）时，用户需要通过更新许可的方式配置本机的许可，使其生效。具体更新操作为：打开 SuperMap 许可中心的"许可更新"页，将目标文件（*.v2c）指定到"文件位置"处，然后单击"更新"按钮即可。

2.3.5 SuperMap Objects 7C 的主要功能

1. 精彩、完美的三维应用

（1）场景可视化与互操作

SuperMap iObjects .NET 7C 提供了三维球体，称为场景，用来模拟现实的地球，并且提供了多种环境渲染，如大气环境、太阳效果、雾环境、宇宙星空、全球影像，更加逼真地贴近现实世界。场景中支持多源数据的显示，包括影像数据、地形数据、三维模型数据、矢量数据，以及二维地图；同时，可以通过丰富的交互操作，实现平移浏览、旋转场景中的球体、改变球体的方位、改变观察球体的视角等，并对这些交互操作，提供鼠标和键盘两种操作形式。

（2）丰富多样的数据支持

在 SuperMap iObjects .NET 7C 的三维场景中可以添加多种类型的三维数据，同时，其二、三维一体化的理念使得三维场景中可以添加所有的二维数据，具体支持的数据内容如下。

① 支持二维矢量数据和栅格数据。

② 支持二维地图数据。

③ 支持 KML、KMZ 数据。

④ 支持三维模型数据，并支持第三方软件制作的三维模型（3DS），也可以是经过缓存处理的三维模型缓存数据。

⑤ 支持海量影像、地形数据，其中影像数据可以是 SIT 数据，也可以是经过缓存处理的影像缓存；地形数据可以为 Grid 数据集，也可以是经过缓存处理的地形缓存。

⑥ 支持三维动画。

（3）绚烂夺目的三维特效

SuperMap iObjects .NET 7C 提供了丰富、炫彩的三维特效，从而使三维表现形式更加贴近现实世界的地理事物。

① 太阳光照随时间的变化，通过设置时间，可以获得该时间点上太阳的照射效果。

② 仿真海洋水体，主要体现水体的波动效果。

③ 支持地下三维场景，如地下油井仿真应用，地下管线应用。

④ 支持海底三维效果。

⑤ 粒子特效，提供火焰、降雨、降雪、喷泉、爆炸、烟火等效果，可以为某种应用提供实况，如火灾发生现场、降雨天气等。

⑥ SuperMap iObjects .NET 7C 提供了立体显示效果，将真空间视觉体验发挥到了极致，使得 GIS 的视觉体验突破了二维屏幕对于真空间显示的限制，用户可以从立体显示中得到前所未有的视觉冲击，三维 GIS 实现了真正的三维可视化。

（4）实用可靠的地理操作

SuperMap iObjects .NET 7C 提供了多种地理量算操作，并且量算可依据地形数据进行，如依地量算距离、量算高程、量算面积等。在场景中，还可以进行地表挖方的仿真操作。此外，还提供了飞行管理功能，通过设置飞行路线上站点的参数，即可实现飞行仿真应用，该功能还可以应用于模拟汽车的行驶过程或者其他与路线行驶相关的过程。

（5）全新的三维空间分析功能

目前，三维空间分析提供通视分析和剖面分析。通视分析可以用于判断三维空间中任意两点之间的通视情况，可广泛应用于工程设计、通信、军事等方面，具有分析结果直观表达的优点。剖面分析支持对地形和模型数据计算剖面线，帮助了解地形的起伏、模型的轮廓形状、分布，以及地形与模型的相对位置等信息，为城市规划、地质勘探、选址分析等分析和应用提供参考。

（6）全新的三维网络分析功能

SuperMap iObjects .NET 7C 提供了一套完整的三维网络分析解决方案，包括三维建模、三维设施网络分析和三维交通网络分析。应用三维网络分析，能够获得更直观、更真实、更多细节的分析效果，提升应用价值。

① 三维网络建模，提供三种方式方便用户灵活构建三维网络数据集，以及为三维网络数据集创建流向。

② 三维设施网络分析，提供数据检查、查找源、查找汇、上下游追踪和上游最近设施查找功能。

③ 三维交通网络分析，目前提供了数据检查和最佳路径分析功能。

2. 二、三维一体化的数据显示与操作

SuperMap iObjects .NET 7C 的理念之一就是实现地理数据的二、三维一体化，二维数据可以显示在场景中，并可以进行风格设置、专题图制作等操作，除此之外，二、三维一体化还体现在以下几个方面。

（1）基于二维数据的快速建模

在三维场景中，除了可以导入通过 3D Max 软件制作的精细三维模型外，还可以基于二维数据进行模型的批量生成，操作流程为：通过对场景中的二维数据图层进行设置，实现对二维几何对象的垂直拉伸获得三维几何对象，然后对三维几何对象进行贴图，从而完成模型的批量构建。快速建模支持对点对象、线对象和面对象进行垂直拉伸，拉伸的高度既可以指定具体的数值也可以依据属性表中某个字段的值。

（2）二、三维数据的联动操作

联动实现了同一地理范围的地图和场景的同步浏览和操作，支持属性表与地图和场景建立同步关系。当地图、场景、属性表三者建立了联动关系，那么在地图中进行漫游操作时，场景的相应范围也会随之移动，如果在地图中选中了某个对象，属性表中也将高亮显示该对象对应的记录，并且场景中该对象也会呈现出选中状态。

(3) 一体化的地图制图操作

SuperMap iObjects .NET 7C 实现了地图制图的二、三维一体化，在地图窗口中可以对二维数据进行风格的设置、进行对象的符号化以及制作各种类型的专题图，在三维场景中同样支持对添加到场景中的二维数据设置显示风格、对几何对象进行符号化，还可以制作各种类型的专题图。

SuperMap iObjects .NET 7C 除了采用一体化的二、三维符号的管理外，显示上也支持二、三维一体化。在对场景中的对象进行符号化时，既可以使用三维符号也可以使用二维符号；在对地图中的点对象进行符号化时，也可以使用三维点符号，此时应用到地图上的三维符号为三维模型的一个快照图片。此外，SuperMap iObjects .NET 7C 提供了丰富的三维符号资源，包括三维点符号和三维线型符号，并且与二维符号实现一体化管理。

3. 全方位的数据处理功能

(1) 生成缓存的数据预处理

地图缓存是快速访问地图服务的有效方式，目前流行的 Google 地图，MapBar 等在线地图均采用缓存地图的方式提高地图访问速度。SuperMap iObjects .NET 7C 产品针对海量数据，特别是三维数据，在客户端高效访问的需求下，为用户提供了一套较为完备的二、三维缓存体系，可以实现对影像数据生成影像缓存、对地形数据生成地形缓存、对矢量数据生成矢量缓存、对三维模型数据生成矢量模型缓存以及生成二、三维地图缓存，还可以对三维场景直接生成场景缓存。

(2) 文本预处理

SuperMap iObjects .NET 7C 提供了对地图中的文本图层、标签专题图层进行文本预处理的功能，可以根据给定的若干比例尺，对地图中的文本图层、标签专题图层进行文本预处理，生成预处理后的地图。当用户要生成地图缓存时，可以直接拿预处理后的地图生成缓存，这样可以避免在生成地图缓存时，由于分块出图导致的文本位置不正确，同时也提高了地图的显示效率。

(3) 数据的基础处理

① 地图配准，不仅能够对栅格数据集、影像数据集和矢量数据集进行配准，还提供对象级别的配准，包括对几何对象和二维坐标点串的配准。

② 地图裁剪，可以对矢量数据和栅格数据进行裁剪。裁剪时，可以选择区域内裁剪或区域外裁剪来确定结果范围，还可以进行精确裁剪或者显示裁剪。

③ 矢量数据融合，用于对矢量线对象和面对象的融合处理，将融合字段值相同的对象合并。

④ 矢量数据和栅格数据重采样，矢量重采样可以简化数据；栅格重采样就是改变数据的分辨率。

⑤ 栅格数据镶嵌，实现将多个栅格数据集或者影像数据集按照地理坐标进行拼接和处理。

(4) 拓扑处理

SuperMap iObjects .NET 7C 提供了一套完整的拓扑处理解决方案，对 GIS 矢量数据依次进行拓扑预处理、拓扑检查、检查后自动修复，之后可进行拓扑处理和进一步的分析。

① 多种拓扑规则，用于检查数据中的拓扑错误。同时，生成的拓扑检查结果报告可以提供详尽的拓扑错误信息，并且对一些拓扑错误进行自动修复。

② 拓扑错误处理，如弧段求交、悬线处理、去除冗余点、合并假节点等；还提供了一系列的拓扑关系处理方法，如提取面边界、查找线的左右多边形等，以满足在实际的应用中经常涉及的一些需要在保证地理要素拓扑关系的基础上进行的要素变更。

③ 拓扑构建功能，在保证空间数据质量的前提下，可实现基于网络模型的高级拓扑分析功能。用于构建拓扑关系的方式包括：线数据集直接构建网络、线数据集构建面数据、多点多线联合构建网络等。

4. 强大、易用的符号制作功能

SuperMap iObjects .NET 7C 提供了崭新的、功能强大、方便易用的多种类型的符号编辑器，包括点符号编辑器、三维点符号编辑器、线型符号编辑器、三维线型符号编辑器、填充符号编辑器，完全满足制作各类符号。

5. 专业的地图制作功能

（1）多样、独具特色的专题图

专题图可以着重显示某一种或某几种自然现象或社会经济现象，从而使地图突出展示某种主题。SuperMap iObjects .NET 7C 可以对一个地图图层制作多种类型的专题图，专题图的类型主要有：单值专题图、分段专题图、统计专题图（多种统计图形式：饼图、柱状图、折线图、玫瑰图、散点图、阶梯图、三维饼图、三维柱状图、三维玫瑰图等）、等级符号专题图、标签专题图，其中最具特色的是矩阵标签专题图；还有对栅格数据制作的单值专题图和分段专题图。

（2）丰富的地图符号资源

SuperMap iObjects .NET 7C 提供了默认符号库资源，其中，提供了丰富的行业应用中的标准地图符号，包括点符号、线型符号以及填充符号，帮助用户进行专业的地图符号化，提高地图制图的效率和质量。

（3）制图表达

SuperMap iObjects .NET 7C 提供了制图表达功能，制图表达是矢量数据集中几何对象所关联的信息，它可以使相应的几何对象在地图窗口中显示时，采用其指定的表现方式，而原来的几何对象不再显示，从而提供一种特殊的地图对象风格化的途径。

（4）标准图幅图框

标准图幅图框的制作，提升了地图制作的专业化水平。利用标准图幅图框功能，可以方便快捷地创建基于国家基本比例尺的各种图幅，在标准图幅内添加具有相同坐标系的居民点、水系、土地利用、等高线、行政区划等国家基础地理信息数据，配以坡度尺、邻接图表、绘制信息等，从而快速创建一幅精美的全要素标准地图。

（5）全面的布局排版

地图的布局排版主要进行地图整饰，为地图添加必要的地图要素，如添加图名、图例、指北针、比例尺等。SuperMap iObjects .NET 7C 为地图的布局排版操作提供了充分的支撑，可以在布局页面上绘制多种布局元素：

① 各种几何形状的地图，支持设置地图边框；
② 提供多种样式的指北针、比例尺，并且可以进行风格样式的个性化设置；
③ 提供为地图添加经纬网的功能，并可以灵活设置经纬网格的样式；
④ 提供在布局页面上绘制各种几何对象的功能以及进行几何对象的风格设置。

此外，SuperMap iObjects .NET 7C 还提供了多种布局排版的辅助工具，方便用户进行布局排版操作。

SuperMap iObjects .NET 7C 支持地图输出为多种格式的影像文件，也支持超大尺寸的地图输出；在地图输出打印方面，支持矢量和栅格两种模式的打印输出，从而满足不同目的的地图打印输出需求。

6. 完善、稳定的企业级数据管理

SuperMap iObjects .NET 7C 内嵌的 SuperMap SDX 加上大型空间数据库引擎为 SuperMap iObjects .NET 7C 使用者提供了强大、稳定的数据管理支撑，也使其具备了优秀的企业级数据管理能力。

① 采用混合多级索引技术——四叉树索引、R 树索引、动态索引（多级网格索引）和图库索引，提高了海量空间数据的访问和查询效率。
② 支持矢量和栅格数据的有损和无损压缩。无损压缩不失真；有损压缩的压缩率较高，失真小。
③ 支持各种空间对象模型，包括点、线、面、文本等简单空间对象；多点、多线、湖中岛、宗地等复合对象；以及 Network（网络模型）、Route（路由模型）、TIN（三角格网模型）、DEM（数字高程模型）、GRID（格网数据）和 Image（影像数据）等复杂数据模型。
④ 支持数据的版本管理和长事务。

7. 便捷的数据转换功能

SuperMap iObjects .NET 7C 数据转换功能支持当今流行的与 GIS 相关的地理数据、影像数据、CAD 数据及属性数据的兼容导入，使得用户在其他平台上的工作成果在 SuperMap 组件产品平台上得以保留，并且能够集成到 SuperMap 数据库中，为后续使用 SuperMap 产品进行数据处理、分析、制图等提供数据基础。同时，也支持将 SuperMap 格式的数据导出为外部数据，保证了用户在数据输出及出图打印时的通用性，也便于多系统之间的交互。

目前，支持 MIF、TAB、SHP、WOR、DXF、DWG、PNG、TIFF、GRD 等多种格式的导入和导出功能。

8. 丰富的数据编辑功能

① 支持多种类型的几何对象。
② 多样的空间对象编辑操作，包括打断、连接、修剪、分割、镜像、旋转和节点编辑等。
③ 编辑捕捉功能，方便地图编辑，提高工作效率和质量。
④ 参数化绘制几何对象，实现对象的精准绘制。

9. 地图可视化与地图互操作

① 满足海量地图数据的高性能显示和操作。
② 支持地图反走样，使得地图表现更加平滑美观。

③ 提供丰富的栅格图层颜色表，也支持用户自定义的颜色表，并允许使用透明色等。

④ 支持地图显示裁剪，通过自定义裁剪范围，只显示地图中的指定范围内的地图内容。

⑤ 丰富的地图互操作功能，可以进行漫游、缩放、自由缩放、选择对象等地图浏览操作，并且支持鼠标、键盘操作。

⑥ 多种地理量算的操作，如量算距离、角度、面积等。

10. 专业的、强大的地理空间分析功能

（1）空间查询功能

① 基于标准 SQL 语句进行属性数据查询，可关联任意的属性表，包括关联非 SuperMap 管理的表格进行查询。

② 支持跨库查询。

③ 基于空间位置的空间查询，支持多种空间查询算子，完全可以满足各种关系的空间对象选取的要求。

（2）基于矢量数据的空间分析

① 多种叠加分析功能，包括：交（intersect）、并（union）、对称差（symmetricdifference）、擦除（erase）、同一（identity）与更新（update），并且对大数据量的叠加分析操作具有高性能的优势，具有高效、准确的特性。

② 邻近分析功能，其中缓冲区分析是应用较为广泛的一种，可以针对不同类型的数据创建多种缓冲区，并支持对数据集中的单个几何对象创建缓冲区。

③ 制图综合功能，目前提供矢量数据融合、碎多边形合并、提取双线数据或面数据的中心线等功能。

（3）全面的网络分析

① 多方面的交通网络分析功能，包括最佳路径分析、旅行商分析、多旅行商分析、服务区分析、资源分配和选址分区分析等，完全可以胜任实现公共交通、物流运输等应用路径分析的领域。

② 设施网络分析功能。它是网络分析功能之一，主要用于进行各类连通性分析和追踪分析。在应用方向上，可以用于市政水网、输电网、天然气管网、电信服务和水流水系等方面，进行建模和分析。

（4）基于栅格数据的空间分析

栅格数据分析是 GIS 空间分析的重要内容，SuperMap iObjects .NET 7C 提供了丰富的基于栅格数据的建模和分析功能。

① 地图代数功能，通过运算符与多种数学函数组合成运算表达式，实现栅格数据间的各种运算。

② 多种栅格数据统计分析功能，如基本统计、常用统计、邻域统计和分带统计。

③ 距离分析功能，可以生成距离栅格、方向栅格和分配栅格，以及计算栅格最短路径。

④ 插值分析功能，可以根据获取的观测值，如土地类型、地面高程等作空间内插生成连续表面模型。

⑤ 地形构建，能够对一个或多个点、线数据集通过数据内插的方法生成 DEM 数据，

还可以通过挖湖操作实现湖泊信息在 DEM 数据上的显示。

⑥ 地形计算，提供对 DEM 数据进行各种地形计算，如生成坡度、坡向图，地形剖面图、三维晕渲图，还可以进行填挖方计算和表面量算等。

⑦ 可视性分析，包括两点之间的可视性分析和给定点的可视域分析。

⑧ 等值线（面）的提取，支持从栅格表面、二维点数据集或记录集、三维点集合中提取等值线或等值面。

⑨ 栅格重分级，对栅格数据的像元值进行重新分类和按照新的分类标准赋值，用新的值取代原像元值。

⑩ 栅格聚合，以整数倍缩小栅格的分辨率，生成一个新的分辨率较粗的栅格。聚合可以通过对数据进行概化，达到清除不需要的信息或者删除微小错误的目的。

11. 全新的公交分析

SuperMap iObjects .NET 7C 公交分析模块是针对公交车、地铁等城市公共交通方式，以公交换乘分析、线路或站点查询为主要功能的分析模块，通过高效、准确和灵活的查找算法，结合以下特性，为使用者提供最优的公交换乘方案。

① 丰富的站点、线路信息设置：如公交票价信息、发车时间和间隔、站点与线路的关系、站点与（轨道交通）出入口的关系设置等。

② 灵活的分析参数设置：站点归并、站点捕捉、最大步行距离、换乘策略、避开/优先站点、线路。

③ 准确的线路导引：通过网络数据集给出准确的步行路线。

12. 易用的海图模块

SuperMap iObjects .NET 7C 海图模块提供基于 S-57、S-52 和 S-58 标准的海图数据转换、显示、查询、编辑和数据验核，以及数据字典管理和环境配置，易于构建符合有关国际标准的 ECDIS 和海图数据生产系统。同时，能够将海图数据和陆地数据进行整合，便于用户在同一个平台中对海图、陆图进行统一的操作和处理，实现海陆一体化的存储、显示和发布，为利用海图进行船舶监控提供更多的资源和信息。

① 基于 IHO S-57 数字海道测量数据传输标准的海图数据的导入、导出，支持导入 001、002……格式的更新文件。

② S-52 标准的海图显示，以及拥有丰富的显示设置，如显示模式、颜色模式、安全水深线、物标的高亮显示风格等。

③ 数据字典管理和环境配置，用于获取符合标准的生产机构、物标属性、物标信息、产品规范物标信息和数据检查信息，以及对显示风格、字典文件路径进行修改。

④ 海图查询，支持通过地图选择查询物标信息，或直接对数据集分组查询符合条件的物标记录。

⑤ 海图编辑功能，可以创建一幅新的海图或修改已有海图，包括海图信息的修改、物标数据集管理、水深管理、物标关联关系管理、拓扑关系构建与维护、物标对象的编辑等。

⑥ 海图数据检查功能，依据 S-58 标准，提供必要的海图数据检查项对海图数据进行检查，有效地保障海图数据符合 S-57 标准和产品规范。

第3章 地图的基本操作

3.1 打开地图

1. 数据

在安装目录\SampleData\World\world.smwu 路径下打开数据文件。

2. 新建文件夹和工程

在【D:\MyProject\】文件中新建一个文件夹，命名为【第3章 地图的基本操作】，并在其目录下建立一个文件夹，命名为【GettingStarted】，在此文件夹里新建一个工程，将此工程命名为 GettingStarted，如图3-1所示。

图3-1 新建工程图

3. 添加控件

添加的各类控件属性见表3-1。

表 3-1　　　　　　　　　　　　控件属性表

控件	Name	Text
toolStrip1	toolStrip1	toolStrip1
Button	toolStripOpen	打开工作空间
mapControl	mapControl1	mapControl1
openFileDialog	openFileDialog1	openFileDialog1
workspace	workspace1	workspace1

4. 窗体设计布局

窗体设计布局如图 3-2 所示。

图 3-2　窗体设计布局图

5. 窗体代码

(1) 添加 using 引用代码

添加 using 引用代码具体如下：

using SuperMap. Data;//添加 data 命名空间,使其相应的类可以用

using SuperMap. Mapping;//添加 mapping 命名空间,使其相应的类可以用

using SuperMap. UI;//添加 UI 工作空间,使其相应的类可以用

(2) 窗体的 FormClose 事件

在退出程序时,要断开空间的关联,关闭地图窗口、工作空间。因此,在属性窗口点击 ⚡ 图标,左键双击 FormClose 事件,添加以下代码：

private void FormDemo_FormClosing(object sender,FormClosingEventArgs e)

```
{mapControl1.Dispose();
    workspace1.Close();//关闭当前工作空间
    workspace1.Dispose();}
```

(3) 添加控件代码

toolStripOpen 控件的 Click 事件,双击 toolStripOpen 控件,添加如下代码,通过此事件打开工作空间:

```
private void toolStripOpen_Click(object sender,EventArgs e)
    {//设置公用打开对话框
        openFileDialog1.Filter="SuperMap 工作空间文件(*.smwu)|*.smwu";
        //判断打开的结果,如果打开就执行下列操作
        if(openFileDialog1.ShowDialog()==DialogResult.OK)
        {//避免连续打开工作空间导致异常信息
          mapControl1.Map.Close();
          workspace1.Close();
        mapControl1.Map.Refresh();
        //定义打开工作空间文件名
        String fileName=openFileDialog1.FileName;
        //打开工作空间文件
            WorkspaceConnectionInfo connectionInfo = new WorkspaceConnectionInfo(fileName);
        //打开工作空间
          workspace1.Open(connectionInfo);
          //建立 MapControl 与 Workspace 的连接
        mapControl1.Map.Workspace=workspace1;
        //判断工作空间中是否有地图
          if(workspace1.Maps.Count==0)
          {MessageBox.Show("当前工作空间中不存在地图!");
          return;}
        //通过名称打开工作空间中的地图
          mapControl1.Map.Open("世界地图_Day");
        //刷新地图窗口
          mapControl1.Map.Refresh();}}
```

6. 运行结果

窗体设计完成后点击 ▶ 按钮,运行该程序,运行后弹出窗体,如图 3-3 所示,如图点击左上角的小图标即可打开地图。

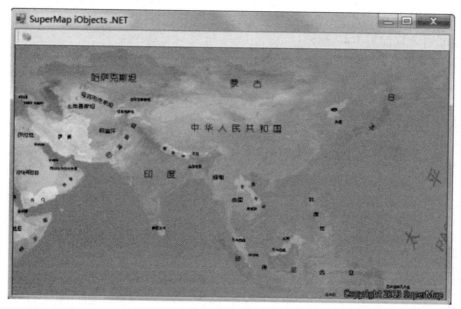

图 3-3　打开地图

3.2　基本操作

在上面程序的基础上添加控件，对地图进行基本操作。

1. 控件属性表

控件属性具体见表 3-2。

表 3-2　　　　　　　　　　　　　　控件属性表

控　件	Name	Text
toolStrip1	toolStrip1	toolStrip1
Button	toolStripOpen	打开工作空间
Button	toolStripPan	漫游
Button	toolStripZoomIn	放大
Button	toolStripZoomOut	缩小
Button	toolStripZoomFree	自由缩放
Button	toolStripViewEntire	全幅显示
Button	toolStripSelect	选择
mapControl	mapControl1	mapControl1
openFileDialog	openFileDialog1	openFileDialog1
workspace	workspace1	workspace1

2. 窗体设计布局

窗体设计布局如图 3-4 所示。

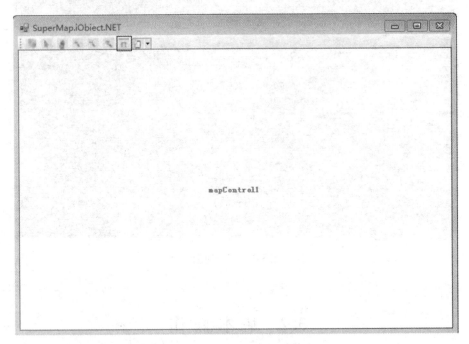

图 3-4　窗体设计布局图

3. 窗体代码

(1) 添加 using 引用代码

添加 using 引用代码具体如下:

using SuperMap. Data;//添加 data 命名空间,使其相应的类可以用

using SuperMap. Mapping;//添加 mapping 命名空间,使其相应的类可以使用

using SuperMap. UI;//添加 UI 命名空间,使其相应的类可以使用

(2) 窗体的 FormClose 事件

窗体的 FormClose 事件同前文中 FormClose 事件代码。

(3) 添加控件代码

① toolStripOpen 控件的 Click 事件。双击 toolStripOpen 控件,添加代码,同 3.1 节中相应内容。

② toolStripSelect 控件的 Click 事件。双击 toolStripSelect 控件,添加如下代码,使选择功能可以使用:

private void toolStripSelect_Click(object sender, EventArgs e)

　　{mapControl1. Action=SuperMap. UI. Action. Select2;}

③ toolStripPan 控件的 Click 事件。双击 toolStripPan 控件,添加如下代码,实现漫游:

private void toolStripPan_Click(object sender, EventArgs e)

　　{mapControl1. Action=SuperMap. UI. Action. Pan;}

④ toolStripZoomIn 控件的 Click 事件。双击 toolStripZoomIn 控件，添加如下代码，实现放大：

private void toolStripZoomIn_Click(object sender, EventArgs e)
　　{mapControl1.Action=SuperMap.UI.Action.ZoomIn;}

⑤ toolStripZoomOut 控件的 Click 事件。双击 toolStripZoomOut 控件，添加如下代码，实现缩小：

private void toolStripZoomOut_Click(object sender, EventArgs e)
　　{mapControl1.Action=SuperMap.UI.Action.ZoomOut;}

⑥ toolStripZoomFree 控件的 Click 事件。双击 toolStripZoomFree 控件，添加如下代码，使自由缩放功能可以使用：

private void toolStripZoomFree_Click(object sender, EventArgs e)
　　{mapControl1.Action=SuperMap.UI.Action.ZoomFree;}

⑦ toolStripViewEntire 控件的 Click 事件。双击 toolStripViewEntire 控件，添加如下代码，实现全幅显示：

private void toolStripViewEntire_Click(object sender, EventArgs e)
　　{mapControl1.Map.ViewEntire();}

4. 运行结果

窗体设计完成后点击按钮▶，运行该程序，运行后弹出窗体，如图3-5所示，如图点击左上角的小图标即可打开地图，点击"选择"、"放大"、"缩小"、"漫游"、"全幅显示"、"自由缩放"按钮，分别对地图进行选择、放大、缩小、漫游、全幅显示、自由缩放的操作。

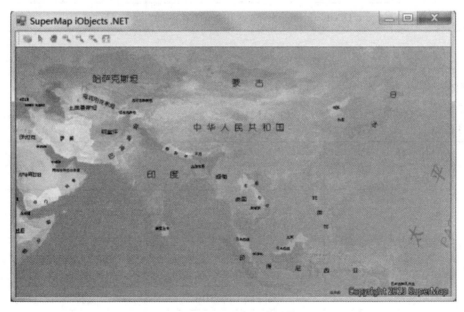

图3-5　地图的基本操作

第4章 空间数据的基本操作

4.1 工作空间管理

1. 新建文件夹和工程

在【D:\MyProject\】文件中新建一个文件夹,命名为【第4章 空间数据的基本操作】,并在其目录下建立一个文件夹,命名为【WorkspaceManage】,在此文件夹里新建一个工程,将此工程命名为 WorkspaceManage。

2. 关键类型/成员

关键类型/成员见表 4-1。

表 4-1　　　　　　　　　　关键类型/成员表

控件/类	方法	属性	事件
WorkspaceConnectionInfo			
Workspace	Open、Create、SavaAs、DeleteWorkspace		

3. 窗体控件属性

窗体控件属性具体见表 4-2 和表 4-3。

表 4-2　　　　　　　　　　窗体控件属性表 1

控件	Name	Text
TextBox	richTextBoxOutput	
tabControl	tabPage1	打开工作空间
	tabPage2	创建工作空间
	tabSave	另存工作空间
	tabDelete	删除工作空间
tabControl	tabPageOpenSDB	文件型工作空间
	tabPageOpenOracle	Oracle 工作空间
	tabPageOpenSQL	SQL 工作空间

表 4-3　　　　　　　　　　　　　　　　窗体控件属性表 2

Text	Text	Label	TextBox	ComboBox	Button
打开工作空间	文件型工作空间	Label1（工作空间路径）	textBoxSDBPath		buttonSDBOK（确定）
		Label2（密码）	textBoxSDBPassword		buttonOpenSDBFile（浏览）
	Oracle工作空间	Label3（服务器名）	textBoxOrcleServer		
		Label4（数据库名）	textBoxOrcleDatabase		
		Label5（用户名）	textBoxOracleUser		
		Label6（密码）	textBoxOraclePassword		
		Label7（工作空间名）	textBoxOracleName		
	SQL工作空间	Label8（工作空间名）	textBoxSQLName		
		Label9（密码）	textBoxSQLPassword		
		Label10（用户名）	textBoxSQLUser		
		Label11（数据库名）	textBoxSQLDatabase		
		Label12（服务器名）	textBoxSQLServer		
创建工作空间	文件型工作空间	Label14（工作空间路径）	textBoxCreateSDBPath		buttonCreateSDBFile（浏览）
		Label13（密码）	textBoxCreateSDBPassword		buttonCreateSDBOK（确定）
		Label（工作空间版本）		comboBoxCreate-SDBVersion	
		Label26（工作空间类型）		comboBoxCreate-SDBType	
		Label50（工作空间名）	textBoxCreateSDBName		
	Oracle工作空间	Label19（服务器名）	textBoxCreateOracleServer		buttonCreateOracleOK（确定）
		Label18（数据库名）	textBoxCreateOracleDatabase		
		Label17（用户名）	textBoxCreateOracleUser		
		Label16（密码）	textBoxCreateOracle-Password		
		Label18（工作空间版本）		comboBoxCreateOracleVersion	
		Label15（工作空间名）	textBoxCreateOracleName		

续表

Text	Text	Label	TextBox	ComboBox	Button
创建工作空间	SQL工作空间	Label30（服务器名）	textBoxCreateSQLServer		buttonCreateSQLOK（确定）
		Label29（数据库名）	textBoxCreateSQLDatabase		
		Label24（用户名）	textBoxCreateSQLUser		
		Label23（密码）	textBoxCreateSQLPassword		
		Label21（工作空间版本）		comboBoxCreateSQLVersion	
		Label22（工作空间名）	textBoxCreateSQLName		
另存工作空间	文件型工作空间	Label34（工作空间路径）	textBoxSaveSDBFilePath		buttonSaveSDBOpenFile（浏览）buttonSaveSDBOK（确定）
		Label33（密码）	textBoxSaveSDBPassword		
		Label32（工作空间版本）		comboBoxSaveSDBVersion	
		Label31（工作空间类型）		comboBoxSaveSDBType	
		Label51（工作空间名）	textBoxSaveSDBName		
	Oracle工作空间	Label41（服务器名）	textBoxSaveOracleServer		buttonSaveOracleOK（确定）
		Label40（数据库名）	textBoxSaveOracleDatabase		
		Label39（用户名）	textBoxSaveOracleUser		
		Label38（密码）	textBoxSaveOraclePassword		
		Label36（工作空间版本）		comboBoxSaveOracleVersion	
		Label37（工作空间名）	textBoxSaveOracleName		
	SQL工作空间	Label48（服务器名）	textBoxSaveSQLServer		buttonSaveSQLOK（确定）
		Label47（数据库名）	textBoxSaveSQLDatabase		
		Label46（用户名）	textBoxSaveSQLUser		
		Label45（密码）	textBoxSaveSQLPassword		
		Label43（工作空间版本）		comboBoxSaveSQLVersion	
		Label44（工作空间名）	textBoxSaveSQLName		

续表

Text	Text	Label	TextBox	ComboBox	Button
删除工作空间		Label49（删除当前的工作空间，确定？）			buttonDeleteOK（确定）
					buttonDeleteCancel（取消）

4. 窗体设计布局

窗体设计布局如图 4-1 所示。

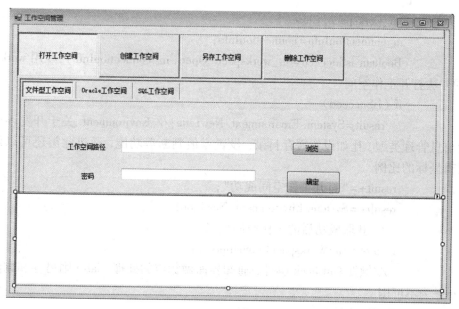

图 4-1　窗体设计布局图

5. 创建 SampleRun 类

（1）添加 using 引用代码

添加 using 引用代码具体如下：

using SuperMap.Data;//添加 Data 命名空间，使其相应的类可以使用。

（2）添加 FormMain()函数前的代码

在 FormMain()函数前添加如下代码：

private Workspace m_workspace;// 定义 Workspace 类对象。Workspace 工作空间类，主要完成数据的组织和管理，包括打开、关闭、创建、保存工作空间

private WorkspaceConnectionInfo m_connectionInfo;//定义 WorkspaceConnectionInfo 类对象。WorkspaceConnectionInfo 工作空间连接信息类，包括工作空间连接的所有信息

public delegate void CheckEventHandler(Boolean isInitialize) ;

```csharp
public event CheckEventHandler OnCheck;//定义 OnCheck 事件
        /// 根据 workspace 构造 SampleRun 对象
public SampleRun(Workspace workspace)
    { try
    { m_workspace = workspace; }
    catch (Exception ex)
    { Trace.WriteLine(ex.Message); } }
//打开工作空间
public String Open(WorkspaceConnectionInfo connectionInfo)
    {String result = String.Empty;
      try
      { m_workspace.Close();
        m_connectionInfo = connectionInfo;
        Boolean isSucceed = m_workspace.Open(m_connectionInfo);//用 workspace.Open 的方法打开工作空间
            if (isSucceed)
            { result = System.Environment.NewLine;// Environment.关于开发环境的一些配置信息管理类型,比如设置缓存目录,设置零值判断等功能,通过此类还可以设置像素与逻辑坐标的比例
            result+="打开工作空间成功";
            result+=System.Environment.NewLine;
            // 获取成功后的工作空间信息
            result+=GetWorkspaceInfomation();
            //触发 OnCheck 事件,通知界面做相应的处理,false 则另存和删除功能可用, true 为不可用
            OnCheck(false); }
            else{ result=System.Environment.NewLine;
            result+="打开工作空间失败";
            result+=System.Environment.NewLine;
            //触发事件,通知界面做相应的处理,false 则另存和删除功能可用, true 为不可用
            OnCheck(true); } }
        catch (Exception ex)
            { Trace.WriteLine(ex.Message); }
        return result; }
    /// 创建工作空间
    public String Create(WorkspaceConnectionInfo connectionInfo)
        {String result = String.Empty;
        try
```

```
            {m_workspace.Close();
            m_connectionInfo=connectionInfo;
                Boolean isSucceed = m_workspace.Create(m_connectionInfo);//通过
workspace.Create方法创建工作空间
                if(isSucceed)
                    {result=System.Environment.NewLine;// Environment.关于开发环境的
一些配置信息管理类型,比如设置缓存目录,设置零值判断等功能,通过此类还可以设置
像素与逻辑坐标的比例。
                    result+="创建工作空间成功";
                    result+=System.Environment.NewLine;
                    result+=GetWorkspaceInfomation();
                    OnCheck(false);}
                else{ result=System.Environment.NewLine;
                    result+="创建工作空间失败";
                    result+=System.Environment.NewLine;
                    OnCheck(true);}}
        catch(Exception ex)
            {Trace.WriteLine(ex.Message);}
        return result;}
    ///另存工作空间
    ///工作空间连接信息
    public String SaveAs(WorkspaceConnectionInfo connectionInfo)
        {String result=String.Empty;
        try
            {m_connectionInfo=connectionInfo;
                Boolean isSucceed = m_workspace.SaveAs(m_connectionInfo);//通过
workspace.SaveAs另存工作空间
                if(isSucceed)
                    {result=System.Environment.NewLine;// Environment.关于开发环境的
一些配置信息管理类型,比如设置缓存目录,设置零值判断等功能,通过此类还可以设置
像素与逻辑坐标的比例。
                    result+="另存工作空间成功";
                    result+=System.Environment.NewLine;
                    result+=GetWorkspaceInfomation();
                    OnCheck(false);// OnCheck 事件}
                else{ result=System.Environment.NewLine;
                    result+="另存工作空间失败";
                    result+=System.Environment.NewLine;
                    OnCheck(false);}}
```

```
        catch ( Exception ex)
         { Trace. WriteLine( ex. Message ); }
        return result; }
///删除工作空间
    public String Delete( )
        { String result = String. Empty;
        try {
            Boolean isSucceed = Workspace. DeleteWorkspace( m_connectionInfo );//通
过 Workspace. Delete 删除工作空间
            if ( isSucceed)
             { result = System. Environment. NewLine;// Environment. 关于开发环境
的一些配置信息管理类型, 比如设置缓存目录, 设置零值判断等功能, 通过此类还可以设
置像素与逻辑坐标的比例。
                result+=" 删除工作空间成功";
                result+=System. Environment. NewLine;
                result+=GetWorkspaceInfomation( );
                OnCheck( true ); }
            else{ result = System. Environment. NewLine;
                result+=" 删除工作空间失败";
                result+=System. Environment. NewLine;
                OnCheck( false ); } }
        catch ( Exception ex)
         { Trace. WriteLine( ex. Message ); }
        return result; }
/// 通过 WorkspaceInfomation 类来获取当前的工作空间信息
    public String GetWorkspaceInfomation( )
        {
        StringBuilder stringBuilder = new StringBuilder( );
        try
            {
            stringBuilder. Append(" 工作空间基本信息");
            stringBuilder. Append( System. Environment. NewLine );
            stringBuilder. Append(" 服务器名或文件路径");
            stringBuilder. Append( m_connectionInfo. Server );
            stringBuilder. Append( System. Environment. NewLine );
            stringBuilder. Append(" 数据库名");
            stringBuilder. Append( m_connectionInfo. Database );
            stringBuilder. Append( System. Environment. NewLine );
            stringBuilder. Append(" 用户名");
```

```
            stringBuilder.Append(m_connectionInfo.User);
            stringBuilder.Append(System.Environment.NewLine);
            stringBuilder.Append("工作空间名称");
            stringBuilder.Append(m_connectionInfo.Name);
            stringBuilder.Append(System.Environment.NewLine);
            stringBuilder.Append("工作空间类型");
            stringBuilder.Append(m_connectionInfo.Type.ToString());
            stringBuilder.Append(System.Environment.NewLine);
            stringBuilder.Append("工作空间版本");
            stringBuilder.Append(m_connectionInfo.Version.ToString());
            stringBuilder.Append(System.Environment.NewLine);
            stringBuilder.Append("*************************************************");
            stringBuilder.Append(System.Environment.NewLine);
            stringBuilder.Append("工作空间中数据源信息");
            Int32 datasourceCount = m_workspace.Datasources.Count;
            if(datasourceCount > 0)
            {for(int i=0;i < datasourceCount;i++)//通过循环来判断是否有数据源,如果有则显示数据源编号,若无数据源则显示工作空间中没有数据源
                {stringBuilder.Append("数据源编号"+i+":");
                stringBuilder.Append(m_workspace.Datasources[i].Alias);
                stringBuilder.Append(System.Environment.NewLine); } }
            else
            { stringBuilder.Append("工作空间中没有数据源");
            stringBuilder.Append(System.Environment.NewLine); }
            stringBuilder.Append(System.Environment.NewLine);
    stringBuilder.Append("*****************************************************");
            stringBuilder.Append(System.Environment.NewLine);
            stringBuilder.Append("工作空间中地图信息");
            Int32 mapCount = m_workspace.Maps.Count;
            if(mapCount > 0)
            { for(int i=0;i < mapCount;i++)//通过循环来判断是否有地图,如果有则显示地图编号,若无数据源则显示工作空间中没有地图
                { stringBuilder.Append("地图编号"+i+":");
                stringBuilder.Append(m_workspace.Maps[i]);
                stringBuilder.Append(System.Environment.NewLine); }
            }
            else
```

```
            { stringBuilder.Append("工作空间中没有地图");
                stringBuilder.Append(System.Environment.NewLine);}}
        catch(Exception ex)
        { Trace.WriteLine(ex.Message);}
        return stringBuilder.ToString();}
```

6. 窗体代码

(1) 添加 using 引用代码

添加 using 引用代码具体如下：

using System.Diagnostics;//添加 System.Diagnostics 命名空间，使其相应的类可以使用

using SuperMap.Data;//添加 Data 命名空间，使其相应的类可以使用

(2) 初始化控件

在 FormMain() 函数前初始化控件及实例化 SampleRun、Workspace、orkspace ConnectionInfo，需添加如下代码：

//定义类对象

private SampleRun m_sampleRun;

private Workspace m_workspace;// 定义 Workspace 类对象。Workspace 工作空间类，主要完成数据的组织和管理，包括打开、关闭、创建、保存工作空间。

private WorkspaceConnectionInfo m_connectionInfo;//定义 WorkspaceConnectionInfo 类对象。WorkspaceConnectionInfo 工作空间连接信息类，包括工作空间连接的所有信息，例如工作空间类型、版本信息、有无数据源和地图。

```
        public FormMain()
        { try
        { InitializeComponent();
    m_workspace=new SuperMap.Data.Workspace();//创建新的 Workspace 对象
        m_connectionInfo = new WorkspaceConnectionInfo( );//创建新的 WorkspaceConnectionInfo 对象
            //实例化 SampleRun
        m_sampleRun=new SampleRun(m_workspace);//创建新的 SampleRun 对象
            //创建一些事件
    m_sampleRun.OnCheck += new SampleRun.CheckEventHandler(m_sampleRun_OnCheck);
            this.comboBoxCreateSDBVersion.SelectedIndexChanged += new EventHandler(comboBoxCreateSDBVersion_SelectedIndexChanged);
            this.comboBoxSaveSDBVersion.SelectedIndexChanged += new EventHandler(comboBoxSaveSDBVersion_SelectedIndexChanged);
            // 初始化控件设置控件的初始值
            comboBoxCreateSQLVersion.SelectedIndex=0;
            comboBoxCreateOracleVersion.SelectedIndex=0;
            comboBoxCreateSDBType.SelectedIndex=0;
```

```
                comboBoxCreateSDBVersion.SelectedIndex=0;
                comboBoxSaveSQLVersion.SelectedIndex=0;
                comboBoxSaveOracleVersion.SelectedIndex=0;
                comboBoxSaveSDBType.SelectedIndex=0;
                comboBoxSaveSDBVersion.SelectedIndex=0;}
        catch(Exception ex)
            {Trace.WriteLine(ex.Message);}}
```

(3) 设置界面

在操作的同时设置界面,需添加如下代码:

```
private void m_sampleRun_OnCheck(Boolean isInitialize)//定义 m_sampleRun_OnCheck
函数的初始化函数
        {try
            {if(isInitialize)
                {//设置控件的有效性
                    tabControlSave.Enabled=false;
                    buttonDeleteOK.Enabled=false;
                    buttonDeleteCancel.Enabled=false;}
                else
                {tabControlSave.Enabled=true;
                    buttonDeleteOK.Enabled=true;
                    buttonDeleteCancel.Enabled=true;}}
        catch(Exception ex)
            {Trace.WriteLine(ex.Message);}}
```

(4) 添加各个控件的代码

① comboBoxSaveSDBVersion 控件的 SelectedIndexChanged 事件。在 comboBoxSaveSDBVersion 的属性窗口中,选择 ![] 按钮后,左键双击 SelectedIndexChanged,即可添加以下代码,通过此事件保存文件型工作空间:

```
private void comboBoxSaveSDBVersion_SelectedIndexChanged(object sender,EventArgs e)
        {try{if(comboBoxSaveSDBVersion.SelectedIndex==0)
                {//文件型添加工作空间类型
                    comboBoxSaveSDBType.Items.Clear();
                    comboBoxSaveSDBType.Items.Add("SMWU");
                    comboBoxSaveSDBType.Items.Add("SXWU");}
                else
                {comboBoxSaveSDBType.Items.Clear();
                    comboBoxSaveSDBType.Items.Add("SMW");
                    comboBoxSaveSDBType.Items.Add("SXW");}
                    comboBoxSaveSDBType.SelectedIndex=0;}
            catch(Exception ex)
```

②comboBoxCreateSDBVersion 控件的 SelectedIndexChanged 事件。在 comboBoxCreateSDBVersion 的属性窗口中，选择 ≠ 按钮后，左键双击 SelectedIndexChanged，即可添加以下代码，通过此事件创建文件型工作空间：

```
private void comboBoxCreateSDBVersion_SelectedIndexChanged(object sender, EventArgs e)
    try
        if(comboBoxCreateSDBVersion.SelectedIndex==0)
            comboBoxCreateSDBType.Items.Clear();
            comboBoxCreateSDBType.Items.Add("SMWU");
            comboBoxCreateSDBType.Items.Add("SXWU");}
        else
            comboBoxCreateSDBType.Items.Clear();
            comboBoxCreateSDBType.Items.Add("SMW");
            comboBoxCreateSDBType.Items.Add("SXW");}
        comboBoxCreateSDBType.SelectedIndex=0;}
    catch(Exception ex)
        Trace.WriteLine(ex.Message);}}
```

③浏览 SDB 文件 Button 控件的 Click 事件。左键双击分析 Button 控件，即可添加如下该按钮的 Click 事件代码，实现浏览：

```
private void buttonOpenSDBFile_Click(object sender, EventArgs e)
    try{ OpenFileDialog openFileDig=new OpenFileDialog();//通过 OpenFileDialog 添加打开的工作空间的路径
        openFileDig.Filter="SXW files (*.sxw)|*.sxw|SMW files (*.smw)|*.smw|SXWU files (*.sxwu)|*.sxwu|SMWU files (*.smwu)|*.smwu";
        if(openFileDig.ShowDialog()==DialogResult.OK)
            textBoxSDBPath.Text=openFileDig.FileName;}}
    catch(Exception ex)
        Trace.WriteLine(ex.Message);}}
```

④打开 SDB 工作空间 Button 控件的 Click 事件。左键双击分析 Button 控件，即可添加如下该按钮的 Click 事件代码，实现打开：

```
private void buttonSDBOK_Click(object sender, EventArgs e)
    try{//通过 workspaceconnectionInfo 获取工作空间的信息，包括服务器名、版本等信息
        m_connectionInfo.Server=textBoxSDBPath.Text;
        m_connectionInfo.Type=this.GetType(System.IO.Path.GetExtension(textBoxSDBPath.Text).ToUpper().Replace(".",String.Empty));
        m_connectionInfo.Name=System.IO.Path.GetFileNameWithoutExtension(textBoxSDBPath.Text);
        m_connectionInfo.Password=textBoxSDBPassword.Text;
```

richTextBoxOutput.Text += m_sampleRun.Open(m_connectionInfo);//运用sampleRun.Open方法打开文件型工作空间}
  catch(Exception ex)
   {Trace.WriteLine(ex.Message);}}

⑤ 打开 Oracle 工作空间 Button 控件的 Click 事件。左键双击分析 Button 控件，即可添加如下该按钮的 Click 事件代码，实现打开：
private void buttonOracleOK_Click(object sender, EventArgs e)
 {try
  {//通过workspaceconnectionInfo获取工作空间的信息，包括服务器名、版本等信息
  m_connectionInfo.Server = textBoxOrcleServer.Text;
  m_connectionInfo.Database = textBoxOrcleDatabase.Text;
  m_connectionInfo.User = textBoxOracleUser.Text;
  m_connectionInfo.Password = textBoxOraclePassword.Text;
  m_connectionInfo.Name = textBoxOrcleName.Text;
  m_connectionInfo.Type = WorkspaceType.Oracle;
  richTextBoxOutput.Text += m_sampleRun.Open(m_connectionInfo);//运用sampleRun.Open方法打开文件型工作空间}
  catch(Exception ex)
   {Trace.WriteLine(ex.Message);}}

⑥ 打开 SQL 工作空间 Button 控件的 Click 事件。左键双击分析 Button 控件，即可添加如下该按钮的 Click 事件代码，实现打开：
private void buttonSQLOK_Click(object sender, EventArgs e)
 {try{//通过workspaceconnectionInfo获取工作空间的信息，包括服务器名、版本等信息
  m_connectionInfo.Server = textBoxSQLServer.Text;
  m_connectionInfo.Database = textBoxSQLDatabase.Text;
  m_connectionInfo.User = textBoxSQLUser.Text;
  m_connectionInfo.Password = textBoxSQLPassword.Text;
  m_connectionInfo.Name = textBoxSQLName.Text;
  m_connectionInfo.Driver = "SQL Server";
  m_connectionInfo.Type = WorkspaceType.SQL;
  richTextBoxOutput.Text += m_sampleRun.Open(m_connectionInfo);}
//运用sampleRun.Open方法打开文件型工作空间
  catch(Exception ex)
   {Trace.WriteLine(ex.Message);}}

⑦ 选择创建的 SDB 路径 Button 控件的 Click 事件。左键双击分析 Button 控件，即可添

加如下该按钮的 Click 事件代码,实现选择创建路径:
```
private void buttonCreateSDBFile_Click(object sender, EventArgs e)
{try{//通过 folderBrowserDig.SelectedPath 选择路径
        FolderBrowserDialog folderBrowserDig=new FolderBrowserDialog();
        if(folderBrowserDig.ShowDialog()==DialogResult.OK)
            {textBoxCreateSDBPath.Text=folderBrowserDig.SelectedPath;}}
    catch(Exception ex)
        {Trace.WriteLine(ex.Message);}}
```

⑧ 创建 SDB 工作空间 Button 控件的 Click 事件。左键双击分析 Button 控件,即可添加如下该按钮的 Click 事件代码,实现创建:
```
private void buttonCreateSDBOK_Click(object sender, EventArgs e)
    {try{//通过 workspaceconnectionInfo 获取工作空间的信息,包括服务器名、版本等信息
        String extension=comboBoxCreateSDBType.SelectedItem as String;
        m_connectionInfo.Version=this.GetVersion(comboBoxCreateSDBVersion.SelectedIndex);
        m_connectionInfo.Type=this.GetType(extension);
        m_connectionInfo.Server=System.IO.Path.Combine(textBoxCreate SDBPath.Text,textBoxCreateSDBName.Text+"."+extension);
        m_connectionInfo.Name=textBoxCreateSDBName.Text;
        m_connectionInfo.Password=textBoxCreateSDBPassword.Text;
        richTextBoxOutput.Text+=m_sampleRun.Create(m_connectionInfo);}//通过 sampleRun.Create 创建工作空间
    catch(Exception ex)
        {Trace.WriteLine(ex.Message);}}
```

⑨ 创建 Oracle 工作空间 Button 控件的 Click 事件。左键双击分析 Button 控件,即可添加如下该按钮的 Click 事件代码:
```
private void buttonCreateOracleOK_Click(object sender, EventArgs e)
    {try{//通过 workspaceconnectionInfo 获取工作空间的信息,包括服务器名、版本等信息
        m_connectionInfo.Server=textBoxCreateOracleServer.Text;
        m_connectionInfo.Database=textBoxCreateOracleDatabase.Text;
        m_connectionInfo.User=textBoxCreateOracleUser.Text;
        m_connectionInfo.Password=textBoxCreateOraclePassword.Text;
        m_connectionInfo.Name=textBoxCreateOracleName.Text;
        m_connectionInfo.Version=GetVersion(comboBoxCreateOracleVersion.SelectedIndex);
```

m_connectionInfo.Type=WorkspaceType.Oracle；

richTextBoxOutput.Text+=m_sampleRun.Create(m_connectionInfo)；}//通过workspaceconnectionInfo 获取工作空间的信息，包括服务器名、版本等信息

catch(Exception ex)

{Trace.WriteLine(ex.Message)；}}

⑩ 创建 SQL 工作空间 Button 控件的 Click 事件。左键双击分析 Button 控件，即可添加如下该按钮的 Click 事件代码：

private void buttonCreateSQLOK_Click(object sender，EventArgs e)

{try{//通过workspaceconnectionInfo 获取工作空间的信息，包括服务器名、版本等信息

m_connectionInfo.Server=textBoxCreateSQLServer.Text；

m_connectionInfo.Database=textBoxCreateSQLDatabase.Text；

m_connectionInfo.User=textBoxCreateSQLUser.Text；

m_connectionInfo.Password=textBoxCreateSQLPassword.Text；

m_connectionInfo.Name=textBoxCreateSQLName.Text；

m_connectionInfo.Version=GetVersion(comboBoxCreateSQLVersion.SelectedIndex)；

m_connectionInfo.Type=WorkspaceType.SQL；

m_connectionInfo.Driver="SQL Server"；

richTextBoxOutput.Text+=m_sampleRun.Create(m_connectionInfo)；}//通过workspaceconnectionInfo 获取工作空间的信息，包括服务器名、版本等信息

catch(Exception ex)

{Trace.WriteLine(ex.Message)；}}

⑪ 另存路径选择 Button 控件的 Click 事件。左键双击分析 Button 控件，即可添加如下该按钮的 Click 事件代码：

private void buttonSaveSDBOpenFile_Click(object sender，EventArgs e)

{try{//通过folderBrowserDig.SelectedPath 选择路径

FolderBrowserDialog folderBrowserDig=new FolderBrowserDialog()；

if(folderBrowserDig.ShowDialog()==DialogResult.OK)

{textBoxSaveSDBFilePath.Text=folderBrowserDig.SelectedPath；}}

catch(Exception ex)

{Trace.WriteLine(ex.Message)；}}

⑫ 另存路径选择 Button 控件的 Click 事件。左键双击分析 Button 控件，即可添加如下该按钮的 Click 事件代码：

private void buttonSaveSDBOpenFile_Click(object sender，EventArgs e)

{try{通过folderBrowserDig.SelectedPath 选择路径

FolderBrowserDialog folderBrowserDig=new FolderBrowserDialog()；

```
                    if (folderBrowserDig.ShowDialog() == DialogResult.OK)
                    {textBoxSaveSDBFilePath.Text = folderBrowserDig.SelectedPath;}}
            catch (Exception ex)
                    {Trace.WriteLine(ex.Message);}}
```

⑬ 另存 SDB 工作空间 Button 控件的 Click 事件。左键双击分析 Button 控件,即可添加如下该按钮的 Click 事件代码:

```
        private void buttonSaveSDBOK_Click(object sender, EventArgs e)
                {try{//通过 workspaceconnectionInfo 获取工作空间的信息,包括服务器名、版本等信息
                    String extension = comboBoxCreateSDBType.SelectedItem as String;
                    m_connectionInfo.Version = GetVersion(comboBoxSaveSDBVersion.SelectedIndex);
                    m_connectionInfo.Type = this.GetType(extension);
                    m_connectionInfo.Server = System.IO.Path.Combine(textBoxSaveSDBFilePath.Text, textBoxSaveSDBName.Text+"."+extension);
                    m_connectionInfo.Name = textBoxSaveSDBName.Text;
                    m_connectionInfo.Password = textBoxSaveSDBPassword.Text;
                    richTextBoxOutput.Text += m_sampleRun.SaveAs(m_connectionInfo);}//通过 sampleRun.SaveAs 方法另存工作空间
            catch (Exception ex)
                    {Trace.WriteLine(ex.Message);}}
```

⑭ 另存 Oracle 工作空间 Button 控件的 Click 事件。左键双击分析 Button 控件,即可添加如下该按钮的 Click 事件代码:

```
        private void buttonSaveOracleOK_Click(object sender, EventArgs e)
                {try{//通过 workspaceconnectionInfo 获取工作空间的信息,包括服务器名、版本等信息
                            m_connectionInfo.Server = textBoxSaveOracleServer.Text;
                            m_connectionInfo.Database = textBoxSaveOracleDatabase.Text;
                            m_connectionInfo.User = textBoxSaveOracleUser.Text;
                            m_connectionInfo.Password = textBoxSaveOraclePassword.Text;
                            m_connectionInfo.Name = textBoxSaveOracleName.Text;
                            m_connectionInfo.Version = GetVersion(comboBoxSaveOracleVersion.SelectedIndex);
                            m_connectionInfo.Type = WorkspaceType.Oracle;
                            richTextBoxOutput.Text += m_sampleRun.SaveAs(m_connectionInfo);}//通过 sampleRun.SaveAs 方法另存工作空间
            catch (Exception ex)
                    {Trace.WriteLine(ex.Message);}}
```

⑮ 另存 SQL 工作空间 Button 控件的 Click 事件。左键双击分析 Button 控件，即可添加如下该按钮的 Click 事件代码：

```
private void buttonSaveSQLOK_Click(object sender, EventArgs e)
    {try{//通过 workspaceconnectionInfo 获取工作空间的信息，包括服务器名、版本等信息
        m_connectionInfo.Server = textBoxSaveSQLServer.Text;
        m_connectionInfo.Database = textBoxSaveSQLDatabase.Text;
        m_connectionInfo.User = textBoxSaveSQLUser.Text;
        m_connectionInfo.Password = textBoxSaveSQLPassword.Text;
        m_connectionInfo.Name = textBoxSaveSQLName.Text;
        m_connectionInfo.Version = GetVersion(comboBoxSaveSQLVersion.SelectedIndex);
        m_connectionInfo.Type = WorkspaceType.SQL;
        m_connectionInfo.Driver = "SQL Server";
        richTextBoxOutput.Text += m_sampleRun.SaveAs(m_connectionInfo);}//通过
sampleRun.SaveAs 方法另存工作空间
    catch (Exception ex)
        {Trace.WriteLine(ex.Message);}}
```

⑯ 删除 Button 控件的 Click 事件。左键双击分析 Button 控件，即可添加如下该按钮的 Click 事件代码：

```
private void buttonDeleteOK_Click(object sender, EventArgs e)
    {richTextBoxOutput.Text += m_sampleRun.Delete();}//通过 sampleRun.Delete 删除工作空间
```

⑰ 取消删除 Button 控件的 Click 事件。左键双击分析 Button 控件，即可添加如下该按钮的 Click 事件代码，实现取消删除：

```
private void buttonDeleteCancel_Click(object sender, EventArgs e)
    {richTextBoxOutput.Text += "删除取消" + System.Environment.NewLine;}
```

(5) 窗体的 FormClose 事件

在退出程序时，要断开空间的关联，关闭地图窗口、工作空间。因此，在属性窗口点击 ⚡ 图标，左键双击 FormClose 事件，添加以下代码：

```
private void FormMain_FormClosing(object sender, FormClosingEventArgs e)
    {ry{m_workspace.Dispose();}//关闭工作空间
    catch (Exception ex)
        {Trace.WriteLine(ex.Message);}}
```

(6) 获取选择的工作空间类型

```
private WorkspaceType GetType(String type)
    {WorkspaceType result = WorkspaceType.Default;
```

```
                switch ( type. ToUpper( ) )
                    { case "SMW" : {
                            result = WorkspaceType. SMW ; }
                            break ;
                        case "SXW" :
                            { result = WorkspaceType. SXW ; }
                            break ;
                        case "SMWU" :
                            { result = WorkspaceType. SMWU ; }
                            break ;
                        case "SXWU" :
                            { result = WorkspaceType. SXWU ; }
                            break ;
                        default :
                            break ; }
                return result ; }
```

（7）获取选择的版本号

获取选择的版本号需添加以下代码：

```
private WorkspaceVersion GetVersion( Int32 selectIndex )
    { WorkspaceVersion version = WorkspaceVersion. SFC50 ;
                switch ( selectIndex )
                    { case 0 :
                            { version = WorkspaceVersion. UGC60 ; }
                            break ;
                        case 1 :
                            { version = WorkspaceVersion. UGC20 ; }
                            break ;
                        case 2 :
                            { version = WorkspaceVersion. SFC60 ;          }
                            break ;
                        default :
                            break ; }
                return version ; }
```

7. 运行结果

窗体设计完成后点击 ▶ 按钮，运行该程序，运行后弹出窗体，如图4-2所示，可以进行文件型、Oracle、SQL 工作空间的打开、另存、创建及删除。

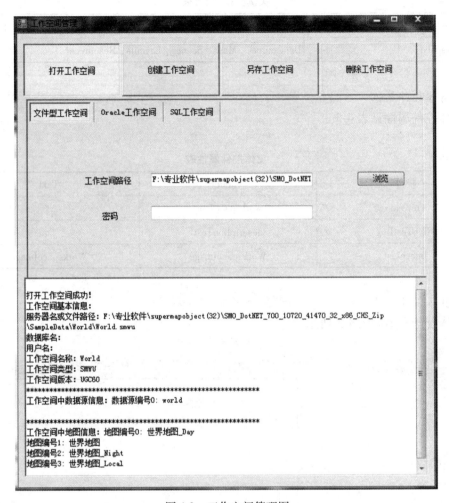

图 4-2 工作空间管理图

4.2 工作树状结构

1. 数据

在安装目录\SampleData\City\Changchun.smwu 路径下打开数据文件。

2. 新建文件夹和工程

在【D:\MyProject\】文件中新建一个文件夹命名为【第 4 章 空间数据的基本操作】，并在其目录下建立一个文件夹命名为【WorkspaceTree】，在此文件夹里新建一个工程，将此工程命名为 WorkspaceTree。

3. 关键类型/成员

关键类型/成员具体见表 4-4。

表 4-4 关键类型/成员表

控件/类	方法	属性	事件
Workspace		Datasources、Maps、Scenes、Layouts、Resources	

4. 窗体控件属性

窗体控件属性见表 4-5。

表 4-5 窗体控件属性表

控件	Name	Text
TreeView	workspaceTree	
RichTextBox	descriptionText	
ImageList	WorkspaceImage	WorkspaceImage

5. 窗体设计布局

窗体设计布局如图 4-3 所示。

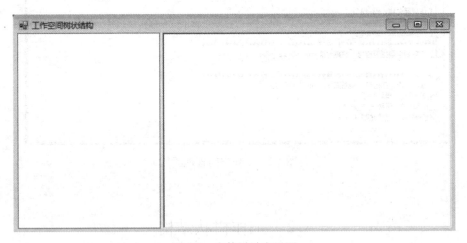

图 4-3 窗体设计布局图

6. 创建 SampleRun 类

(1) 添加 using 引用代码

添加 using 引用代码具体如下：

using SuperMap.Data; //添加 Data 命名空间，使其对应的类可以使用
using SuperMap.Mapping; //添加 Mapping 命名空间，使其对应的类可以使用
using SuperMap.Layout; //添加 Layout 命名空间，使其对应的类可以使用
using SuperMap.Realspace; //添加 Realspace 命名空间，使其对应的类可以使用

(2) 在 public class SampleRun 中添加代码

4.2 工作树状结构

在 public class SampleRun 中添加如下代码：
//定义类对象

```csharp
private Workspace m_workspace;
private TreeView m_workspaceTree;// TreeView 显示标记项的分层集合
///根据 Yworkspace 构造 SampleRun 对象
public SampleRun(Workspace workspace, TreeView workspaceTree)
    {m_workspace = workspace;
    m_workspaceTree = workspaceTree;
    SetWorkspaceTree(m_workspaceTree, m_workspace);}
private void SetWorkspaceTree(TreeView workspaceTree, Workspace workspace)//定义
{SetWorkspaceTree 初始化函数
    { try { workspaceTree.Nodes.Clear();
        // 工作空间节点
        TreeNode workspaceNode = this.CreateNode(workspace.Caption, "Workspace", "Workspace");
        TreeNode datasourcesNode = this.CreateNode("数据源", "Datasources", "Datasources");
        this.SetDatasourcesNode(datasourcesNode, workspace.Datasources);
        TreeNode mapsNode = this.CreateNode("地图", "Maps", "Maps");
        this.SetMapsNode(mapsNode, workspace.Maps);
        TreeNode scenesNode = this.CreateNode("三维场景", "Scenes", "Scenes");
        this.SetScenesNode(scenesNode, workspace.Scenes);
        TreeNode layoutsNode = this.CreateNode("布局", "Layouts", "Layouts");
        this.SetLayoutsNode(layoutsNode, workspace.Layouts);
        TreeNode resourcesNode = this.CreateNode("资源", "Resources", "Resources");
        this.SetResourcesNode(resourcesNode, workspace.Resources);
        // 添加节点
        workspaceNode.Nodes.Add(datasourcesNode);
        workspaceNode.Nodes.Add(mapsNode);
        workspaceNode.Nodes.Add(scenesNode);
        workspaceNode.Nodes.Add(layoutsNode);
        workspaceNode.Nodes.Add(resourcesNode);
        workspaceNode.Expand();
        workspaceTree.Nodes.Add(workspaceNode); }
    catch (Exception ex)
        { Trace.WriteLine(ex.Message); }}
/// 获取数据源子节点
private void SetDatasourcesNode(TreeNode datasourcesNode, Datasources datasources)//
定义{//SetDatasourcesNode 初始化函数
```

```
        {try{//通过判断数据源集合中数据源的个数来决定是否将先前创建的树节点添
加到树节点集合
            if(datasources.Count！=0)
            {for(int i=0; i<datasources.Count; i++)
                {TreeNode sourceNode=this.CreateNode(datasources[i].Alias,"Datasource","
Datasource");
                sourceNode.Tag=datasources[i];
                this.SetDatsourceNode(sourceNode,datasources[i]);
                datasourcesNode.Nodes.Add(sourceNode);//添加树节点}}}
            catch(Exception ex)
                {Trace.WriteLine(ex.Message);}}
        /// 获取maps子节点
    private void SetMapsNode(TreeNode mapsNode, Maps maps)//定义SetMapsNode初始化
函数
        {try{//通过判断地图集合中地图对象的个数来决定是否将先前创建的树节点添
加到树节点集合
            for(int i=0; i<maps.Count; i++)
                {TreeNode mapNode=this.CreateNode(maps[i],"Map","Map");
                mapNode.Tag=maps[i];
                mapsNode.Nodes.Add(mapNode);//添加树节点}}
            catch(Exception ex)
            {Trace.WriteLine(ex.Message);}}
        /// 获取三维场景子节点
    private void SetScenesNode(TreeNode scenesNode, Scenes scenes)//定义SetScenesNode
初始化函数
        {try//通过判断三维场景集合中三维场景的个数来决定是否将先前创建的树节点
添加到树节点集合
            for(int i=0; i<scenes.Count;++i)
                {TreeNode sceneNode=this.CreateNode(scenes[i],"Scene","Scene");
                sceneNode.Tag=scenes[i];
                scenesNode.Nodes.Add(sceneNode);//添加树节点}}
            catch(Exception ex)
                {Trace.WriteLine(ex.Message);}}
        /// 获取布局子节点
    private void SetLayoutsNode(TreeNode layoutsNode, Layouts layouts)//定义SetLayouts
Node 初始化函数
        {try//通过判断布局集合中布局的个数来决定是否将先前创建的树节点添加到树
节点集合
                {for(int i=0; i<layouts.Count;++i)
```

```
            {TreeNode layoutNode=this.CreateNode(layouts[i],"Layout","Layout");
             layoutNode.Tag=layouts[i];
             layoutsNode.Nodes.Add(layoutNode);//添加树节点}}
             catch(Exception ex)
             {Trace.WriteLine(ex.Message);}}
             ///设置资源子节点
    private void SetResourcesNode(TreeNode resourcesNode,Resources resources)//定义
SetResourcesNode 初始化函数
           {try
             {TreeNode pointResNode=this.CreateNode("点符号库","PointLibaray","resource");
             pointResNode.Tag=resources.MarkerLibrary;
             TreeNode lineResNode = this.CreateNode("线符号库"," LineLibaray"," 
resource");
             lineResNode.Tag=resources.LineLibrary;
             TreeNode fillResNode = this.CreateNode("填充符号库"," FillLibaray"," 
resource");
             fillResNode.Tag=resources.FillLibrary;
             resourcesNode.Nodes.Add(pointResNode);//添加树节点
             resourcesNode.Nodes.Add(lineResNode);//添加树节点
             resourcesNode.Nodes.Add(fillResNode);//添加树节点}
             catch(Exception ex)
             {Trace.WriteLine(ex.Message);}}
             ///添加遍历数据源下的数据集,将它们变成节点加入 sourcesNode 之下
    private void SetDatasourceNode(TreeNode sourceNode,Datasource datasource)//定义
SetDatasourceNode 初始化函数
          {try{Datasets datasets=datasource.Datasets;
             foreach(Dataset dataset in datasets)
                {TreeNode datasetNode = this.CreateNode(dataset.Name,GetDatasetTypeImg
(dataset),"Dataset");
             datasetNode.Tag=dataset;
             SetDatasetNode(datasetNode,dataset);//遍历数据集
             sourceNode.Nodes.Add(datasetNode);//添加树节点}    }
             catch(Exception ex)
             {Trace.WriteLine(ex.Message);}}
             ///遍历数据集添加数据集节点
    private void SetDatasetNode(TreeNode datasetNode,Dataset dataset)//定义 SetDatasetNode
初始化函数
             {try
                {if(dataset.Type==DatasetType.Network)
```

```
            {DatasetVector network=dataset as DatasetVector;
             TreeNode childDatasetNode=this.CreateNode(network.ChildDataset.Name,"
Dataset_Point","Dataset");
             childDatasetNode.Tag=network.ChildDataset;
             datasetNode.Nodes.Add(childDatasetNode);//添加树节点}}
         catch(Exception ex)
            {Trace.WriteLine(ex.Message);}}
      ///设置节点的图标
     private void SetNodeImage(TreeNode node,String imageKey)//定义 SetNodeImage 初始化函数
        {node.ImageKey=imageKey;//设置数据节点未选中时的图像关联键
         node.SelectedImageKey=imageKey;//设置数据节点选中是的图像关联键}
         /// 根据数据集的类型确定它们的图标
        private String GetDatasetTypeImg(Dataset dataset)
        {switch(dataset.Type)
         {case DatasetType.Region:
             return "Dataset_Polygon";
          case DatasetType.Text:
             return "Dataset_Text";
          case DatasetType.Line:
             return "Dataset_Line";
          case DatasetType.Point:
             return "Dataset_Point";
          case DatasetType.Network:
             return "Network";
          default:
             return "Adorn";}}
         /// 创建一个节点同时设置节点的 text,imageKey,selectImageKey
        private TreeNode CreateNode(String nodeText,String imageKey,String nodeName)
           {TreeNode node=new TreeNode(nodeText);
            node.ImageKey=imageKey;
            node.SelectedImageKey=imageKey;
            node.Name=nodeName;
         return node;}
         /// 通过 GetNodeInfo 获取节点的相关信息
     public String GetNodeInfo(TreeNode node)
         {String result="";
            try{ switch(node.Name)
             {case "Dataset":
```

```
                result = this.GetDatasetInfo(node.Tag as Dataset);
                break;
            case "Map":
                Map map = new Map(m_workspace);
                map.Open(node.Tag as String);
                result = this.GetMapInfo(map);
                break;
            case "Layout":
                MapLayout layout = new MapLayout(m_workspace);
                layout.Open(node.Tag as String);
                result = this.GetLayoutInfo(layout);
                break;
            case "Scene":
                result = this.GetSceneInfo(node.Tag as String);
                break;
            case "resource":
                result = this.GetResourceInfo(node);
                break;
            default:
                break; } }
        catch (Exception ex)
            { Trace.WriteLine(ex.Message); }
            return result; }
```
/// 获取数据集信息
```
private String GetDatasetInfo(Dataset dataset)
    { StringBuilder result = new StringBuilder(); //创建 StringBuilder 的一个新对象
        try {
            result.AppendLine("所属数据源" + dataset.Datasource.Alias);
            result.AppendLine("数据集名" + dataset.Name);
            result.AppendLine("描述信息" + dataset.Description);
            result.AppendLine("数据表名" + dataset.TableName);
            result.AppendLine("类型" + dataset.Type.ToString());
            result.AppendLine("最小外接矩形" + dataset.Bounds.ToString());
            result.AppendLine("编码方式" + dataset.EncodeType.ToString());
            result.AppendLine("是否已经打开" + dataset.IsOpen.ToString());
            result.AppendLine("是否只读" + dataset.IsReadOnly.ToString()); }
        catch (Exception ex)
            { Trace.WriteLine(ex.Message); }
        return result.ToString();
```

```csharp
// 通过GetMapInfo获取地图信息
private String GetMapInfo(Map map)
    {StringBuilder result = new StringBuilder();
        try{ result.AppendLine("关联的工作空间"+map.Workspace.Caption);
        result.AppendLine("地图名"+map.Name);
        result.AppendLine("描述信息"+map.Description);
        result.AppendLine("空间范围"+map.Bounds.ToString());
        result.AppendLine("中心点"+map.Center.ToString());
        result.AppendLine("颜色模式"+map.ColorMode.ToString());
        result.AppendLine("坐标单位"+map.CoordUnit.ToString());
        result.AppendLine("比例尺"+map.Scale.ToString());
        result.AppendLine("是否反走样地图"+map.IsAntialias.ToString());
            result.AppendLine("自定义地图边界是否有效"+map.IsCustomBoundsEnabled.ToString());
            result.AppendLine("是否动态投影显示"+map.IsDynamicProjection.ToString());
        result.AppendLine("是否被修改"+map.IsModified.ToString());
        result.AppendLine("是否绘制地图背景"+map.IsPaintBackground.ToString());
        result.AppendLine("文本角度是否固定"+map.IsTextAngleFixed.ToString());
        result.AppendLine(" ************************************* ");
        result.AppendLine("图层信息");
        Layers layers = map.Layers;
        result.AppendLine("图层总数"+layers.Count.ToString());
        for(int i=0; i<layers.Count; i++)//通过判断图层集合中图层的个数来判断显示的相关信息
            {Layer layer = layers[i];
        result.AppendFormat("图层编号{0}:{1}\n", i, layer.Caption);
        result.AppendLine("图层名"+layer.Name);
        result.AppendLine("描述信息"+layer.Description);
        result.AppendLine("对应数据集"+layer.Dataset.Name);
        result.AppendLine("过滤条件"+layer.DisplayFilter.ToString());
        result.AppendLine("不透明度"+layer.OpaqueRate.ToString());
        result.AppendLine("制图表达字段"+layer.RepresentationField);
        if(layer.Theme != null)
            {result.AppendLine("专题图"+layer.Theme.ToString());}
        result.AppendLine("是否可见"+layer.IsVisible.ToString());
        result.AppendLine("图层是否可编辑"+layer.IsEditable.ToString());
            result.AppendLine("是否采用制图表达"+layer.IsRepresentationEnabled.ToString());
```

```
            result.AppendLine("图层中对象是否可选择"+layer.IsSelectable.ToString());
            result.AppendLine("是否可捕捉"+layer.IsSnapable.ToString());}}
         catch(Exception ex)
            {Trace.WriteLine(ex.Message);}
            return result.ToString();}
///通过GetSceneInfo获取三维场景信息,目前只是场景名
private String GetSceneInfo(String sceneName)
      {String result=sceneName;
         return result;}
///通过GetSceneInfo获取布局信息
private String GetLayoutInfo(MapLayout layout)
      {StringBuilder result=new StringBuilder();//创建新的StringBuilder对象
         try{ result.AppendLine("关联的工作空间"+layout.Workspace.Caption);
            result.AppendLine("布局名"+layout.Name);
            result.AppendLine("空间范围"+layout.Bounds.ToString());
            result.AppendLine("中心点"+layout.Center.ToString());
            result.AppendLine(" ******************************* \n");
            result.AppendLine("布局元素信息");
            LayoutElements layoutElements=layout.Elements;
            result.AppendLine("布局元素总数"+layoutElements.Count);
            int index=-1;
            while(! layoutElements.IsEOF)
              {index++;
             result.AppendFormat("布局元素编号",index,layoutElements.GetGeometry().Type.ToString());
               layoutElements.MoveNext();//移动指针到下一条布局元素}}
         catch(Exception ex)
            {Trace.WriteLine(ex.Message);}
            return result.ToString();}
///通过GetResourceInfo获取资源信息
private String GetResourceInfo(TreeNode resourceNode)
      {StringBuilder result=new StringBuilder();//创建新的StringBuilder对象
               try{ SymbolLibrary marker=resourceNode.Tag as SymbolLibrary;//
SymbolLibrary:符号库基类。点状符号库类、线型符号库类和填充符号库类都继承自该抽象
类。用来管理符号对象,包括符号对象的添加、删除。
                  if(marker! =null)
                  {SymbolGroup group=marker.RootGroup;// SymbolGroup:符号库分组类。该
类主要为管理符号库中的子分组和符号对象提供一种逻辑化的组织结构形式,从而实现结
构化、层次化的清晰管理模式,该类不能获取系统符号也并不负责符号对象的添加与删除
```

以及符号对象的导入与导出操作，这类操作主要由符号库基类(SymbolLibrary 类)提供。

```
        Int32 count=group.Count;
        for(int i=0;i < count;++i)//根据符号个数来判断符号索引
    {result.AppendFormat("符号索引", i, group[i].Name);}}}
        catch(Exception ex)
    {Trace.WriteLine(ex.Message);}
        return result.ToString();}
```

7. 窗体代码

(1) 添加 using 引用代码

添加 using 引用代码具体如下：

using System.Diagnostics;//添加 Diagnostics 命名空间，使其对应的类可以使用

using SuperMap.Data;//添加 Data 命名空间，使其对应的类可以使用

(2) 实例化代码

实例化 SampleRun、Workspace，定义 m_workspacePath 变量以及 public FormMain() 函数部分的代码如下：

```
//定义类对象
private SampleRun m_sampleRun;
private Workspace m_workspace;
private String m_workspacePath;
public FormMain()
    {try{InitializeComponent();
        m_workspacePath="../../SampleData/City/Changchun.smwu";
        m_workspace=new Workspace();
        m_workspace.Open(new WorkspaceConnectionInfo(m_workspacePath));
        m_sampleRun=new SampleRun(m_workspace, workspaceTree);//创建新的 SampleRun 对象
         workspaceTree.NodeMouseClick+=new TreeNodeMouseClickEventHandler(workspaceTree_NodeMouseClick);
        this.Load+=new EventHandler(FormMain_Load);}
        catch(Exception ex)
    {Trace.WriteLine(ex.Message);}}
```

(3) 其他控件的代码

workspaceTree 的 NodeMouseClick 事件。在控件的属性窗口单击 ⚡ 图标，左键双击 NodeMouseClick，添加如下代码，来显示相应对象的信息：

```
private void workspaceTree_NodeMouseClick(object sender, TreeNodeMouseClickEventArgs e)
        {try{descriptionText.Text=m_sampleRun.GetNodeInfo(e.Node);//通过 sampleRun.GetNodeInfo 来获取信息}
```

catch(Exception ex)

{Trace.WriteLine(ex.Message);}}

(4)窗体的 FormClose 事件

在退出程序时,要断开空间的关联,释放对象。因此,在属性窗口中双击 FormClose 事件,添加以下代码:

private void FormMain_FormClosing(object sender,FormClosingEventArgs e)

　　{try

　　{if(m_workspace!=null)

　　{m_workspace.Dispose();//关闭工作空间释放资源}}

catch(Exception ex)

{Trace.WriteLine(ex.Message);}}

8. 运行结果

窗体设计完成后点击▶按钮,运行该程序,运行后弹出窗体,如图4-4所示。图中显示的是工作空间的树状结构,包括数据源、地图、三维场景、布局、符号库。

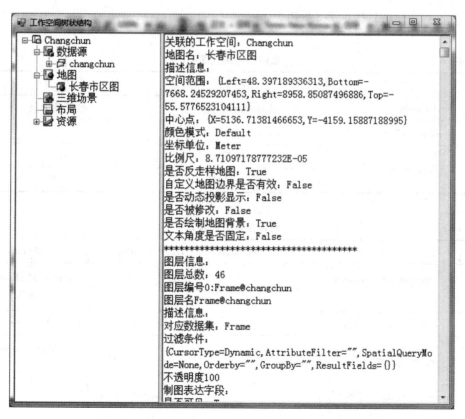

图4-4　工作空间树状结构

4.3 数据源管理

1. 新建文件夹和工程

在【D:\MyProject\】文件中新建一个文件夹，命名为【第 4 章　空间数据的基本操作】，并在其目录下建立一个文件夹，命名为【DatasourceManage】，在此文件夹里新建一个工程，将此工程命名为 DatasourceManage。

2. 关键类型/成员

关键类型/成员见表 4-6。

表 4-6　关键类型/成员表

控件/类	方法	属性	事件
Datasources	Open、Create		

3. 窗体控件属性

窗体控件的属性具体见表 4-7 和表 4-8。

表 4-7　窗体控件属性表 1

控件	Name	Text
GroupBox	datasourceGroup	数据源信息
RichTextBox	datasourceText	
TabControl	operationTabCtrl	打开
	operationTabCtrl	新建

表 4-8　窗体控件属性表 2

Text	Label	ComboBox	TextBox	Button
打开	openEngineTypeLabel（引擎类型：）	openEngineTypeCombox		openFolderBrowser（浏览）
	openPathLabel（路径：）		openPathText	openFileSource（打开）
	openKeyText（密匙：）		openKeyTable	openDatabaseSource（打开）
	openServerLabel（服务器名）		openServerText	
	openDatabaseLabel（数据库名）		openDatabaseText	
	openUserNameLabel（用户名）		openUserNameText	
	openPasswordLabel（密码）		openPasswordText	

续表

Text	Label	ComboBox	TextBox	Button
新建	createTypeLabel（引擎类型：）	createEngineTypeCombox		createFileFolderBrowser（浏览）
	createFilePathLabel（路径：）		createSourceText	createFileSource（创建）
	createSourceNameLabel（数据源名：）		createSourceText	createDatabaseSource（创建）
	createServerLabel（服务器名：）		createServerText	
	createDatabaseLabel（数据库名）		createDatabaseText	
	createUserLabel（用户名）		createUserText	
	createPasswordLabel（密码）		createPasswordText	

4. 窗体设计布局

窗体设计布局如图 4-5 所示。

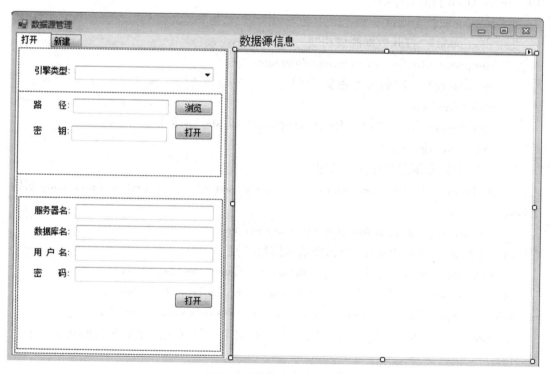

图 4-5　窗体设计布局图

5. 创建 SampleRun 类

（1）添加 using 引用代码

添加 using 引用代码具体如下：
using SuperMap.Data;//添加 Data 命名空间，使其对应的类可以使用
（2）在 public class SampleRun 中添加代码
在 public class SampleRun 中添加代码如下：
//定义类对象
private Workspace m_workspace;
///根据 Yworkspace 构造 SampleRun 对象
public SampleRun(Workspace workspace)
　　{m_workspace=workspace;}
///根据数据源连接信息打开数据源，返回相应的描述信息
public String OpenDatasource(DatasourceConnectionInfo datasourceInfo)
　　{m_workspace.Datasources.CloseAll();
String discription="";
if(m_workspace!=null)
　{try
　{Datasource datasource=m_workspace.Datasources.Open(datasourceInfo);//通过 Datasources.Open 打开数据源
　if(datasource!=null)
　{discription="打开数据源成功";
　discription+=GetDatasourceString(datasource);}
　else{discription="打开数据源失败";}}
　catch(Exception e)
　{discription="打开数据源失败！异常信息"+e.Message;}}
　return discription;}
　/// 获取数据源的描述字符串
private String GetDatasourceString(Datasource datasource)//定义 GetDatasourceString 初始化函数
　　{StringBuilder dsString=new StringBuilder();
try{dsString.AppendLine("数据源基本信息:");
dsString.AppendLine("实例名称:"+datasource.ConnectionInfo.Server);
dsString.AppendLine("引擎类型:"+datasource.EngineType.ToString());
dsString.AppendLine("***");
if(datasource.Datasets.Count!=0)//根据数据源中数据集的个数来判断所显示的数据集信息
　{dsString.AppendLine("数据源中数据集信息:");
foreach(Dataset dataset in datasource.Datasets)
　{dsString.AppendFormat("数据集名称:{0}\t{1}\n",dataset.Name,dataset.Type.ToString());}}
　else{dsString.AppendLine("数据源中没有数据集");}}

```
    catch(Exception ex)
        {Trace.WriteLine(ex.Message);}
            return dsString.ToString();}
            /// 根据数据源连接信息创建数据源,返回相应的描述信息
    public String CreateDatasource(DatasourceConnectionInfo datasourceInfo)
            {String discription="";
        try{ m_workspace.Datasources.CloseAll();
            Datasource datasource=m_workspace.Datasources.Create(datasourceInfo);通过
Datasources.Create 创建数据源
            if(datasource! =null)
            {discription="创建数据源成功!";
            discription+=GetDatasourceString(datasource);}
            else
            {discription="创建数据源失败!";}}
        catch(Exception e)
            {discription="创建数据源失败! 异常信息"+e.Message;}}
            return discription;}
```

6. 窗体代码

(1)添加 using 引用代码

添加 using 引用代码具体如下:

```
using System.Diagnostics;//添加 Diagnostics 命名空间,使对应的类可以使用
using SuperMap.Data;//添加 Data 命名空间,使对应的类可以使用
using System.IO;//添加 IO 命名空间,使对应的类可以使用
```

(2)实例化代码

实例化 SampleRun、Workspace,public FormMain 函数部分的代码如下:

```
//定义类对象
private SampleRun m_sampleRun;
private Workspace m_workspace;
private DatasourceConnectionInfo m_datasourceInfo;
private String[] m_engineType;
public FormMain()
    {try{ InitializeComponent();
    Initialize();}
    catch(Exception ex)
    {Trace.WriteLine(ex.Message);}}}
```

(3)初始化控件

初始化控件需添加如下代码:

```
private void Initialize()
    {try
```

```csharp
{m_workspace = new Workspace();//定义相应类的新对象
    m_datasourceInfo = new DatasourceConnectionInfo();
    m_sampleRun = new SampleRun(m_workspace);
    if(m_sampleRun != null)
    {//设置panel的位置和是否可见并初始化控件
        openDatabasePanel.Location = new Point(0,45);
        openFilePanel.Location = new Point(0,45);
        openFilePanel.Visible = true;
        openDatabasePanel.Visible = false;
        m_engineType = new String[11];
        m_engineType[0] = "ImagePlus 引擎";
        m_engineType[1] = "UDB 引擎";
        m_engineType[2] = "OraclePlus 引擎";
        m_engineType[3] = "SQLPlus 引擎";
        m_engineType[4] = "PostgreSQL 引擎";
        m_engineType[5] = "DB2 引擎";
        m_engineType[6] = "OGC 引擎(WMS)";
        m_engineType[7] = "GoogleMaps 引擎";
        m_engineType[8] = "SuperMapCloud 引擎";
        m_engineType[9] = "REST 引擎";
        m_engineType[10] = "MapWorld 引擎";
        InitializeOpenEngineType();
        this.Resize += new EventHandler(FormMain_Resize);
        // 打开界面的相关事件
        this.openFileSource.Enabled = false;
        this.openEngineTypeCombox.SelectedIndexChanged += new
    EventHandler(openEngineTypeCombox_SelectedIndexChanged);
        this.openFolderBrowser.Click += new EventHandler(openFolderBrowser_Click);
    this.openFileSource.Click += new EventHandler(openFileSource_Click);
    this.openDatabaseSource.Click += new EventHandler(openDatabaseSource_Click);
    this.openPathText.TextChanged += new EventHandler(openPathText_TextChanged);
    // 创建界面的相关事件
    this.operationTabCtrl.SelectedIndexChanged + = new EventHandler(operationTabCtrl_SelectedIndexChanged);
        this.createEngineTypeCombox.SelectedIndexChanged += new
    EventHandler(createEngineTypeCombox_SelectedIndexChanged);
        this.createFileFolderBrowser.Click += new EventHandler(createFileFolderBrowser_Click);
        this.createSourceText.TextChanged += new EventHandler(createSourceText_TextChanged);
        this.createFileSource.Click += new EventHandler(createFileSource_Click);
```

```
       this.createDatabaseSource.Click+=new EventHandler(createDatabaseSource_Click);}
       else{ MessageBox.Show("初始化失败!");}}
    catch(Exception ex)
       {Trace.WriteLine(ex.Message);}}
```

(4)窗体的 Resize 事件。在窗体的属性窗口单击 图标,左键双击 Resize,添加如下代码,重新设置窗口大小:

```
    private void FormMain_Resize(object sender,EventArgs e)
        {try
{datasourceGroup.Width=this.Width - operationTabCtrl.Width - 10;//重新设置窗口的大小
        }
    catch(Exception ex)
       {Trace.WriteLine(ex.Message);}}
```

(5)定义 InitializeOpenEngineType 初始化函数

定义 InitializeOpenEngineType 初始化函数需添加如下代码:

```
    private void InitializeOpenEngineType()//定义 InitializeOpenEngineType 初始化函数
        {try{//设置控件的状态
            openEngineTypeCombox.Items.AddRange(m_engineType);
            openEngineTypeCombox.SelectedIndex=1;
            openKeyTable.Visible=false;
            openKeyText.Visible=false;}
        catch(Exception ex)
           { Trace.WriteLine(ex.Message);}}
```

(6)根据 typename 获取对应的 EngineType

根据 typename 获取对应的 EngineType,需添加如下代码:

```
    private EngineType GetEngineType(Int32 engineIndex)
        {EngineType type=EngineType.UDB;
        switch(engineIndex)
            { case 0:
               { type=EngineType.ImagePlugs; }
              break;
              case 1:
               { type=EngineType.UDB; }
              break;
              case 2:
               { type=EngineType.OraclePlus;}
               break;
              case 3:
               { type=EngineType.SQLPlus; }
```

```
            break;
            case 4:
            {type=EngineType.PostgreSQL;}
            break;
            case 5:
            {type=EngineType.DB2;}
            break;
            case 6:
            { type=EngineType.OGC;}
            break;
            case 7:
            { type=EngineType.GoogleMaps;}
            break;
            case 8:
             { type=EngineType.SuperMapCloud;}
               break;
               case 9:
               { type=EngineType.iServerRest;}
            break;
            default:
              { type=EngineType.MapWorld;}
                break;}
               return type;}
```

(7) 其他控件的代码

① openEngineTypeCombox 的 SelectedIndexChanged 事件。在控件的属性窗口单击 图标，左键双击 SelectedIndexChanged，添加如下代码：

```
//设置不同引擎下界面上应该显示的东西
            private void openEngineTypeCombox_SelectedIndexChanged(object sender, EventArgs e)
                {try{
            Boolean isVisible=(openEngineTypeCombox.Text.CompareTo(m_engineType[2])==0
                    || openEngineTypeCombox.Text.CompareTo(m_engineType[3])==0
                    || openEngineTypeCombox.Text.CompareTo(m_engineType[4])==0
                    || openEngineTypeCombox.Text.CompareTo(m_engineType[5])==0);
                    openDatabasePanel.Visible=isVisible;
                    openFilePanel.Visible=! isVisible;
                    if(openFilePanel.Visible)
                        {   openPathText.Text="";
                         if(openEngineTypeCombox.Text.CompareTo(m_engineType[6])==0
                          || openEngineTypeCombox.Text.CompareTo(m_engineType[9])==0)
```

```
                    { openPathLabel.Text="网址";
                      openKeyText.Visible=false;
                      openKeyTable.Visible=false;
                      openFolderBrowser.Visible=false;}
                    else if(openEngineTypeCombox.Text.CompareTo(m_engineType
[7])==0)
                    { openPathLabel.Text="服务地址";
                      openPathText.Text="http://maps.google.com";
                      openPathText.Enabled=false;
                      openKeyText.Visible=true;
                      openKeyTable.Visible=true;

                      openFolderBrowser.Visible=false;}
                    else if(openEngineTypeCombox.Text.CompareTo(m_engineType
[8])==0
                        || openEngineTypeCombox.Text.CompareTo(m_engineType
[10])==0)
                    { openPathLabel.Text="服务地址";
                      openPathText.Enabled=false;
                      openKeyText.Visible=false;
                      openKeyTable.Visible=false;
                      openFolderBrowser.Visible=false;
                      if(openEngineTypeCombox.Text.CompareTo(m_engineType
[8])==0)
                      { openPathText.Text="http://beijing.supermapcloud.com";}
                      else
                      { openPathText.Text="http://www.tianditu.com";}}
                    else
                    {  openPathLabel.Text="路径";
                      openPathText.Enabled=true;
                      openKeyText.Visible=false;
                      openKeyTable.Visible=false;
                      openFolderBrowser.Visible=true;}
                }
                else
                {
                    openServerText.Text="";
                    openDatabaseText.Text="";
```

```
                    openUserNameText.Text="";
                    openPasswordText.Text="";
                    if(openEngineTypeCombox.Text.CompareTo(m_engineType[5])==0)
                    {
                        openServerText.Enabled=false;
                    }
                    else
                    {
                        openServerText.Enabled=true;
                    }
                }
                //根据所选 type 设置连接信息的 EngineType
                m_datasourceInfo=new DatasourceConnectionInfo();
                m_datasourceInfo.EngineType=GetEngineType(openEngineTypeCombox.SelectedIndex);
            }
            catch(Exception ex)
            {
                Trace.WriteLine(ex.Message);
            }
        }//设置不同引擎下界面上应该显示的东西
```

② openFolderBrowser 的 click 事件。在控件的属性窗口单击 图标。左键双击 click，添加如下代码，添加打开路径：

```
private void openFolderBrowser_Click(object sender,EventArgs e)
{ try{ OpenFileDialog fileDlg=new OpenFileDialog();//创建 OpenFileDialog 的新对象
    EngineType type=GetEngineType(openEngineTypeCombox.SelectedIndex);
    switch(type)
    {case EngineType.ImagePlugins:
        fileDlg.Filter="支持的影像文件(*.tif,*.sit,*.bmp,*.png,*.sct,*.tga,*.raw)|*.tif;*.sit;*.bmp;*.png;*.sct;*.tga;*.raw";
        break;
    case EngineType.UDB:
        fileDlg.Filter="UDB 数据文件(*.udb)|*.udb";
        break;
    default:
        fileDlg.Filter="所有文件|*.*";
        break;}
```

// 该按钮只在文件型数据源时有效
```
    if ( fileDlg. ShowDialog( ) = = DialogResult. OK)
    { openPathText. Text = fileDlg. FileName; } }
    catch ( Exception ex )
    { Trace. WriteLine( ex. Message) ; } }
```

③ openFileSource 的 click 事件。在控件的属性窗口单击 ≠ 图标,左键双击 click,添加如下代码,实现打开文件型数据源:

```
private void openFileSource_Click( object sender, EventArgs e)
    { try
{ openFileSource. Enabled = false;
datasourceText. Text = "数据源打开中.... \n";
datasourceText. Update( );
if ( m_datasourceInfo. EngineType = = EngineType. OGC)
{// 这里只允许设置 WMS 和 WFS 两种类型具体看发布的网络数据源类型
    m_datasourceInfo. Database = "WMS"; }
    m_datasourceInfo. Server = openPathText. Text;
    datasourceText. Text = m_sampleRun. OpenDatasource ( m_datasourceInfo); 通过 sampleRun. OpenDatasource 打开数据源
    openFileSource. Enabled = true; }
    catch ( Exception ex )
        { Trace. WriteLine( ex. Message) ; }
        m_workspace. Datasources. CloseAll( );}
```

④ openDatabaseSource 的 click 事件。在控件的属性窗口单击 ≠ 图标,左键双击 click,添加如下代码,实现打开数据库型数据源:

```
private void openDatabaseSource_Click( object sender, EventArgs e)
    {try { openDatabaseSource. Enabled = false;
    datasourceText. Text = "数据源打开中.... \n";
    datasourceText. Update( );
//数据源的各种信息如用户名、密码等
    m_datasourceInfo. Server = openServerText. Text;
    m_datasourceInfo. Database = openDatabaseText. Text;
    m_datasourceInfo. User = openUserNameText. Text;
    m_datasourceInfo. Password = openPasswordText. Text;
    if ( m_datasourceInfo. EngineType = = EngineType. SQLPlus)
        { m_datasourceInfo. Driver = "SQL Server"; }
    datasourceText. Text = m_sampleRun. OpenDatasource ( m_datasourceInfo); 通过 sampleRun. OpenDatasource 创建数据源
        openDatabaseSource. Enabled = true; }
        catch ( Exception ex )
```

```
        { Trace. WriteLine( ex. Message ) ; }
        m_workspace. Datasources. CloseAll( ) ;}
```

⑤ openPathText 的 TextChanged 事件。在控件的属性窗口单击 图标,左键双击 TextChanged,添加如下代码:

```
private void openPathText_TextChanged( object sender, EventArgs e)
    {//设置控件的有效性
        if ( openPathText. Text. Length ! =0)
         {openFileSource. Enabled = true ; }
        else
         { openFileSource. Enabled = false ; } }
```

⑥ operationTabCtrl 的 SelectedIndexChanged 事件。在控件的属性窗口单击 图标,左键双击 TextChanged,添加如下代码:

```
//当切换到创建页的时候,才初始化该界面上的元素即初始化控件
private void operationTabCtrl_SelectedIndexChanged( object sender, EventArgs e)
    { try
    { if ( operationTabCtrl. SelectedIndex = =1 & createEngineTypeCombox. Items. Count = =0 )
     { createEngineTypeCombox. Items. Add( m_engineType[1] );
       createEngineTypeCombox. Items. Add( m_engineType[2] );
       createEngineTypeCombox. Items. Add( m_engineType[3] );
       createEngineTypeCombox. Items. Add( m_engineType[4] );
       createEngineTypeCombox. Items. Add( m_engineType[5] );
       createEngineTypeCombox. SelectedIndex = 0 ;
       createFileSourcePanel. Location = new Point( 0, 45 ) ;
       createFileSourcePanel. Visible = true ;
       createDatabasePanel. Location = new Point( 0, 45 ) ;
       createDatabasePanel. Visible = false ; } }
    catch ( Exception ex )
    { Trace. WriteLine( ex. Message ) ; } }
```

⑦ createEngineTypeCombox 的 SelectedIndexChanged 事件。在控件的属性窗口单击 图标,左键双击 TextChanged,添加如下代码:

```
//创建数据源根据选择不同的引擎,显示不同的界面
private void createEngineTypeCombox_SelectedIndexChanged( object sender, EventArgs e)
    { try
    {Boolean isVisible = ( createEngineTypeCombox. Text. CompareTo( m_engineType[2] ) = =0
       || createEngineTypeCombox. Text. CompareTo( m_engineType[3] ) = =0
       || createEngineTypeCombox. Text. CompareTo( m_engineType[4] ) = =0
       || createEngineTypeCombox. Text. CompareTo( m_engineType[5] ) = =0 ) ;
       createDatabasePanel. Visible = isVisible ;
       createFileSourcePanel. Visible = ! isVisible ;
```

```
            if(createDatabasePanel.Visible)
            {createServerText.Text="";
        createDatabaseText.Text="";
        createUserText.Text="";
        createPasswordText.Text="";
            if(createEngineTypeCombox.Text.CompareTo(m_engineType[5])==0)
            {createServerText.Enabled=false;}
            else
            {createServerText.Enabled=true;}}
            else
            {createFilePathText.Text="";
        createSourceText.Text="";}
            //根据所选 type 设置连接信息 EngineType
        m_datasourceInfo=new DatasourceConnectionInfo();
        m_datasourceInfo.EngineType = GetEngineType( createEngineTypeCombox.
SelectedIndex+1);
            createFileSource.Enabled=false;}
            catch(Exception ex)
            {Trace.WriteLine(ex.Message);}}
```

⑧ createFileFolderBrowser 的 Click 事件。在控件的属性窗口单击 图标,左键双击 Click,添加如下代码:

```
//通过 FolderBrowserDialog 设置创建数据源的路径
private void createFileFolderBrowser_Click(object sender, EventArgs e)
    {try{
    FolderBrowserDialog folderDlg=new FolderBrowserDialog();
        if(folderDlg.ShowDialog()==DialogResult.OK)
        {   createFilePathText.Text=folderDlg.SelectedPath;}}
        catch(Exception ex)
        {Trace.WriteLine(ex.Message);}}
```

⑨ createSourceText 的 TextChanged 事件。在控件的属性窗口单击 图标,左键双击 TextChanged,添加如下代码:

```
//控制创建按钮是否可用,如果有创建路径则按钮可见,否则为不可见
private void createSourceText_TextChanged(object sender, EventArgs e)
    {if(createSourceText.Text.Length!=0)
    {createFileSource.Enabled=true;}
else
    {createFileSource.Enabled=false;}}
```

⑩ createDatabaseSourc 的 Click 事件。在控件的属性窗口单击 图标，左键双击 Click，添加如下代码：

//通过 sampleRun. CreateDatasource 创建数据库数据源

```
private void createDatabaseSource_Click(object sender, EventArgs e)
    {try{ createDatabaseSource.Enabled=false;
        datasourceText.Text="数据源创建中....\n";
        datasourceText.Update();
        //数据源的各种信息
        m_datasourceInfo.Server=createServerText.Text;
        m_datasourceInfo.Database=createDatabaseText.Text;
        m_datasourceInfo.User=createUserText.Text;
        m_datasourceInfo.Password=createPasswordText.Text;
        if(m_datasourceInfo.EngineType==EngineType.SQLPlus)
          { m_datasourceInfo.Driver="SQL Server";}
        datasourceText.Text=m_sampleRun.CreateDatasource(m_datasourceInfo);
        createDatabaseSource.Enabled=true; }
    catch(Exception ex)
        { Trace.WriteLine(ex.Message); } }
```

⑪ createFileSource 的 Click 事件。在控件的属性窗口单击 图标，左键双击 Click，添加如下代码：

//创建文件型数据源

```
private void createFileSource_Click(object sender, EventArgs e)
    {try{ createFileSource.Enabled=false;
        datasourceText.Text="数据源创建中....\n";
        datasourceText.Update();
         m_datasourceInfo.Server=createFilePathText.Text+"\\"+createSourceText.Text;
        datasourceText.Text=m_sampleRun.CreateDatasource(m_datasourceInfo);
```
通过 sampleRun. CreateDatasource 创建数据源

```
        createFileSource.Enabled=true; }
    catch(Exception ex)
        { Trace.WriteLine(ex.Message); }}
```

（8）窗体的 FormClose 事件

在退出程序时，要断开空间的关联，关闭地图窗口、工作空间。因此在属性窗口点击 图标，左键双击 FormClose 事件，添加代码如前文相应内容。

7. 运行结果

窗体设计完成后点击 ▶ 按钮，运行该程序，运行后弹出窗体，如图 4-6 所示，图中显示的是打开和新建不同类型的数据源。

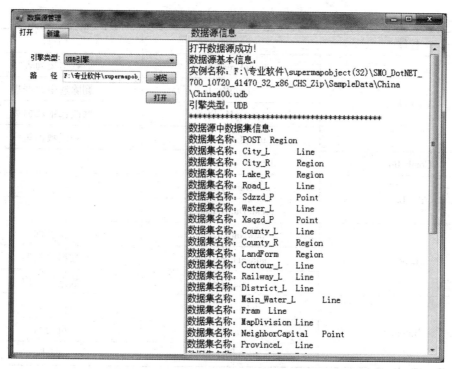

图 4-6　数据源管理

4.4　数据集管理

1. 数据

在安装目录\SampleData\World\World.smwu 路径下打开数据文件。

2. 新建文件夹和工程

在【D:\MyProject\】文件中新建一个文件夹，命名为【第 4 章　空间数据的基本操作】，并在其目录下建立一个文件夹，命名为【DatasetManage】，在此文件夹里新建一个工程，将此工程命名为 DatasetManage。

3. 关键类型/成员

关键类型/成员具体见表 4-9。

表 4-9　　　　　　　　　　　　　　关键类型/成员表

控件/类	方法	属性	事件
Datasets	Delete、Rename、Create		

4. 窗体控件属性

窗体控件属性见表 4-10。

表4-10　　　　　　　　　　　　　窗体控件属性表

控件	Name	Text
GroupBox	groupCreateNew	数据集列表
	groupBox1	删除选中数据集
	groupModify	修改选中数据集
	groupCreateNew	新建数据集
ComboBox	datasetTypeCombox	
TextBox	newNameTextBox	
	createDatasetName	
Label	newNameLabel	新名称
	datasetTypeLabel	数据集类型
	dataName	数据集名称
Button	deleteCurrent	删除选中
	changeName	重命名
	createDataset	创建

5. 窗体设计布局

窗体设计布局如图4-7所示。

6. 创建 SampleRun 类

(1) 添加 using 引用代码

添加 using 引用代码具体如下：

using SuperMap. Data;//添加 Data 命名空间，使其对应的类可以使用

using System. IO；//添加 IO 命名空间，使其对应的类可以使用

using System. Windows. Forms；//添加 System. Windows. Forms 命名空间，使其对应的类可以使用

(2) 在 public class SampleRun 中添加代码

在 public class SampleRun 中添加如下代码：

private Workspace m_workspace；//定义类对象

//保存拷贝后的数据源

private Datasource m_datasource；

//构造函数，需要传入可用的工作空间和数据源连接信息

　　public SampleRun(Workspace workspace，DatasourceConnectionInfo dsInfo)
　　　　｛try
　　　　　｛this. Initialize(workspace)；
　　　　　　if (m_workspace. Datasources. Open(dsInfo) == null)
　　　　　　　｛MessageBox. Show("数据打开失败")；｝

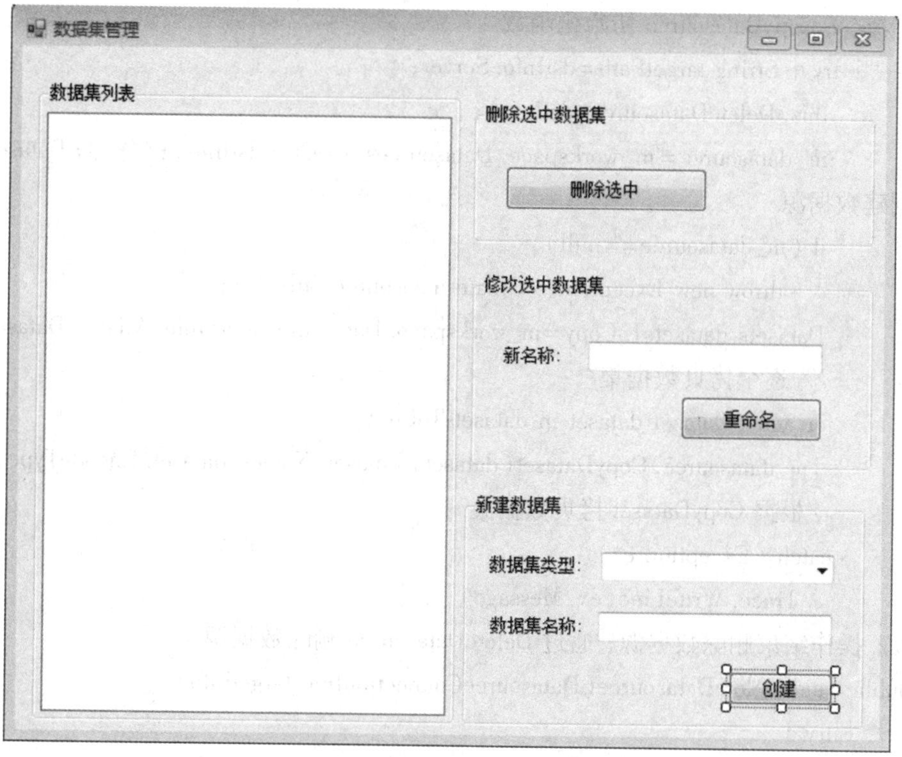

图 4-7 窗体设计布局图

```
                }
            catch（Exception ex）
            ｛Trace.WriteLine（ex.Message）；｝｝
/// 初始化成员
private void Initialize（Workspace workspace）
    ｛m_datasource = null；
        m_workspace = workspace；｝
/// 防止破坏数据，此处拷贝一份新的
public void CopyDatasource（DatasourceConnectionInfo dstInfo）
    ｛try｛if（m_workspace.Datasources.Count == 0）
            ｛m_datasource = null；
        return；｝
        CopyDatasource（m_workspace.Datasources[0].ConnectionInfo，dstInfo）｝；
        //通过 CopyDatasource 拷贝
        catch（Exception ex）
            ｛Trace.WriteLine（ex.Message）；｝｝
/// 拷贝数据源
```

```csharp
private void CopyDatasource(DatasourceConnectionInfo srcInfo, DatasourceConnectionInfo dstInfo)//定义 CopyDatasource 初始化函数
{ try { String targetPath = dstInfo.Server;
    this.DeleteDatasource(dstInfo);
    m_datasource = m_workspace.Datasources.Create(dstInfo);//根据 Datasources.Create 创建数据源
    if(m_datasource == null)
    { throw new Exception("Create datasource failed"); }
    Datasets datasetsToCopy = m_workspace.Datasources[srcInfo.Alias].Datasets;
    // 逐个拷贝数据集
    foreach(Dataset dataset in datasetsToCopy)
    {m_datasource.CopyDataset(dataset, dataset.Name, dataset.EncodeType);}}
    //根据 CopyDataset 拷贝数据集
    catch(Exception ex)
    { Trace.WriteLine(ex.Message); } }
/// 操作结束删除数据源，通过 DeleteDatasource 删除数据源
public void DeleteDatasource(DatasourceConnectionInfo targetInfo)
{ try {
    this.CloseDatasource(targetInfo);
    String targetPath = targetInfo.Server;
    File.Delete(targetPath);
    String sddPath = Path.ChangeExtension(targetPath, "udd");
    File.Delete(sddPath); }
    catch(Exception ex)
    { Trace.WriteLine(ex.Message); } }
/// 关闭数据源
private void CloseDatasource(DatasourceConnectionInfo targetInfo)
{ try
    {if(m_datasource != null)
    {m_workspace.Datasources.Close(targetInfo.Alias);}}
    //通过 Datasources.Close 关闭数据源
    catch(Exception ex)
    { Trace.WriteLine(ex.Message); } }
/// 获取数据源中所有数据集的名称
public String[] GetDatasetsNames()
{ try
    { if(m_datasource != null)
```

```
            { String[ ] datasetNames=new String[m_datasource.Datasets.Count];
              Int32 index=0;
              foreach (Dataset dataset in m_datasource.Datasets)
              { datasetNames[index++]=dataset.Name;}
                return datasetsNames;} }
              catch (Exception ex)
              { Trace.WriteLine(ex.Message);}
                return null;}
```
// 删除数据集
```
public Boolean DeleteDataset(String datasetName)
    { try {   if (m_datasource ！=null)
             { return m_datasource.Datasets.Delete(datasetName);} }
              //通过 Datasets.Delete 删除数据集
              catch (Exception ex)
                 { Trace.WriteLine(ex.Message);  }
                return false;}
```
/// 数据集重命名
```
public Boolean RenameDataset(String srcName, String targetName)
      { try{ if (m_datasource ！=null)
              { if (! m_datasource.Datasets.IsAvailableDatasetName(targetName))
                  { MessageBox.Show("该名字已经存在或不合法");   }
                   else
                     { return m_datasource.Datasets.Rename(srcName, targetName);//
通过 Datasets.Rename 重命名数据集} } }
              catch (Exception ex)
                 { Trace.WriteLine(ex.Message);}
            return false;}
```
/// 通过 CreateDataset 创建数据集
```
public Boolean CreateDataset(DatasetType datasetType, String datasetName)
       { Boolean result=false;
            if (m_datasource==null)
              { return result;}
        // 首先要判断输入的名字是否可用
         if (! m_datasource.Datasets.IsAvailableDatasetName(datasetName))
              { MessageBox.Show("该名字已经存在或不合法");
                   return result;}
        Datasets datasets=m_datasource.Datasets;// Datasets:所有数据集类型(如矢量
```

数据集、栅格数据集等)的基类。提供各数据集共有的属性、方法和事件。

  DatasetVectorInfo vectorInfo = new DatasetVectorInfo( );//DatasetVectorInfo:矢量数据集信息类。包括了矢量数据集的信息,如矢量数据集的名称、数据集的类型、编码方式、是否选用文件缓存等。文件缓存只针对图库索引而言。

  vectorInfo. Name = datasetName;

  try { //point 等为 Vector 类型,类型是一样的,可以统一处理

  switch (datasetType)// DatasetType:该枚举定义了数据集类型常量。

//数据集一般为存储在一起的相关数据的集合;根据数据类型的不同,分为矢量数据集、栅格数据集和影像数据集,以及为了处理特定问题而设计的如拓扑数据集,网络数据集等。根据要素的空间特征的不同,矢量数据集又分为点数据集、线数据集、面数据集、CAD 数据集、文本数据集、纯属性数据集等。

  { case DatasetType. Point:

  case DatasetType. Line:

  case DatasetType. CAD:

  case DatasetType. Region:

  case DatasetType. Text:

  case DatasetType. Tabular://纯属性数据集

  { vectorInfo. Type = datasetType;

  if (datasets. Create(vectorInfo)！ = null)

  result = true; }

  break;

  case DatasetType. Grid://栅格数据

  { DatasetGridInfo datasetGridInfo = new DatasetGridInfo( );//DatasetGridInfo:栅格数据集信息类。包括了返回和设置栅格数据集的相应的属性信息等,例如栅格数据集的名称、宽度、高度、像素格式、编码方式、存储分块大小和空值等。

  datasetGridInfo. Name = datasetName;

  datasetGridInfo. Height = 200;

  datasetGridInfo. Width = 200;

  datasetGridInfo. NoValue = 1.0;

  datasetGridInfo. PixelFormat = PixelFormat. Single;// PixelFormat:该枚举定义了栅格与影像数据存储的像素格式类型常量。Sigel:每个像元用 4 个字节来表示。可表示 $-3.402823E+38$ 到 $3.402823E+38$ 范围内的单精度浮点数。

  datasetGridInfo. EncodeType = EncodeType. LZW;// EncodeType:该枚举定义了数据集存储时的压缩编码方式类型常量。

  //对矢量数据集,支持四种压缩编码方式,即单字节、双字节、三字节和四字节编码方式,这四种压缩编码方式采用相同的压缩编码机制,但是压缩的比率不同。其均为有损压缩。需要注意的是点数据集、纯属性数据集以及 CAD 数据集不可压缩编码。对光栅数

据,可以采用四种压缩编码方式,即 DCT、SGL、LZW 和 COMPOUND。其中 DCT 和 COMPOUND 为有损压缩编码方式,SGL 和 LZW 为无损压缩编码方式。

LZW:LZW 是一种广泛采用的字典压缩方法,其最早是用在文字数据的压缩方面。LZW 的编码原理是用代号来取代一段字符串,后续的相同的字符串就使用相同代号,所以,该编码方式不仅可以对重复数据起到压缩作用,还可以对不重复数据进行压缩操作。适用于索引色影像的压缩方式,这是一种无损压缩编码方式。(适用于栅格和影像数据)

  if (datasets. Create(datasetGridInfo) ! = null)

 result = true ; } break ;

 case DatasetType. Image ://影像数据集

 { DatasetImageInfo datasetImageInfo = new DatasetImageInfo( ) ;//DatasetImageInfo:影像数据集信息类。该类用于设置影像数据集的创建信息,包括名称、宽度、高度、波段数和存储分块大小等。

  datasetImageInfo. Name = datasetName ;

  datasetImageInfo. BlockSizeOption = BlockSizeOption. BS_128 ;// BlockSizeOption:该枚举定义了栅格数据集的栅格块大小的类型常量。BS-128:表示 128 像素 * 128 像素的分块。

  datasetImageInfo. Height = 200 ;

  datasetImageInfo. Width = 200 ;

  //datasetImageInfo. Palette = Colors. MakeRandom(10) ;

  datasetImageInfo. EncodeType = EncodeType. None ;// EncodeType:该枚举定义了数据集存储时的压缩编码方式类型常量。

//对矢量数据集,支持四种压缩编码方式,即单字节、双字节、三字节和四字节编码方式,这四种压缩编码方式采用相同的压缩编码机制,但是压缩的比率不同。其均为有损压缩。需要注意的是,点数据集、纯属性数据集以及 CAD 数据集不可压缩编码。对光栅数据,可以采用四种压缩编码方式,即 DCT、SGL、LZW 和 COMPOUND。其中 DCT 和 COMPOUND 为有损压缩编码方式,SGL 和 LZW 为无损压缩编码方式。None:不使用编码方式。

  if (datasets. Create(datasetImageInfo) ! = null)

  result = true ; } break ;

  default : break ; } }

 catch (Exception ex)

 { Trace. WriteLine(ex. Message) ; }

 return result ; }

7. 窗体代码

(1)添加 using 引用代码

添加 using 引用代码具体如下:

```csharp
using System.Diagnostics;//添加Diagnostics命名空间，使其对应的类可以使用
using SuperMap.Data;//添加Data命名空间，使其对应的类可以使用
```
（2）实例化代码

实例化SampleRun、Workspace、DatasourceConnectionInfo，public FormMain()函数部分的代码如下：

```csharp
//定义类对象
private SampleRun m_sampleRun;
private Workspace m_workspace;
private DatasourceConnectionInfo m_srcDatasourceInfo;
private DatasourceConnectionInfo m_dstDatasourceInfo;
public FormMain()
{ try
{   InitializeComponent();
    InitializeDatasetTypeCombox();
    m_workspace = new Workspace();
    //定义相应类的新对象
   m_srcDatasourceInfo = new DatasourceConnectionInfo("../../SampleData/World/world.udb", "world", "");
    m_dstDatasourceInfo = new DatasourceConnectionInfo
   ("../../SampleData/World/copyworld.udb", "copyworld", "");
   m_sampleRun = null; }
   catch(Exception ex)
{   Trace.WriteLine(ex.Message); } }
```

（3）窗体的Load事件

窗体的Load事件，在窗体的属性窗口单击 图标，左键双击Load，即可添加以下代码：

```csharp
//窗口加载时，绑定一些事件，创建一些事件
private void FormMain_Load(object sender, EventArgs e)
{try { m_sampleRun = new SampleRun(m_workspace, m_srcDatasourceInfo);
    m_sampleRun.CopyDatasource(m_dstDatasourceInfo);//通过sampleRun.CopyDatasource拷贝数据源
      InitializeDatasetView();
      this.FormClosing += new FormClosingEventHandler(FormMain_FormClosing);
      this.deleteCurrent.Click += new EventHandler(deleteCurrent_Click);
      this.changeName.Enabled = false;
      this.changeName.Click += new EventHandler(changeName_Click);
     this.newNameTextBox.TextChanged += new EventHandler(newNameTextBox_
```

TextChanged);
            this.createDataset.Enabled=false;
            this.createDataset.Click+=new EventHandler(createDataset_Click);
             this.createDatasetName.TextChanged+=new EventHandler(createDatasetName_TextChanged);
              this.datasetView.NodeMouseClick += new TreeNodeMouseClickEventHandler(datasetView_NodeMouseClick);
            this.datasetView.LostFocus+=new EventHandler(datasetView_LostFocus);}
        catch(Exception ex)
            {Trace.WriteLine(ex.Message);} }

（4）初始化 Dataset 的名称列表
//初始化 Dataset 的名称列表,将数据集名称添加到 Dataset 中
private void InitializeDatasetView()
    { try{if(m_sampleRun ！=null)
    { String[] datasetNames = m_sampleRun.GetDatasetsNames();//sampleRun.GetDatasetsNames:获取数据源名称
        foreach(String name in datasetNames)
        {datasetView.Nodes.Add(name);}} }
        catch(Exception ex)
        {Trace.WriteLine(ex.Message);} }

（5）选择不同类型的数据集
初始化数据集类型列表在创建数据集时需选择不同的类型,实现代码如下：
private void InitializeDatasetTypeCombox()
        {try{ datasetTypeCombox.Items.Add("点数据集");
            datasetTypeCombox.Items.Add("线数据集");
            datasetTypeCombox.Items.Add("面数据集");
            datasetTypeCombox.Items.Add("文本数据集");
            datasetTypeCombox.Items.Add("CAD 数据集");
            datasetTypeCombox.Items.Add("数据表数据集");
            datasetTypeCombox.Items.Add("栅格数据集");
            datasetTypeCombox.Items.Add("影像数据集");
            datasetTypeCombox.SelectedIndex=0;}
        catch(Exception ex)
        {Trace.WriteLine(ex.Message);} }

（6）获取数据类型 Type
通过数据集类型字符可获取对应的数据类型 Type,添加代码如下：
private DatasetType GetDatasetType(String typeName)

```csharp
{DatasetType result = DatasetType.Point;
try{ switch(typeName)
    { case "点数据集":
        result = DatasetType.Point;
        break;
      case "线数据集":
        result = DatasetType.Line;
        break;
      case "面数据集":
        result = DatasetType.Region;
        break;
      case "文本数据集":
        result = DatasetType.Text;
        break;
      case "CAD 数据集":
        result = DatasetType.CAD;
        break;
      case "数据表数据集":
        result = DatasetType.Tabular;
        break;
      case "栅格数据集":
        result = DatasetType.Grid;
        break;
      case "影像数据集":
        result = DatasetType.Image;
        break;
      default:
        result = DatasetType.Line;
        break; } }
catch(Exception ex)
    { Console.WriteLine(ex.Message); }
    return result; }
```

(7) 窗体的 FormClose 事件

在退出程序时，要断开空间的关联，关闭地图窗口、工作空间。因此在属性窗口点击 图标，左键双击 FormClose 事件，添加代码同前文中相应内容。

(8) 其他控件的代码

① datasetView 的 LostFocus 事件。在控件的属性窗口单击 图标，左键双击

LostFocus，添加如下代码：
　　//此函数为了实现节点在失去焦点时依然可以看到选中状态
　　private void datasetView_LostFocus(object sender, EventArgs e)
　　　　{try{//设置节点的前景色和背景色
　　　　　　datasetView.SelectedNode.BackColor=Color.DarkBlue;
　　　　　　datasetView.SelectedNode.ForeColor=Color.White;}
　　　　catch(Exception ex)
　　　　　　{Trace.WriteLine(ex.Message);}}
　　② datasetView 的 NodeMouseClick 事件。在控件的属性窗口单击图标，左键双击 click，添加如下代码：
　　//设置鼠标选中的文本颜色
　　private void datasetView_NodeMouseClick(object sender, TreeNodeMouseClickEventArgs e)
　　　　{try { if(datasetView.SelectedNode != null)
　　　　　　 datasetView.SelectedNode.BackColor=Color.White;
　　　　　　 datasetView.SelectedNode.ForeColor=Color.Black;}}
　　　　catch(Exception ex)
　　　　　　{Trace.WriteLine(ex.Message);}}
　　③ deleteCurrent 的 click 事件。在控件的属性窗口单击图标，左键双击 click，添加如下代码：
　　//删除选中的数据集
　　private void deleteCurrent_Click(object sender, EventArgs e)
　　　　{try { if(datasetView.SelectedNode==null)
　　 {MessageBox.Show("没有选中任何数据集");
　　　　return;}
　　　　if(MessageBox.Show("确定要删除数据集？不可恢复","警告",
MessageBoxButtons.OKCancel)==DialogResult.OK)
　　　　{m_sampleRun.DeleteDataset(datasetView.SelectedNode.Text);// 通过 sampleRun.DeleteDataset 删除数据集
　　　　datasetView.Nodes.Remove(datasetView.SelectedNode);}}
　　catch(Exception ex) { Trace.WriteLine(ex.Message);}}
　　④ changeName 的 click 事件。在控件的属性窗口单击图标，左键双击 click，添加如下代码：
　　//修改名称响应函数 利用 sampleRun.RenameDataset 修改名称
　　private void changeName_Click(object sender, EventArgs e)
　　　　{try{ if(datasetView.SelectedNode==null)

```
            {MessageBox.Show("没有选中任何数据集!");
            return;}
            if(m_sampleRun.RenameDataset(datasetView.SelectedNode.Text,
newNameTextBox.Text))
            {datasetView.SelectedNode.Text=newNameTextBox.Text;
            newNameTextBox.Text="";}}
        catch(Exception ex)
        {Console.WriteLine(ex.Message);}}
```

⑤ newNameTextBox 的 TextChanged 事件。在控件的属性窗口单击 ⚡ 图标，左键双击 TextChanged，添加如下代码：

```
//判断修改按钮是否可见 当输入新名字时，修改名字按钮才可用
private void newNameTextBox_TextChanged(object sender, EventArgs e)
        {try{ if(newNameTextBox.Text.Length!=0)
        {changeName.Enabled=true;}
        else{changeName.Enabled=false;}}
        catch(Exception ex)
        {Console.WriteLine(ex.Message);}}
```

⑥ createDatasetName 的 TextChanged 事件。在控件的属性窗口单击 ⚡ 图标，左键双击 TextChanged，添加如下代码：

```
//判断新建按钮是否可见 当输入名字时，新建数据集按钮才可用
private void createDatasetName_TextChanged(object sender, EventArgs e)
        {try{ if(createDatasetName.Text.Length!=0)
        {createDataset.Enabled=true;}
        else{createDataset.Enabled=false;}}
        catch(Exception ex)
          {Console.WriteLine(ex.Message);}}
```

⑦ createDataset 的 Click 事件。在控件的属性窗口单击 ⚡ 图标，左键双击 Click，添加如下代码：

```
//创建数据集
private void createDataset_Click(object sender, EventArgs e)
        {try{if(m_sampleRun.CreateDataset(GetDatasetType(datasetTypeCombox.Text),
createDatasetName.Text))  通过 sampleRun.CreateDataset 创建数据集
        {datasetView.Nodes.Add(createDatasetName.Text);
        MessageBox.Show("添加数据集成功");}}
        catch(Exception ex)
          {Trace.WriteLine(ex.Message);}}
```

8. 运行结果

窗体设计完成后,点击▶按钮,运行该程序,运行后弹出窗体如图4-8所示,可对数据集进行删除、重命名、新建。

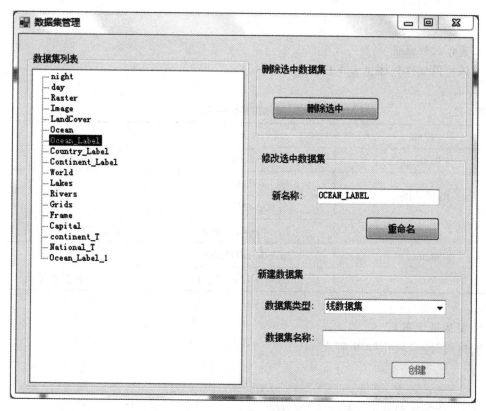

图4-8　数据集管理

4.5　投 影 转 换

1. 数据

在安装目录\SampleData\China\China400.smwu 路径下打开数据文件。

2. 新建文件夹和工程

在【D:\MyProject\】文件中新建一个文件夹,命名为【第4章　空间数据的基本操作】,并在其目录下建立一个文件夹,命名为【PrjCoordSysTransform】,在此文件夹里新建一个工程,将此工程命名为 PrjCoordSysTransform。

3. 关键类型/成员

关键类型/成员见表4-11。

表 4-11　　　　　　　　　　　　　关键类型/成员表

控件/类	方法	属性	事件
CoordSysTranslator	Convert		
PrjCoordSys			

4. 窗体控件属性

窗体控件属性具体见表 4-12。

表 4-12　　　　　　　　　　　　　窗体控件属性表

控件	Name	Text
Toolstrip	Toolstrip1	Toolstrip1
RichTexBox	srcPrjText	
	targetPrjText	
Button	transformGaussBut	高斯-克吕格投影转换
	transformUTMBut	UTM 投影转换
	transformLambertBut	兰勃托投影转换

5. 窗体设计布局

窗体设计布局如图 4-9 所示。

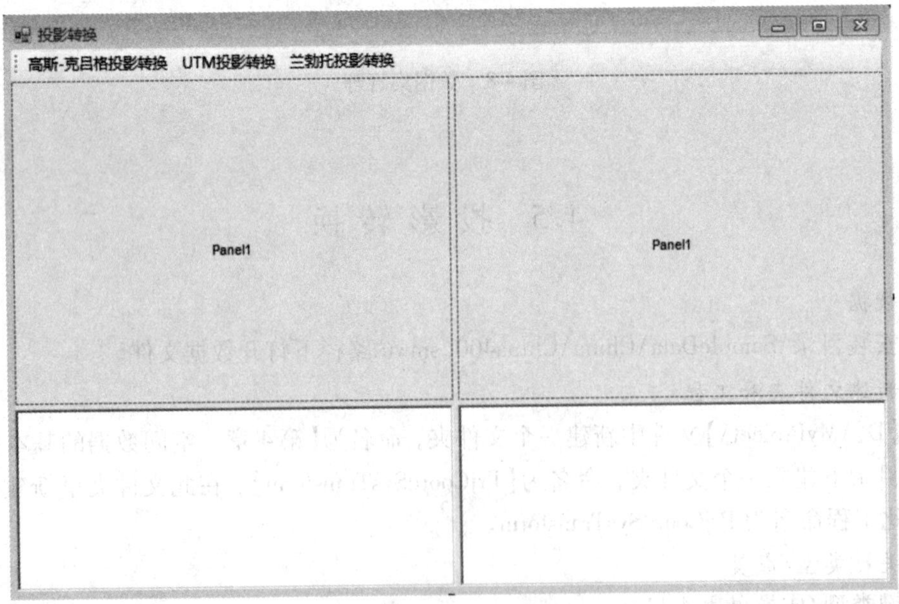

图 4-9　窗体设计布局图

6. 创建 SampleRun 类

（1）添加 using 引用代码

添加 using 引用代码具体如下：

using SuperMap.Data;//添加 SuperMap.Data 命名空间，使其对应的类可以使用

using SuperMap.UI; //添加 SuperMap.UI 命名空间，使其对应的类可以使用

（2）在 public class SampleRun 中添加代码

在 public class SampleRun 中添加如下代码：

```
//定义类对象和控件
    private Workspace m_workspace;
    private MapControl m_srcMapControl;
    private MapControl m_targMapControl;
    private Dataset m_dataset;
    private Dataset m_processDataset;
    private String m_bufPrjDataName;
    // 根据 workspace 和 map 构造 SampleRun 对象
    public SampleRun（Workspace workspace, MapControl srcMapControl, MapControl targMapControl）
        { try { m_workspace=workspace;
        m_srcMapControl=srcMapControl;
        m_targMapControl=targMapControl;
        Initialize();}
        catch（Exception ex）
        { Trace.WriteLine(ex.Message);}}
    /// 打开需要的工作空间文件及地图
    private void Initialize()
        { if（m_workspace ！=null）
            { try {    m_workspace.Open(new
        WorkspaceConnectionInfo（@"../../SampleData/China/China400.smwu"));//
通过 workspace.Open 打开工作空间
                m_dataset=m_workspace.Datasources[0].Datasets["County_R_Prj"];
                m_srcMapControl.Map.Layers.Add(m_dataset,true);//添加图层
                m_srcMapControl.Map.ViewEntire();//全幅显示地图
                m_bufPrjDataName="bufPrj"; }
            catch（Exception ex）
            {Trace.Write(ex.Message); } } }
    /// 删除临时数据集并创建新数据
    private void CopyDataset(String name)//定义 CopyDataset 初始化函数
        { try {
        m_dataset.Datasource.Datasets.Delete(name);
```

```
                m_processDataset = m_dataset. Datasource. CopyDataset(m_dataset, name, m_dataset.
EncodeType);}
            catch(Exception ex)
            { Trace. WriteLine(ex. Message);}}
    /// 获取指定数据集的投影描述信息
    public String GetPrjStr(Dataset dataset)
        { StringBuilder prjStrBuilder = new StringBuilder();//定义 StringBuilder 新对象
                if (dataset == null)
                { MessageBox. Show("数据打开失败:");
                 return null;}
                PrjCoordSys crtPrjSys = dataset. PrjCoordSys;//获取数据集投影信息
                prjStrBuilder. AppendLine("当前投影信息:");
                prjStrBuilder. Append(this. GetEarthPrjStr(crtPrjSys));//获取投影描述信息
                return prjStrBuilder. ToString();}
    /// 获取投影坐标系的描述信息
    private String GetEarthPrjStr(PrjCoordSys crtPrj)
            { StringBuilder prjStrBuilder = new StringBuilder();
                try{ PrjParameter crtPrjParm = crtPrj. PrjParameter;
                    prjStrBuilder. AppendLine("投影类型:"+crtPrj. Name);
                    prjStrBuilder. AppendLine("投影方式:"+crtPrj. Projection. Name);
                    prjStrBuilder. AppendLine("中央经线:"+
                    crtPrjParm. CentralMeridian. ToString("0. 000000"));
                    prjStrBuilder. AppendLine("原点纬线:"+rtPrjParm. CentralParallel. ToString
());
                    prjStrBuilder. AppendLine("标准纬线(1):"+crtPrjParm. StandardParallel1.
ToString());
                    prjStrBuilder. AppendLine("标准纬线(2):"+crtPrjParm. StandardParallel2.
ToString());
                    prjStrBuilder. AppendLine("水平偏移量:"+crtPrjParm. FalseEasting. ToString
(".0000"));
                     prjStrBuilder. AppendLine("垂直偏移量:" + crtPrjParm. FalseNorthing.
ToString(".0000"));
                    prjStrBuilder. AppendLine("比例因子:"+crtPrjParm. ScaleFactor. ToString
());
                     prjStrBuilder. AppendLine("方位角:" + crtPrjParm. Azimuth. ToString
(".0000"));
                    prjStrBuilder. AppendLine("第一点经线:"+crtPrjParm. FirstPointLongitude.
ToString();
```

prjStrBuilder. AppendLine(" 第 二 点 经 线:" + crtPrjParm. SecondPointLongitude. ToString());

prjStrBuilder. AppendLine("地理坐标系:"+crtPrj. GeoCoordSys. Name);

prjStrBuilder. AppendLine(" 大地参考系:"+crtPrj. GeoCoordSys. GeoDatum. Name);

prjStrBuilder. AppendLine("参考椭球体:"+crtPrj. GeoCoordSys. GeoDatum. GeoSpheroid. Name);

prjStrBuilder. AppendLine(" 椭球长半轴:"+crtPrj. GeoCoordSys. GeoDatum. GeoSpheroid. Axis. ToString(".00"));

prjStrBuilder. AppendLine(" 椭球扁率:"+crtPrj. GeoCoordSys. GeoDatum. GeoSpheroid. Flatten. ToString());

prjStrBuilder. AppendLine(" 本 初 子 午 线:" + crtPrj. GeoCoordSys. GeoPrimeMeridian. LongitudeValue. ToString("0.000000"));}

catch(Exception ex)

{Trace. WriteLine(ex. Message);}

return prjStrBuilder. ToString();}

/// 对应不同的投影类型转换后投影的描述信息

public String TransformPrj(int type)

{try{m_targMapControl. Map. Layers. Clear();

this. CopyDataset(m_bufPrjDataName);

PrjCoordSys gaussPrjSys = this. GetTargetPrjCoordSys(type);// PrjCoordSys:投影坐标系类型。投影坐标系由地图投影方式、投影参数、坐标单位和地理坐标系组成。SuperMap iObjects .NET 中提供了很多预定义的投影系统,用户可以直接使用,此外,用户还可以定制自己的投影系统。

Boolean result = CoordSysTranslator. Convert(m_processDataset, gaussPrjSys, new CoordSysTransParameter(), CoordSysTransMethod. GeocentricTranslation);// CoordSysTransMethod. :投影转换类型。主要用于投影坐标之间及投影坐标系之间的转换。

m_targMapControl. Map. Layers. Add(m_processDataset, true);//添加图层

m_targMapControl. Map. Center = m_targMapControl. Map. Bounds. Center;//获取中心点

m_targMapControl. Map. Scale = m_srcMapControl. Map. Scale;//获取比例尺

m_targMapControl. Map. Refresh();//刷新

return this. GetPrjStr(m_processDataset);}

catch(Exception ex)

{Trace. WriteLine(ex. Message);}

return null;}

// 按照不同的投影类型,初始化投影坐标系

private PrjCoordSys GetTargetPrjCoordSys(int type)

```
            { PrjCoordSys targetPrjCoordSys = null;
                PrjParameter parameter = null;// PrjParameter:地图投影参数类型。地图投影
的参数，比如中央经线、原点纬度、双标准纬线的第一和第二条纬线等。
                Projection projection = null;// Projection:投影坐标系地图投影类型。
                switch（type）
                    {case 1:
                        {targetPrjCoordSys = new PrjCoordSys（PrjCoordSysType.UserDefined）;
                        projection = new Projection（ProjectionType.GaussKruger）;
                        targetPrjCoordSys.Projection = projection;
                        parameter = new PrjParameter（）;
                        parameter.CentralMeridian = 110;
                        parameter.StandardParallel1 = 20;
                        parameter.StandardParallel2 = 40;
                        targetPrjCoordSys.PrjParameter = parameter;}
                        break;
                    case 2:
                    {targetPrjCoordSys = new PrjCoordSys（PrjCoordSysType.UserDefined）;
                    projection = new Projection（ProjectionType.TransverseMercator）;
                    targetPrjCoordSys.Projection = projection;
                    parameter = new PrjParameter（）;
                    parameter.CentralMeridian = 110;
                    parameter.StandardParallel1 = 0;
                    targetPrjCoordSys.PrjParameter = parameter;}
                    break;
                    case 3:
                    {targetPrjCoordSys = new PrjCoordSys（PrjCoordSysType.UserDefined）;
                    projection = new Projection（ProjectionType.LambertConformalConic）;
                    targetPrjCoordSys.Projection = projection;
                    parameter = new PrjParameter（）;
                    parameter.CentralMeridian = 110;
                    parameter.StandardParallel1 = 30;
                    targetPrjCoordSys.PrjParameter = parameter;}
                    break;
                    default:
                    break;}
                return targetPrjCoordSys;}
```

7. 窗体代码

（1）添加 using 引用代码

添加 using 引用代码具体如下：

using System. Diagnostics；//添加 System. Diagnostics 命名空间，使其对应的类可以使用
using SuperMap. Data；//添加 SuperMap. Data 命名空间，使其对应的类可以使用
using SuperMap. UI；//添加 SuperMap. UI 命名空间，使其对应的类可以使用

（2）实例化代码

实例化 SampleRun、Workspace、MapControl，public FormMain（）函数部分所需添加的代码：

```
//定义类对象
private SampleRun m_sampleRun；
private Workspace m_workspace；
private MapControl m_srcMapControl；
private MapControl m_targetMapControl；
public FormMain（）
    ｛try｛
InitializeComponent（）；
//创建相应类的新对象
m_workspace = new Workspace（）；
//初始化控件
m_srcMapControl = new MapControl（m_workspace）；
m_srcMapControl. Dock = DockStyle. Fill；
srcSplit. Panel1. Controls. Add（m_srcMapControl）；
m_targetMapControl = new MapControl（m_workspace）；
m_targetMapControl. Dock = DockStyle. Fill；
targetSplit. Panel1. Controls. Add（m_targetMapControl）；｝
catch（Exception ex）
    ｛Trace. WriteLine（ex. Message）；｝｝
```

（3）窗体的 Load 事件

窗体的 Load 事件，在窗体的属性窗口单击图标，左键双击 Load，即可添加如下代码，初始化源数据投影信息：

```
private void FormMain_Load（object sender，EventArgs e）
    ｛try
    ｛m_sampleRun = new SampleRun（m_workspace，m_srcMapControl，m_targetMapControl）；//创建新对象
    if（m_srcMapControl. Map. Layers. Count！=0）
    ｛srcPrjText. Text = m_sampleRun. GetPrjStr（m_srcMapControl. Map. Layers[0]. Dataset）；//获取指定数据集的投影信息｝
    transformGaussBut. Click+= new EventHandler（transformGaussBut_Click）；
```

transformUTMBut. Click+=new EventHandler(transformUTMBut_Click);
transformLambertBut. Click+=new EventHandler(transformLambertBut_Click);}
catch(Exception ex)
{Trace. WriteLine(ex. Message);}}

(4)其他控件的代码

① transformLambertBut 的 Click 事件。在控件的属性窗口单击 图标，左键双击 click，添加如下代码，实现兰勃托投影转换：

private void transformLambertBut_Click(object sender, EventArgs e)
{try
{this. Cursor=Cursors. WaitCursor;
targetPrjText. Text = m_sampleRun. TransformPrj(3);// 通过 sampleRun. TransformPrj(3)实现投影转换
this. Cursor=Cursors. Arrow;}
catch(Exception ex)
{Trace. WriteLine(ex. Message);}}

② transformUTMBut 的 Click 事件。在控件的属性窗口单击 图标，左键双击 Click，添加如下代码，实现 UTM 投影转换：

private void transformUTMBut_Click(object sender, EventArgs e)
{try
{this. Cursor=Cursors. WaitCursor;
targetPrjText. Text=m_sampleRun. TransformPrj(2);// 通过 sampleRun. TransformPrj(2)实现投影转换
this. Cursor=Cursors. Arrow;}
catch(Exception ex)
{Trace. WriteLine(ex. Message);}}

③ transformGaussBut 的 Click 事件。在控件的属性窗口单击 图标，左键双击 Click，添加如下代码，实现高斯-克吕格投影转换：

private void transformGaussBut_Click(object sender, EventArgs e)
{try { this. Cursor=Cursors. WaitCursor;
targetPrjText. Text = m_sampleRun. TransformPrj(1);//通过 sampleRun. TransformPrj(1)实现投影转换
this. Cursor=Cursors. Arrow;}
catch(Exception ex)
{Trace. WriteLine(ex. Message); }}

(5)窗体的 FormClose 事件

在退出程序时，要断开空间的关联，关闭地图窗口、工作空间。因此，在属性窗口点击 图标，左键双击 FormClose 事件，添加以下代码：

```
private void FormMain_FormClosing(object sender, FormClosingEventArgs e)
    {try{ m_targetMapControl.Dispose();
    m_srcMapControl.Dispose();
    m_workspace.Dispose();//关闭工作空间}
    catch(Exception ex)
    {Trace.WriteLine(ex.Message);}}
```

8. 运行结果

窗体设计完成后点击▶按钮，运行该程序，运行后弹出窗体，如图4-10所示,可对数据进行高斯-克吕格投影转换、UTM投影转换、兰勃托投影转换。图中结果为高斯-克吕格投影转换。

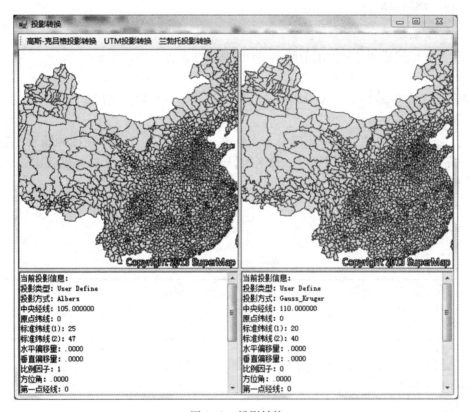

图 4-10 投影转换

第 5 章　地图制图及专题图制作

5.1　地图制图

5.1.1　图层添加

1. 添加数据集

实例简介如何添加各种类型的数据集到地图，并设置风格。

2. 数据

在安装目录\SampleData\World\World.smwu 路径下打开数据文件。

3. 新建文件夹和工程

在【D:\MyProject\】文件中新建一个文件夹，命名为【第 5 章　地图制图及专题图制作】，并在其目录下建立一个文件夹，命名为【LayerAdding】，在此文件夹里新建一个工程，将此工程命名为 LayerAdding。

4. 关键类型/成员

关键类型/成员见表 5-1。

表 5-1　　　　　　　　　　　　**关键类型/成员表**

控件/类	方法	属性	事件
Workspace	Open		
Layers	Add		
LayerSettingVector		Style	
LayerSettingImage		OpaqueRate	
LayerSettingGrid		IsSpecialValueTransparent	

5. 窗体控件属性

窗体控件属性具体见表 5-2。

6. 窗体设计布局

窗体设计布局如图 5-1 所示。

表 5-2　　　　　　　　　　　　窗体控件属性表

控件	Name	Text
toolStrip	m_toolStrip	toolStrip1
Botton	m_toolStripButtonPoint	添加点数据集
Botton	m_toolStripButtonLine	添加线数据集
Botton	m_toolStripButtonRegion	添加面数据集
Botton	m_toolStripButtonGrid	添加栅格数据集
Botton	m_toolStripButtonImage	添加影像数据集
Botton	m_toolStripButtonClear	清空图层
ComboBox	m_toolStripComboBoxIsCustom	

图 5-1　窗体设计布局图

7. 创建 SampleRun 类

（1）添加 using 引用代码

添加 using 引用代码具体如下：

using SuperMap.Data; //添加 SuperMap.Data 命名空间，使其对应的类可以使用
using SuperMap.UI; //添加 Map.UI 命名空间，使其对应的类可以使用
using SuperMap.Mapping；//添加 SuperMap.Mapping 命名空间，使其对应的类可以使用

（2）在 public class SampleRun 中添加代码

在 public class SampleRun 中添加如下代码：

//定义类对象和控件

private Workspace m_workspace；

private MapControl m_mapControl；

private Datasource m_datasource；// Datasource：数据源类。该类管理投影信息、数据源与数据库的连接信息和对其中的数据集的相关操作，如通过已有数据集复制生成新的数据集等。

//添加的各种类型数据集的名称
private readonly String m_datasetPoint = "Capital";
private readonly String m_datasetLine = "Grids";
private readonly String m_datasetRegion = "World";
private readonly String m_datasetImage = "day";
private readonly String m_datasetGrid = "Raster";
/// 根据 workspace 和 map 构造 SampleRun 对象
public SampleRun(Workspace workspace, MapControl mapControl)
    {try {m_workspace = workspace;
        m_mapControl = mapControl;
        m_mapControl.Map.Workspace = workspace;
        catch (Exception ex)
        {Trace.WriteLine(ex.Message);}
        Initialize();} }
/// 打开需要的工作空间文件及地图
private void Initialize()
    {try{ //打开工作空间及地图
WorkspaceConnectionInfo conInfo = new WorkspaceConnectionInfo(@"..\..\SampleData\World\World.smwu");
    m_workspace.Open(conInfo);//根据 m-workspace.Open 打开工作空间
        m_datasource = m_workspace.Datasources[0];
        this.m_mapControl.Map.Open(m_workspace.Maps[0]);
        // 调整 mapControl 的状态
        m_mapControl.Action = SuperMap.UI.Action.Pan;
        m_mapControl.Map.ViewEntire();}
        catch (Exception ex)
        { Trace.WriteLine(ex.Message); } }
/// 清空图层
public void ClearLayers()
    { try{ m_mapControl.Map.Layers.Clear();
        m_mapControl.Map.Refresh();//刷新地图}
        catch (Exception ex)
        { Trace.WriteLine(ex.Message); }}
/// 向地图中添加点数据集,是否自定义风格
public void AddPoint(Boolean isWithStyle)
    { try{ DatasetVector dataset = m_datasource.Datasets[m_datasetPoint] as DatasetVector;// DatasetVector:矢量数据集信息类。包括了矢量数据集的信息,如矢量数据集的名称、数据集的类型、编码方式、是否选用文件缓存等。文件缓存只针对图库索引而言。

//设置风格并添加数据集
Layer layer=null;
if(isWithStyle)
｛LayerSettingVector setting=new LayerSettingVector();//创建新对象 LayerSettingVector:矢量图层设置类。该类主要用来设置矢量图层的显示风格。矢量图层用单一的符号或风格绘制所有的要素。当用户只想可视化地显示空间数据,只关心空间数据中各要素在什么位置,而不关心各要素在数量或性质上的不同时,可以用普通图层来显示要素数据。
//设置风格
setting.Style.LineColor=Color.SeaGreen;
setting.Style.MarkerSize=new Size2D(4,4);
setting.Style.MarkerSymbolID=12;
layer=m_mapControl.Map.Layers.Add(dataset,setting,true);｝
else｛layer=m_mapControl.Map.Layers.Add(dataset,true);/｝//添加图层
//全幅显示添加的图层
m_mapControl.Map.EnsureVisible(layer);//居中显示
m_mapControl.Map.Refresh();｝//刷新
catch(Exception ex)
｛Trace.WriteLine(ex.Message);｝｝
///向地图中添加线数据集,是否自定义风格
public void AddLine(Boolean isWithStyle)
｛try｛DatasetVector dataset=m_datasource.Datasets[m_datasetLine] as DatasetVector;
Layer layer=null;
if(isWithStyle)
｛LayerSettingVector setting=new LayerSettingVector();//创建新对象
//设置风格
setting.Style.LineColor=Color.Gray;
setting.Style.LineSymbolID=2;
setting.Style.LineWidth=0.3;
layer=m_mapControl.Map.Layers.Add(dataset,setting,true);｝//添加图层
else｛ layer=m_mapControl.Map.Layers.Add(dataset,true);｝
//全幅显示添加的图层
m_mapControl.Map.EnsureVisible(layer);//居中显示
m_mapControl.Map.Refresh();//刷新｝
catch(Exception ex)
｛Trace.WriteLine(ex.Message);｝ ｝
///向地图中添加面数据集,是否自定义风格
public void AddRegion(Boolean isWithStyle)

```csharp
try{ DatasetVector dataset = m_datasource.Datasets[m_datasetRegion] as DatasetVector;// DatasetVector：矢量数据集类。描述矢量数据集，并提供相应的管理和操作。对矢量数据集的操作主要包括数据查询、修改、删除、建立索引等。
    Layer layer = null;
    if(isWithStyle)
    {LayerSettingVector setting = new LayerSettingVector();//创建新对象
        //设置风格
        setting.Style.LineColor = Color.Teal;
        setting.Style.LineSymbolID = 11;
        setting.Style.LineWidth = 0.5;
        setting.Style.FillForeColor = Color.FromArgb(2,138,226);
        setting.Style.FillBackColor = Color.FromArgb(232,245,254);
        setting.Style.FillGradientMode = FillGradientMode.Radial;
        layer = m_mapControl.Map.Layers.Add(dataset,setting,true);}//添加图层
    else{ layer = m_mapControl.Map.Layers.Add(dataset,true); }
    //全幅显示添加的图层
    m_mapControl.Map.EnsureVisible(layer);//居中显示
    m_mapControl.Map.Refresh(); }
catch(Exception ex)
{ Trace.WriteLine(ex.Message);} }
///向地图中添加影像数据集，是否自定义风格
public void AddImage(Boolean isWithStyle)
{ try{ DatasetImage dataset = m_datasource.Datasets[m_datasetImage] as DatasetImage;// DatasetImage：影像数据集类，用于描述影像数据，不具备属性信息，例如影像地图、多波段影像和实物地图等。
    Layer layer = null;
    if(isWithStyle)
    { LayerSettingImage setting = new LayerSettingImage();
        setting.OpaqueRate = 50;//设置不透明度
        layer = m_mapControl.Map.Layers.Add(dataset,setting,true); }
    else
    { layer = m_mapControl.Map.Layers.Add(dataset,true); }
    //全幅显示添加的图层
    m_mapControl.Map.EnsureVisible(layer);
    m_mapControl.Map.Refresh();}
catch(Exception ex)
{ Trace.WriteLine(ex.Message); }}
///向地图中添加栅格数据集，是否自定义风格
public void AddGrid(Boolean isWithStyle)
```

```csharp
{try {DatasetGrid dataset = m_datasource.Datasets[m_datasetGrid] as DatasetGrid;//
DatasetGrid:栅格数据集类,用于描述栅格数据,例如高程数据集和土地利用图。
            Layer layer = null;
            if(isWithStyle)
            { LayerSettingGrid setting = new LayerSettingGrid();
                //设置风格
        setting.ColorTable = Colors.MakeGradient(20, ColorGradientType.BlueWhite, false);
        setting.SpecialValue = -9999;
        setting.IsSpecialValueTransparent = true;
        layer = m_mapControl.Map.Layers.Add(dataset, setting, true); }
            else
            { layer = m_mapControl.Map.Layers.Add(dataset, true); }
        //全幅显示添加的图层
            m_mapControl.Map.EnsureVisible(layer);
        m_mapControl.Map.Refresh();}
        catch(Exception ex)
        { Trace.WriteLine(ex.Message); } }
```

8. 窗体代码

(1)添加 using 引用代码

添加 using 引用代码具体如下:

using System.Diagnostics;//添加 System.Diagnostics 命名空间,使其对应的类可以使用

(2)实例化代码

实例化 SampleRun, public FormMain() 函数部分的代码如下:

```csharp
//定义对应类的对象
private SampleRun m_sampleRun;
private SuperMap.Data.Workspace m_workspace;
private SuperMap.UI.MapControl m_mapControl;
private Boolean m_isCustomStyle = true;
public FormMain()
    { try { InitializeComponent();
        this.Load += new EventHandler(FormMain_Load);
        //定义相应类的新对象
        this.m_workspace = new SuperMap.Data.Workspace();
        this.m_mapControl = new SuperMap.UI.MapControl();
        //初始化控件
    m_mapControl.Dock = DockStyle.Fill;
```

```
        base.Controls.Add(m_mapControl);
        base.Controls.SetChildIndex(m_mapControl, 0);  }
    catch(Exception ex)
    { Trace.WriteLine(ex.Message); } }
```

(3) 窗体的 Load 事件

在窗体的属性窗口中,选择 按钮后,左键双击 Load,即可添加以下代码:

```
private void FormMain_Load(object sender, EventArgs e)
    { try{//实例化 SampleRun
        m_sampleRun = new SampleRun(m_workspace, m_mapControl);
        //初始化 m_toolStripComboBoxIsCustom
        m_toolStripComboBoxIsCustom.Items.Add(自定义风格);
        m_toolStripComboBoxIsCustom.Items.Add(默认风格);
        m_toolStripComboBoxIsCustom.SelectedItem = m_toolStripComboBoxIsCustom.Items[0]; }
        catch(Exception ex)
        { Trace.WriteLine(ex.Message); } }
```

(4) 窗体的 FormClose 事件

在退出程序时,要断开空间的关联,关闭地图窗口、工作空间。因此,在属性窗口点击 图标,左键双击 FormClose 事件,添加 m_mapControl.Dispose();和 m_workspace.Dispose();/关闭,其他部分同前文相应内容。

(5) 添加各个控件的代码

① 添加点数据集 Button 控件的 Click 事件。左键双击添加点数据集 Button 控件,即可添加如下该按钮的 Click 事件代码:

```
//添加点数据集
private void m_toolStripButtonPoint_Click(object sender, EventArgs e)
    { try{m_sampleRun.AddPoint(m_isCustomStyle);//通过 sampleRun.AddPoint 添加点}
        catch(Exception ex)
        { Trace.WriteLine(ex.Message); } }
```

② 添加线数据集 Button 控件的 Click 事件。左键双击添加线数据集 Button 控件,即可添加如下该按钮的 Click 事件代码:

```
//添加线数据集
private void m_toolStripButtonLine_Click(object sender, EventArgs e)
    { try{m_sampleRun.AddLine(m_isCustomStyle);//通过 sampleRun.AddLine 添加线}
        catch(Exception ex)
        { Trace.WriteLine(ex.Message); } }
```

③ 添加面数据集 Button 控件的 Click 事件。左键双击添加面数据集 Button 控件，即可添加如下该按钮的 Click 事件代码：

//添加面数据集
private void m_toolStripButtonRegion_Click(object sender, EventArgs e)
　　｛try｛m_sampleRun. AddRegion(m_isCustomStyle)；//通过 sampleRun. AddRegion 添加面｝
catch（Exception ex）
　｛Trace. WriteLine(ex. Message)；｝｝

④ 添加影像数据集 Button 控件的 Click 事件。左键双击添加影像数据集 Button 控件，即可添加如下该按钮的 Click 事件代码：

//添加影像数据集
private void m_toolStripButtonImage_Click(object sender, EventArgs e)
　　｛try｛m_sampleRun. AddImage(m_isCustomStyle)；//通过 sampleRun. AddImage 添加影像｝
catch（Exception ex）
　｛Trace. WriteLine(ex. Message)；｝｝

⑤ 添加栅格数据集 Button 控件的 Click 事件。左键双击添加栅格数据集 Button 控件，即可添加如下该按钮的 Click 事件代码：

//添加栅格数据集
private void m_toolStripButtonGrid_Click(object sender, EventArgs e)
　　｛try｛ m_sampleRun. AddGrid(m_isCustomStyle)；｝//通过 sampleRun. AddGrid 添加栅格
　　catch（Exception ex）
　｛Trace. WriteLine(ex. Message)；｝｝

⑥ 清空图层 Button 控件的 Click 事件。左键双击清空图层 Button 控件，即可添加如下该按钮的 Click 事件代码：

//清空图层
private void m_toolStripButtonClear_Click(object sender, EventArgs e)
　　｛try｛ m_sampleRun. ClearLayers( )；｝//通过 sampleRun. ClearLayers 清空图层
　catch（Exception ex）
　｛Trace. WriteLine(ex. Message)；｝｝

9. 运行结果

窗体设计完成后点击 ▶ 图标，运行该程序，运行后弹出窗体，如图 5-2 所示。分别点击添加"点数据集"、"添加线数据集"、"添加面数据集"、"添加栅格数据集"、"添加影像数据集"各不同按钮将添加不同的数据集。点击"清空图层"按钮会将添加的数据集清空。

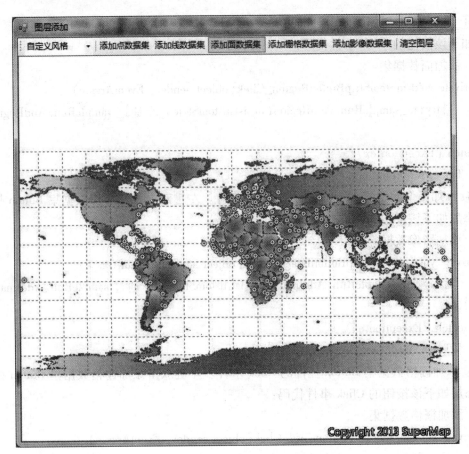

图 5-2　运行结果

5.1.2　地图属性

1. 数据

在安装目录\SampleData\City\Changchun.smwu 路径下打开数据文件。

2. 新建文件夹和工程

在【D:\MyProject\】文件中新建一个文件夹,命名为【第 5 章　地图制图及专题图制作】,并在其目录下建立一个文件夹,命名为【MapProperty】,在此文件夹里新建一个工程,将此工程命名为 MapProperty。

3. 关键类型/成员

关键类型/成员见表 5-3。

表 5-3　　　　　　　　　　关键类型/成员表

控件/类	方法	属　性	事件
Map		BackgroundStyle、IsAntialias、Angle、IsTextAngleFixed、ColorMode、CustomBounds、IsCustomBoundsEnabled、IsClipRegionEnabled	

4. 窗体控件属性
窗体控件属性具体见表 5-4。

表 5-4　　　　　　　　　　　　窗体控件属性表

控件	Name	Text
statusStrip	statusStrip1	statusStrip1
Label	Label1	显示模式
Label	Label2	自定义地图边界
Label	Label3	旋转角度
ComboBox	comboBoxDisplyMode	
ComboBox	comboBoxMapBorder	
CheckBox	checkTextAngleFixed	文本角度固定
CheckBox	checkAntialias	反走样
CheckBox	checkClipView	裁剪当前视图
statusStrip	statusStrip1	statusStrip1
ColorDialog	backColorDialog	backColorDialog

4. 窗体设计布局
窗体设计布局如图 5-3 所示。

图 5-3　窗体设计布局图

5. 创建 SampleRun 类
（1）添加 using 引用代码

添加 using 引用代码具体如下：

using SuperMap.Data; //添加 SuperMap.Data 命名空间，使其对应的类可以使用

using SuperMap. UI; //添加 SuperMap. UI 命名空间,使其对应的类可以使用
using SuperMap. Mapping; //添加 SuperMap. Mapping 命名空间,使其对应的类可以使用

(2)在 public class SampleRun 中添加代码

在 public class SampleRun 中添加如下代码:

```csharp
//定义类对象
private Workspace m_workspace;
private MapControl m_mapControl;
private ToolStripStatusLabel m_statusLabel;
private Map m_map;
private Datasource m_datasource;
/// 根据 Yworkspace 和 map 构造 SampleRun 对象
public SampleRun( Workspace workspace, MapControl mapControl, ToolStripStatusLabel statusLabel)
    { m_workspace = workspace;
      m_mapControl = mapControl;
      m_statusLabel = statusLabel;
      m_mapControl. Map. Workspace = workspace;
      Initialize( );}
///打开工作空间及地图
private void Initialize( )
 {try
  {//打开工作空间及地图
      WorkspaceConnectionInfo conInfo = new WorkspaceConnectionInfo( @"..\..\SampleData\City\Changchun. smwu");
      m_workspace. Open( conInfo);//通过 workspace. Open 打开工作空间
      m_datasource = m_workspace. Datasources[0];
      m_mapControl. Map. Open( m_workspace. Maps[0]);//通过 Map. Open 打开地图
      m_map = m_mapControl. Map;
      m_map. Drawn + = new MapDrawnEventHandler( MapDrawnHandler);}
      catch( Exception ex)
       { Trace. WriteLine( ex. Message);}}
/// 地图刷新后事件
private void MapDrawnHandler( object sender, MapDrawnEventArgs e)
    { String info = String. Empty;
    //获取当前比例尺
    Double scale = 1 / m_map. Scale;
    //获取中心点坐标
    Double x = m_map. Center. X;
```

```csharp
            Double y = m_map.Center.Y;
            //获取坐标单位
            Unit unit = m_map.CoordUnit;
              info + = "比例尺:1" + scale.ToString() + "    ";
              info + = "中心点:" + x + "  " + unit.ToString() + "   " + y + "  " + unit.ToString();
              m_statusLabel.Text = info;}
/// 设置背景颜色
public Color BackColor
     {set{ try
         {m_map.BackgroundStyle.FillForeColor = value;
              m_map.Refresh();//刷新地图}
         catch(Exception ex)
         {Trace.WriteLine(ex.Message);}}}
/// 设置反走样
public Boolean IsAntialias
     {set { try { m_map.IsAntialias = value;
              m_map.Refresh();}
          catch(Exception ex)
            { Trace.WriteLine(ex.Message);} } }
/// 设置地图旋转角度
public Int32 Angle
     { set { try
         {m_map.Angle = value;
          m_map.Refresh();}
      catch(Exception ex)
      { Trace.WriteLine(ex.Message);}}}
/// 设置文本角度固定
public Boolean TestAngleFixed
     { set{ try { m_map.IsTextAngleFixed = value;
                 m_map.Refresh();}
              catch(Exception ex)
     {Trace.WriteLine(ex.Message);}}}
/// 裁剪当前显示区域
public Boolean IsClipRegionEnable
      {set{ try
        { if(value)
          {//获取当前视图范围
         Rectangle2D viewBounds = m_map.ViewBounds;
         GeoRectangle geoRectangle = new GeoRectangle(viewBounds, 0);
```

```
            m_map.ClipRegion = geoRectangle.ConvertToRegion();}
         //设置裁剪有效
            m_map.IsClipRegionEnabled = value;}
         catch(Exception ex)
         {Trace.WriteLine(ex.Message);}}}
/// 设置自定义全幅范围,下拉框选择的索引
public void SetCustomBounds(Int32 index)
    {try{Rectangle2D bounds = m_map.Bounds;
    switch(index)
        {case 0:
            {m_map.IsCustomBoundsEnabled = false;}
             break;
         case 1:
         {// 宽城区
         bounds = new Rectangle2D(2189.5034307084,-3525.4051686629,
                       7839.8477808474,-83.0480041011);
         //设置自定义范围
            m_map.CustomBounds = bounds;
            m_map.IsCustomBoundsEnabled = true;}
             break;
         case 2:
         {// 绿园区
         bounds = new Rectangle2D(-2029.3482323858,-6469.0219094643,
                       6893.0400156267,9.7634455669);
         //设置自定义范围
            m_map.CustomBounds = bounds;
            m_map.IsCustomBoundsEnabled = true;}
             break;
         case 3:
         {// 朝阳区
         bounds = new Rectangle2D(1191.8948102088,-7653.3705467338,
                       7102.3318954094,-3361.6436945694);
         //设置自定义范围
            m_map.CustomBounds = bounds;
            m_map.IsCustomBoundsEnabled = true;}
             break;
         case 4:
         {// 南关区
         bounds = new Rectangle2D(2867.0454279395,-7472.1969118647,
```

```
                        8777.4825131401, -3180.4700597003);
            //设置自定义范围
            m_map.CustomBounds = bounds;
            m_map.IsCustomBoundsEnabled = true;}
            break;
            case 5:
            //=道区
          { bounds = new Rectangle2D(2613.0322255848, -7630.794392856,
                13041.020644187, -58.7523822514);
            // 设置自定义范围
            m_map.CustomBounds = bounds;
            m_map.IsCustomBoundsEnabled = true;}
            break;
            case 6:
           {// 当前视图
              bounds = m_map.ViewBounds;
              // 设置自定义范围
              m_map.CustomBounds = bounds;
              m_map.IsCustomBoundsEnabled = true;}
            break;
            default:
            break;}
            m_map.ViewEntire();//全幅显示
            m_map.Refresh();}
       catch(Exception ex)
         { Trace.WriteLine(ex.Message);}}
/// 设置地面的彩色模式,下拉框选择的索引
public void SetDisplayMode(Int32 index)
     {MapColorMode colorMode;
        switch(index)
       { case 0:
           { colorMode = MapColorMode.Default;}
            break;
            case 1:
           {colorMode = MapColorMode.Gray;}
            break;
            case 2:
           { colorMode = MapColorMode.BlackWhite;}
            break;
```

```
                    case 3:
                        {colorMode = MapColorMode.BlackWhiteReverse;}
            break;
            case 4:
                {colorMode = MapColorMode.OnlyBlackWhiteReverse;}
            break;
            default:
                {colorMode = MapColorMode.Default;}
            break;}
            m_map.ColorMode = colorMode;
        m_map.Refresh();}
```

6. 窗体代码

(1) 添加 using 引用代码

添加 using 引用代码具体如下：

using System.Diagnostics; //添加 System.Diagnostics 命名空间，使其对应的类可以使用

(2) 实例化代码

实例化 SampleRun，以及 public FormMain() 函数部分的代码如下：

```
//定义类对象
private SampleRun m_sampleRun;
private SuperMap.Data.Workspace m_workspace;
private SuperMap.UI.MapControl m_mapControl;
public FormMain()
        {try{InitializeComponent();
        // 创建一个新的类对象
    this.m_workspace = new SuperMap.Data.Workspace();
    this.m_mapControl = new SuperMap.UI.MapControl();
    //初始化控件
    m_mapControl.Dock = DockStyle.Fill;
    //实例化 SampleRun
    m_sampleRun = new SampleRun(m_workspace, m_mapControl, statusLabel);
    base.Controls.Add(m_mapControl);
    base.Controls.SetChildIndex(m_mapControl, 0);
    // 填充显示模式
    comboBoxDisplyMode.Items.Add("彩色模式");
    comboBoxDisplyMode.Items.Add("灰度模式");
    comboBoxDisplyMode.Items.Add("黑白模式");
    comboBoxDisplyMode.Items.Add("黑白反色");
    comboBoxDisplyMode.Items.Add("仅黑白反色");
    comboBoxDisplyMode.SelectedIndex = 0;
```

// 填充地图边界
comboBoxMapBorder.Items.Add("未定义");
comboBoxMapBorder.Items.Add("宽城区");
comboBoxMapBorder.Items.Add("绿园区");
comboBoxMapBorder.Items.Add("朝阳区");
comboBoxMapBorder.Items.Add("南关区");
comboBoxMapBorder.Items.Add("二道区");
comboBoxMapBorder.Items.Add("当前视图");
comboBoxMapBorder.SelectedIndex=0;
//定义默认值
checkAntialias.Checked=true;}
catch(Exception ex)
{Trace.WriteLine(ex.Message);}}

(3)窗体的 FormClose 事件

在退出程序时,要断开空间的关联,关闭地图窗口、工作空间。因此,在属性窗口点击 图标,左键双击 FormClose 事件,添加代码同前文中相应内容。

(4)添加各个控件的代码

① 背景颜色 Button 控件的 Click 事件。左键双击背景颜色 Button 控件,即可添加如下该按钮的 Click 事件代码,设定背景颜色。

private void bttonBackColor_Click(object sender, EventArgs e)
{ if (backColoeDialog.ShowDialog()==DialogResult.OK)
{m_sampleRun.BackColor=backColoeDialog.Color;//通过 sampleRun.BackColor 实现}}

② 显示模式 ComboBox 控件的 SelectedIndexChanged 事件。在 ComboBoxDisplyMode 的属性窗口中,选择 按钮后,左键双击 SelectedIndexChanged,即可添加以下代码,选择显示模式:

private void comboBoxDisplyMode_SelectedIndexChanged(object sender, EventArgs e)
{ m_sampleRun.SetDisplayMode(comboBoxDisplyMode.SelectedIndex);
//通过 sampleRun.SetDisplayMode 实现}

③ 反走样 CheckAntialias 控件的 CheckedChanged 事件。在 CheckAntialias 的属性窗口中,选择 按钮后,左键双击 CheckedChanged,即可添加以下代码,实现反走样:

private void checkAntialias_CheckedChanged(object sender, EventArgs e)
{m_sampleRun.IsAntialias=checkAntialias.Checked;//通过 sampleRun.IsAntialias 实现}

④ 裁剪当前视图 CheckClipView 控件的 CheckedChanged 事件。在 CheckClipView 的属性窗口中,选择 按钮后,左键双击 CheckedChanged,即可添加以下代码,实现裁剪:

private void checkClipView_CheckedChanged(object sender, EventArgs e)
{m_sampleRun.IsClipRegionEnable=checkClipView.Checked;
//通过 sampleRun.IsClipRegionEnable 实现}

⑤ 定义地图边界 ComboBoxMapBorder 控件的 SelectedIndexChanged 事件。在 ComboBoxMapBorder 的属性窗口中，选择 按钮后，左键双击 SelectedIndexChanged，即可添加以下代码：

private void comboBoxMapBorder_SelectedIndexChanged(object sender, EventArgs e)
　　{m_sampleRun.SetCustomBounds(comboBoxMapBorder.SelectedIndex);
//通过 sampleRun.SetCustomBounds 实现}

⑥ 设置地图旋转角度 upDownAngle 控件的 ValueChanged 事件。在 upDownAngle 的属性窗口中，选择 按钮后，左键双击 ValueChanged，即可添加以下代码，设置地图旋转角度：

private void upDownAngle_ValueChanged(object sender, EventArgs e)
　　{m_sampleRun.Angle=(Int32)upDownAngle.Value;//通过 sampleRun.Angle 实现}

⑦ 设置文本是否固定 CheckTextAngleFixed 控件的 CheckedChanged 事件。在 CheckTextAngleFixed 的属性窗口中，选择 按钮后，左键双击 CheckedChanged，即可添加以下代码，设置文本是否固定：

private void checkTextAngleFixed_CheckedChanged(object sender, EventArgs e)
　　{m_sampleRun.TestAngleFixed=checkTextAngleFixed.Checked;
//通过 sampleRun.TestAngleFixed 实现}

7. 运行结果

窗体设计完成后点击 按钮，运行该程序，运行后弹出窗体，如图 5-4 所示，可以设置地图背景颜色、设置显示模式、设置是否反走样、设置是否剪裁当前视图、设置自定义地图边界、设置地图旋转角度、设置文本角度固定。

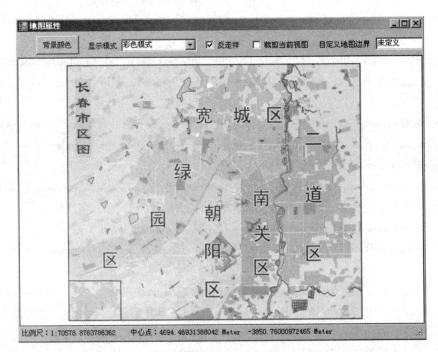

图 5-4　运行结果

5.1.3 地图输出

1. 数据

在安装目录\SampleData\City\Changchun.smwu 路径下打开数据文件。

2. 新建文件夹和工程

在【D:\MyProject\】文件中新建一个文件夹,命名为【第 5 章　地图制图及专题图制作】,并在其目录下建立一个文件夹,命名为【MapOutput】,在此文件夹里新建一个工程,将此工程命名为 MapOutput。

3. 关键类型/成员

关键类型/成员见表 5-5。

表 5-5　　　　　　　　　　关键类型/成员表

控件/类	方法	属性	事件
Map	OutputMapToBMP、OutputMapToEPS、OutputMapToPNG		
Workspace	Open		

4. 窗体控件属性

窗体控件属性具体见表 5-6。

表 5-6　　　　　　　　　　窗体控件属性表

控件	Name	Text
toolStrip	toolStrip1	toolStrip1
Button	m_toolStripButtonOutToBMP	BMP 出图
	m_toolStripButtonOutToEPS	EPS 出图
	m_toolStripButtonOutToPNGBackTransparent	PNG 出图(背景透明)
	m_toolStripButtonOutToPNGBackNotTransparent	PNG 出图(背景不透明)

5. 窗体设计布局

窗体设计布局如图 5-5 所示。

6. 创建 SampleRun 类

(1) 添加 using 引用代码

添加 using 引用代码具体如下:

using System.Diagnostics；//添加 System.Diagnostics 命名空间,使其对应的类可以使用

using System.Drawing；//添加 System.Drawing 命名空间,使其对应的类可以使用

using System.Windows.Forms；//添加 System.Windows.Forms 命名空间,使其对应的类可以使用

using SuperMap.Data；//添加 SuperMap.Data 命名空间,使其对应的类可以使用

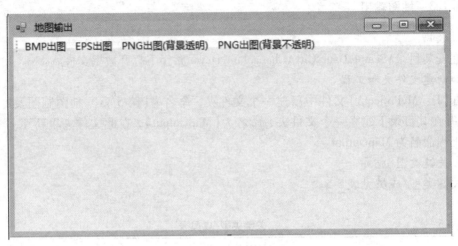

图 5-5　窗体设计布局图

using SuperMap.UI;　//添加 SuperMap.UI 命名空间，使其对应的类可以使用

（2）在 public class SampleRun 中添加代码

在 public class SampleRun 中添加如下代码：

//定义类对象

private Workspace m_workspace;

private MapControl m_mapControl;

/// 根据 workspace 和 map 构造 SampleRun 对象

public SampleRun(Workspace workspace, MapControl mapControl)

　　{m_workspace = workspace;

　m_mapControl = mapControl;

　m_mapControl.Map.Workspace = workspace;

　　Initialize();}

private void Initialize()

　　{try{//打开工作空间及地图

　　　WorkspaceConnectionInfo conInfo = new WorkspaceConnectionInfo(@"..\..\SampleData\City\Changchun.smwu");

　　　m_workspace.Open(conInfo);//根据 workspace.Open 打开工作空间

　　　this.m_mapControl.Map.Open(m_workspace.Maps[0]);//根据 Map.Open 打开地图

　　// 调整 mapControl 的状态

　　m_mapControl.Map.ViewEntire();}

　　catch (Exception ex)

　　{Trace.WriteLine(ex.Message);}}

/// 根据文件后缀名选择要保存的文件路径和格式，并返回选择的路径
```
private String GetFilePath(String postfix)
    { String result = String.Empty;
      SaveFileDialog dialog = new SaveFileDialog();//通过SaveFileDialog保存路径
      dialog.Title = "请选择要保存的文件路径";
      dialog.Filter = String.Format("{0}格式(*.{0})|*.{0}", postfix);
      if(dialog.ShowDialog() == DialogResult.OK)
        {result = dialog.FileName;}
      return result;}
```
/// 出 BMP 图
```
public void OutputBMP()
    {try{ String fileName = GetFilePath("bmp");
      m_mapControl.Map.OutputMapToBMP(fileName);//通过 Map.OutputMapToBMP 出 bmp}
     catch(Exception ex)
     {Trace.WriteLine(ex.Message);}}
```
/// 出 EPS 图
```
public void OutputEPS()
    {try{ String fileName = GetFilePath("eps");
      m_mapControl.Map.OutputMapToEPS(fileName);//通过 Map.OutputMapToEPS 出 EPS}
     catch(Exception ex)
     {Trace.WriteLine(ex.Message);}}
```
/// 出 PNG 图
```
public void OutputPNG(Boolean isBackTransparent)
    {try{ String fileName = GetFilePath("png");
        m_mapControl.Map.OutputMapToPNG(fileName, isBackTransparent);//通过 Map.OutputMapToPNG 出 PNG}
     catch(Exception ex)
     { Trace.WriteLine(ex.Message);}}
```

7. 窗体代码

(1) 添加 using 引用代码

添加 using 引用具体代码如下：

using System.Diagnostics;//添加 System.Diagnostics 命名空间，使其对应的类可以使用

(2) 实例化代码

实例化 SampleRun，以及 public FormMain() 函数部分的代码如下：

//定义类对象和控件

```
private SampleRun m_sampleRun;
private SuperMap. Data. Workspace m_workspace;
private SuperMap. UI. MapControl m_mapControl;
public FormMain( )
    {try{ InitializeComponent( );
  创建相应类的新对象
this. m_workspace = new SuperMap. Data. Workspace( );
this. m_mapControl = new SuperMap. UI. MapControl( );
//初始化控件
m_mapControl. Dock = DockStyle. Fill;
base. Controls. Add( m_mapControl);
base. Controls. SetChildIndex( m_mapControl, 0);
m_sampleRun = new SampleRun( m_workspace, m_mapControl); }
catch (Exception ex)
    { Trace. WriteLine( ex. Message); } }
```

（3）窗体的 FormClose 事件

在退出程序时，要断开空间的关联，关闭地图窗口、工作空间。因此，在属性窗口点击 图标，左键双击 FormClose 事件，添加 m_mapControl. Dispose( );以及 m_workspace. Dispose( );来关闭工作空间，其他部分同前文中的相应部分。

（4）添加各个控件的代码

① BMP 出图 Button 控件的 Click 事件。左键双击 BMP 出图 Button 控件，即可添加该按钮的 Click 事件代码：

```
//出 BMP 图
private void m_toolStripButtonOutToBMP_Click( object sender, EventArgs e)
    {try{ m_sampleRun. OutputBMP( );}//根据 sampleRun. OutputBMP 出 BMP 图
    catch(Exception ex)
    { Trace. WriteLine( ex. Message); } }
```

② EPS 出图 Button 控件的 Click 事件。左键双击 EPS 出图 Button 控件，即可添加该按钮的 Click 事件在 try 后添加代码 m_sampleRun. OutputEPS( )，其他同上。

③ PNG 出图（背景透明）Button 控件的 Click 事件。左键双击 PNG 出图（背景透明）Button 控件，即可添加该按钮的 Click 事件代码 m_sampleRun. OutputPNG(true)，其他同上，true 代表设置背景为透明，如果是 false，则背景不透明。

8. 运行结果

窗体设计完成后点击 按钮，运行该程序，运行后弹出窗体，如图 5-6 所示，可以设置地图输出的格式包括 BMP、EPS、PNG（背景透明或者背景不透明）。

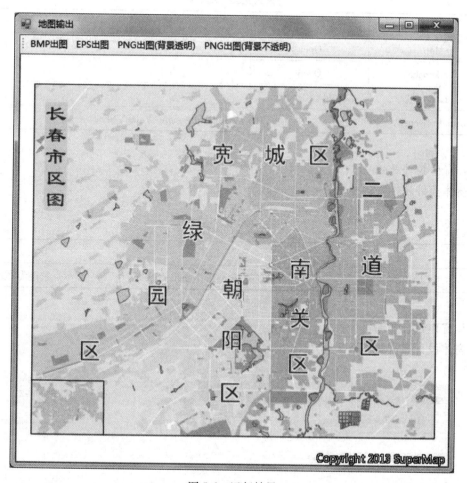

图 5-6 运行结果

5.2 专题图制作

5.2.1 点密度专题图

1. 数据

在安装目录\SampleData\ThematicMaps\ThematicMaps.smwu 路径下打开数据文件。

2. 新建文件夹和工程

在【D:\MyProject\】文件中新建一个文件夹，命名为【第 5 章　地图制图及专题图制作】，并在其目录下建立一个文件夹，命名为【ThemeDotDensityDisplay】，在此文件夹里新建一个工程，将此工程命名为 ThemeDotDensityDisplay。

3. 关键类型/成员

关键类型/成员见表 5-7。

表 5-7　关键类型/成员表

控件/类	方法	属性	事件
Map	Open		
Workspace	Open		
Layers	Add、Remove		
ThemeDotDensity		DotExpression、Value、Style	

4. 窗体控件属性

窗体控件属性具体见表 5-8。

表 5-8　窗体控件属性表

控件	Name	Text
CheckBox	m_checkBoxThemeDotDensityDisplay	显示点密度专题图

5. 窗体设计布局

窗体设计布局如图 5-7 所示。

图 5-7　窗体设计布局图

6. 创建 SampleRun 类

（1）添加 using 引用代码

添加 using 引用代码具体如下：

using System.Diagnostics;　//添加 System.Diagnostics 命名空间，使其对应的类可以使用

using System.Drawing;　//添加 System.Drawing a 命名空间，使其对应的类可以使用

using System.Windows.Forms;　//添加 System.Windows.Forms 命名空间，使其对应的类可以使用

using SuperMap.Data；//添加 SuperMap.Data 命名空间，使其对应的类可以使用
using SuperMap.UI；//添加 SuperMap.UI 命名空间，使其对应的类可以使用
using SuperMap.Mapping；//添加 SuperMap.Mapping 命名空间，使其对应的类可以使用
（2）在 public class SampleRun 中添加代码
在 public class SampleRun 中添加如下代码：
//定义类对象和控件
private Workspace m_workspace；
private MapControl m_mapControl；
private DatasetVector m_dataset；// DatasetVector：矢量数据集类。描述矢量数据集，并提供相应的管理和操作。对矢量数据集的操作主要包括数据查询、修改、删除、建立索引等。
private String m_themeLayerName；
private Boolean m_isThemeLayerDisplay；
public Boolean IsThemeLayerDisplay
    ｛set｛m_isThemeLayerDisplay = value；｝｝
/// 根据 workspace 和 mapControl 构造 SampleRun 对象
public SampleRun（Workspace workspace，MapControl mapControl）
    ｛m_workspace = workspace；
m_mapControl = mapControl；
m_mapControl.Map.Workspace = m_workspace；
Initialize（）；｝
private void Initialize（）
    ｛try｛//打开工作空间及地图
WorkspaceConnectionInfo conInfo = new WorkspaceConnectionInfo
（@"..\..\SampleData\ThematicMaps\ThematicMaps.smwu"）；
m_workspace.Open（conInfo）；//根据 workspace.Open 打开工作空间
m_dataset = m_workspace.Datasources［"Thematicmaps"］.Datasets［"BaseMap_R"］ as DatasetVector；
m_mapControl.Map.Open（"京津地区地图"）；//根据 Map.Open 打开地图
m_mapControl.Map.Refresh（）；
// 调整 mapControl 的状态
m_mapControl.Action = SuperMap.UI.Action.Pan；｝
catch（Exception ex）
｛Trace.WriteLine（ex.Message）；｝｝
/// 通过属性 m_isThemeLayerDisplay 来控制是否显示点密度专题图
public void CheckThemeDotDensityDisplay（）
    ｛try｛if（m_isThemeLayerDisplay）
    ｛AddThemeDotDensityLayer（）；｝

else{m_mapControl.Map.Layers.Remove(m_themeLayerName);}//删除指定图层
m_mapControl.Map.Refresh();//刷新地图}}
catch(Exception ex)
{Trace.WriteLine(ex.Message);}}
/// 设置ThemeDotDensity的属性,添加点密度专题图图层到地图
private void AddThemeDotDensityLayer()
{try{ThemeDotDensity dotDensity = new ThemeDotDensity();//ThemeDotDensity:点密度专题图类型。SuperMap iObjects.NET的点密度专题图用一定大小、形状相同的点表示现象分布范围、数量特征和分布密度。点的多少和所代表的意义由地图的内容确定。
dotDensity.DotExpression = "Pop_Density99";//设置点密度专题图是根据哪个属性制作
dotDensity.Value = 0.00030;//设置点代表的最低值
GeoStyle geostyle = new GeoStyle();//GeoStyle:几何风格类。用于定义点状符号、线状符号、填充符号风格及其相关属性。
//设置风格
geostyle.LineColor = Color.Red;
geostyle.MarkerSize = new Size2D(0.8,0.8);
dotDensity.Style = geostyle;
//将制作好的专题图添加到地图中显示
Layer themeLayer = m_mapControl.Map.Layers.Add(m_dataset,dotDensity,true);//Layer:图层类。该类提供了图层显示和控制等便于地图管理的一系列属性。
m_mapControl.Map.Layers.MoveDown(0);
m_themeLayerName = themeLayer.Name;
m_mapControl.Map.Refresh();}
catch(Exception ex)
{Trace.WriteLine(ex.Message);}}

7. 窗体代码

(1)添加using引用代码

添加using引用代码具体如下:

using System.Diagnostics;//添加System.Diagnostics命名空间,使其对应的类可以使用

(2)实例化代码

实例化SampleRun,以及public FormMain()函数部分的代码如下:

//定义类对象
private SampleRun m_sampleRun;
private SuperMap.Data.Workspace m_workspace;
private SuperMap.UI.MapControl m_mapControl;
public FormMain()
{try{InitializeComponent();

this. m_workspace = new SuperMap. Data. Workspace( );//创建 Workspace 新对象

this. m_mapControl = new SuperMap. UI. MapControl( );//创建 MapControl 新对象

//初始化控件

m_mapControl. Dock = DockStyle. Fill;

base. Controls. Add(m_mapControl);

m_sampleRun = new SampleRun(m_workspace, m_mapControl);

m_checkBoxThemeDotDensityDisplay. Checked = true;}

catch (Exception ex)

{Trace. WriteLine(ex. Message); }}

(3) 窗体的 FormClose 事件

在退出程序时,要断开空间的关联,关闭地图窗口、工作空间。因此,在属性窗口点击图标,左键双击 FormClose 事件,添加代码同前文中的相应部分。

(4) 添加各个控件的代码

显示点密度专题图 CheckBox 控件的 CheckedChanged 事件,在 m_checkBox Theme DotDensityDisplay 的属性窗口中,选择 按钮后,左键双击 CheckedChanged,即可添加以下代码:

private void m_panelCheckBoxe_CheckedChanged(object sender, EventArgs e)

{ m _ sampleRun. IsThemeLayerDisplay = m _ checkBoxThemeDotDensityDisplay. Checked;

m _ sampleRun. CheckThemeDotDensityDisplay ( );//通过 sampleRun. CheckTheme DotDensityDisplay 显示点密度专题图

}//在 CheckBox 的选中事件触发时,运行 SampleRun 的功能操作

8. 运行结果

窗体设计完成后点击 按钮,运行该程序,运行后弹出窗体,如图 5-8 所示,可以通过勾选"显示点密度专题图"进行显示专题图。点密度专题图是通过点的密集程度来表示数据的分布情况的。

5.2.2 内存数据专题图

1. 数据

在安装目录\SampleData\ThematicMaps\ThematicMap. smwu 路径下打开数据文件。

2. 新建文件夹和工程

在【D:\MyProject\】文件中新建一个文件夹,命名为【第 5 章　地图制图及专题图制作】,并在其目录下建立一个文件夹,命名为【ThemeMemoryDataDisplay】,在此文件夹里新建一个工程,将此工程命名为 ThemeMemoryDataDisplay。

3. 关键类型/成员

关键类型/成员见表 5-9。

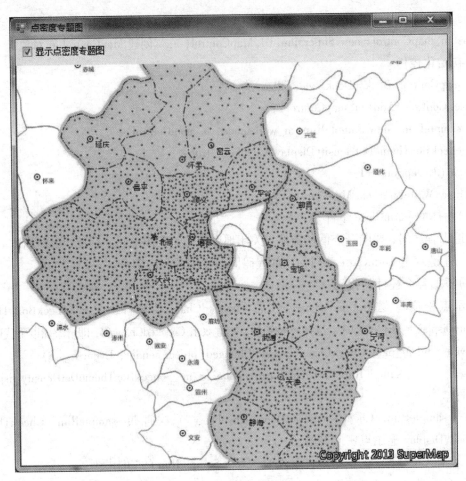

图 5-8 运行结果

表 5-9 关键类型/成员表

控件/类	方法	属性	事件
ThemeGraph	SetMemoryKeys		
ThemeGraphItem	SetMemoryDoubleValues		

4. 窗体控件属性

窗体控件属性见表 5-10。

表 5-10 窗体控件属性表

控件	Name	Text
CheckBox	m_checkBoxThemeGraduatedSymbolDisplay	显示等级符号专题图

5. 窗体设计布局

窗体设计布局如图 5-9 所示。

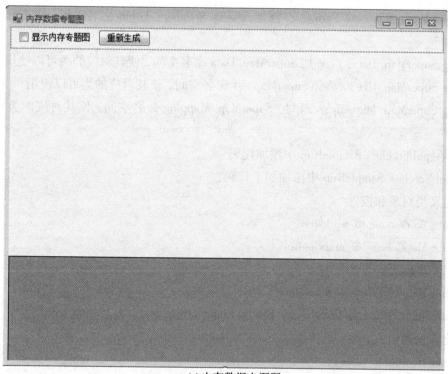

(a)内存数据专题图

(b)等级符号专题图

图 5-9　窗体设计布局图

6. 创建 SampleRun 类

(1)添加 using 引用代码

添加 using 引用代码具体如下：

using System.Diagnostics；//添加 System.Diagnostics 命名空间，使其对应的类可以使用
using System.Drawing；//添加 System.Drawing 命名空间，使其对应的类可以使用
using System.Windows.Forms；//添加 System.Windows.Forms 命名空间，使其对应的类可以使用
using SuperMap.Data；//添加 SuperMap.Data 命名空间，使其对应的类可以使用
using SuperMap.UI；//添加 SuperMap.UI 命名空间，使其对应的类可以使用
using SuperMap.Mapping；//添加 SuperMap.Mapping 命名空间，使其对应的类可以使用

（2）在 public class SampleRun 中添加代码

在 public class SampleRun 中添加如下代码：

//定义类对象和控件

private Workspace m_workspace；

private MapControl m_mapControl；

private DatasetVector m_dataset；

private DataGridView m_dataGrid；

private static String m_layerName = String.Empty；

//定义变量

private Double[ ] value1；

private Double[ ] value2；

private Double[ ] value3；

private Double[ ] value4；

/// 根据 workspace 和 mapControl 构造 SampleRun 对象

public SampleRun (Workspace workspace, MapControl mapControl, DataGridView dataGridView)

｛try｛m_workspace = workspace；

m_mapControl = mapControl；

m_dataGrid = dataGridView；

m_mapControl.Map.Workspace = workspace；

Initialize( )；｝

catch (Exception ex)

｛Trace.WriteLine(ex.Message)；｝｝

private void Initialize( )

｛try｛value1 = new Double[15]；

value2 = new Double[15]；

value3 = new Double[15]；

```
            value4 = new Double[15];
    //打开工作空间及地图
    WorkspaceConnectionInfo conInfo = new
    WorkspaceConnectionInfo(@"..\..\SampleData\ThematicMaps\ThematicMaps.smwu");
        m_workspace.Open(conInfo);//根据 workspace.Open 打开工作空间
        m_mapControl.Map.Open("京津地区地图");//根据 Map.Open 打开地图
        m_mapControl.Action = SuperMap.UI.Action.Pan;
        m_dataset = m_workspace.Datasources[0].Datasets["BaseMap_R"] as DatasetVector;
        InitializeDataGrid();}
        catch(Exception ex)
        {Trace.WriteLine(ex.Message);}}
    private void InitializeDataGrid()//定义 InitializeDataGrid()初始化函数 填充 dataGrid
        {m_dataGrid.Columns.Add("AreaName","AreaName");
    m_dataGrid.Columns[0].ValueType = typeof(String);
        m_dataGrid.Columns.Add("Key","Key");
        m_dataGrid.Columns[1].ValueType = typeof(Int32);
        m_dataGrid.Columns.Add("value1","value1");
        m_dataGrid.Columns[2].ValueType = typeof(Double);
        m_dataGrid.Columns.Add("value2","Value2");
        m_dataGrid.Columns[3].ValueType = typeof(Double);
        m_dataGrid.Columns.Add("value3","Value3");
        m_dataGrid.Columns[4].ValueType = typeof(Double);
        m_dataGrid.Columns.Add("value4","Value4");
        m_dataGrid.Columns[5].ValueType = typeof(Double);
        Recordset recordset = m_dataset.GetRecordset(false, CursorType.Static);
        m_dataGrid.Rows.Add(recordset.RecordCount);
        for(int index = 0; index < recordset.RecordCount; recordset.MoveNext(), ++index)
        {m_dataGrid.Rows[index].Cells[0].Value = recordset.GetFieldValue("Name");//
recordset:记录集类。通过此类,可以实现对矢量数据集中的数据进行操作。
        m_dataGrid.Rows[index].Cells[1].Value = recordset.GetFieldValue("SmID");}
        recordset.Close();
        DisplayMemoryDataTheme(true);}
    private void GetRandomNums()//定义 GetRandomNums()初始化函数
        {Random r = new Random((Int32)DateTime.Now.Ticks);// Random:表示为随机数
生成器,一种能够产生满足某些随机性统计要求的数字序列的设备。
            for(int i = 0; i < 15; ++i)
```

```
            {value1[i] = (r.NextDouble()+0.1) * 1000;
             value2[i] = (r.NextDouble()+0.1) * 1000;
             value3[i] = (r.NextDouble()+0.1) * 1000;
             value4[i] = (r.NextDouble()+0.1) * 1000;
             m_dataGrid.Rows[i].Cells["value1"].Value = value1[i];
             m_dataGrid.Rows[i].Cells["value2"].Value = value2[i];
             m_dataGrid.Rows[i].Cells["value3"].Value = value3[i];
             m_dataGrid.Rows[i].Cells["value4"].Value = value4[i];}}
```

/// 显示内存数据专题图,并判断是否显示

public void DisplayMemoryDataTheme(Boolean isDisplay)
　　{try{m_mapControl.Map.Layers.Remove(m_layerName);
　　　if(isDisplay)
{//构造统计专题图
　　ThemeGraph themeMemoryData = new ThemeGraph();//ThemeGraph:统计专题图类。统计专题图通过为每个要素或记录绘制统计图来反映其对应的专题值的大小。统计专题图可以基于多个变量,反映多种属性,即可以将多个专题变量的值绘制在一个统计图上。

　　hemeMemoryData.SetMemoryKeys(new Int32[]{1, 2, 3, 4, 5, 6, 7, 8, 9, 10, 11, 12, 13, 14, 15});

　　GetRandomNums();

　　// 初始化子项,并设置子项的风格

　　GeoStyle geoStyle = new GeoStyle();//GeoStyle:几何风格类。用于定义点状符号、线状符号、填充符号风格及其相关属性。

　　//设置风格

　　geoStyle.LineWidth = 0.1;

　　geoStyle.FillGradientMode = FillGradientMode.Linear;// FillGradientMode:该枚举定义了渐变填充模式的渐变类型常量。所有渐变类型都是两种颜色之间的渐变,即从渐变起始色到渐变终止色之间的渐变。渐变风格的计算都是以填充区域的边界矩形,即最小外接矩形作为基础的,因而以下提到的填充区域范围即为填充区域的最小外接矩形。Linear:线性渐变。从水平线段的起始点到终止点,其颜色从起始色均匀渐变到终止色,垂直于该线段的直线上颜色不发生渐变。

　　ThemeGraphItem item0 = new ThemeGraphItem();//新建 ThemeGraphItem 对象

　　//设置风格

　　geoStyle.FillForeColor = Color.FromArgb(115, 178, 255);

　　item0.UniformStyle = geoStyle;

　　item0.SetMemoryDoubleValues(value1);

　　//通过 geoStyle 设置 item1 风格

```
ThemeGraphItem item1 = new ThemeGraphItem();
geoStyle.FillForeColor = Color.FromArgb(255,85,0);
item1.UniformStyle = geoStyle;
item1.SetMemoryDoubleValues(value2);
//通过 geoStyle 设置 item2 风格
ThemeGraphItem item2 = new ThemeGraphItem();
geoStyle.FillForeColor = Color.FromArgb(255,255,115);
item2.UniformStyle = geoStyle;
item2.SetMemoryDoubleValues(value3);
//通过 geoStyle 设置 item3 风格
ThemeGraphItem item3 = new ThemeGraphItem();
geoStyle.FillForeColor = Color.FromArgb(222,113,255);
item3.UniformStyle = geoStyle;
item3.SetMemoryDoubleValues(value4);
// 添加项目
themeMemoryData.Add(item0);
themeMemoryData.Add(item1);
themeMemoryData.Add(item2);
themeMemoryData.Add(item3);
// 设置统计专题图属性
themeMemoryData.IsOffsetFixed = false;
themeMemoryData.IsFlowEnabled = false;
themeMemoryData.IsOverlapAvoided = false;
themeMemoryData.MaxGraphSize = 120000;
themeMemoryData.MinGraphSize = 50000;
themeMemoryData.IsGraphTextDisplayed = true;
themeMemoryData.GraphTextFormat = ThemeGraphTextFormat.Value;
themeMemoryData.GraphType = ThemeGraphType.Pie;
//设置文本风格
TextStyle textStyle = themeMemoryData.GraphTextStyle;
textStyle.FontHeight = 9000;
textStyle.IsSizeFixed = false;
// 添加内存数据换题图图层
Layer layer = this.m_mapControl.Map.Layers.Add(m_dataset, themeMemoryData, true);
            m_layerName = layer.Name;}// 刷新地图
m_mapControl.Map.Refresh();}
```

catch(Exception ex)
{Trace.WriteLine(ex.Message);}}

7.窗体代码

(1)添加 using 引用代码

添加 using 引用代码具体如下：

using System.Diagnostics;//添加 System.Diagnostics 命名空间，使其对应的类可以使用

(2)初始化代码

初始化 SampleRun，以及 public FormMain()函数部分的代码如下：

//定义类对象及控件

private SampleRun m_sampleRun;

private SuperMap.Data.Workspace m_workspace;

private SuperMap.UI.MapControl m_mapControl;

public FormMain()

{try{ InitializeComponent();

this.m_workspace = new SuperMap.Data.Workspace(this.components);//创建 Workspace 新对象

this.m_mapControl=new SuperMap.UI.MapControl();

//初始化控件

m_mapControl.Dock=DockStyle.Fill;

m_sampleRun=new SampleRun(m_workspace,m_mapControl);//创建 SampleRun 新对象

base.Controls.Add(m_mapControl);

base.Controls.SetChildIndex(m_mapControl,0);

m_mapControl.Map.ViewEntire();

m_checkBoxThemeGraduatedSymbolDisplay.Checked=true;}

catch(Exception ex)

{Trace.WriteLine(ex.Message);}}

(3)窗体的 FormClose 事件

在退出程序时，要断开空间的关联，关闭地图窗口、工作空间。因此，在属性窗口点击 图标，左键双击 FormClose 事件，添加代码同前文相应部分。

(4)添加各个控件的代码

① 内存数据专题图 CheckBox 控件的 CheckedChanged 事件。在 checkBoxShowThemeLayer 的属性窗口中，选择 按钮后，左键双击 CheckedChanged，即可添加以下代码，实现显示内存数据专题图：

private void checkBoxShowThemeLayer_CheckedChanged(object sender, EventArgs e)

{ m_sampleRun.DisplayMemoryDataTheme(checkBoxShowThemeLayer.Checked);}

5.2 专题图制作

//通过 DisplayMemoryDataTheme 显示内存数据专题图

② 重新生成 button 的 click 事件。左键双击 button，即可添加以下代码，实现重新生成：

private void makeNewData_Click(object sender, EventArgs e)
{m_sampleRun. DisplayMemoryDataTheme(checkBoxShowThemeLayer. Checked);//通过 DisplayMemoryDataTheme 重新生成}

8. 运行结果

窗体设计完成后点击 ▶ 按钮，运行该程序，运行后弹出窗体如图 5-10 所示，可以通过勾选"显示内存专题图"进行显示专题图，通过圆形的大小来表示数据的分布。

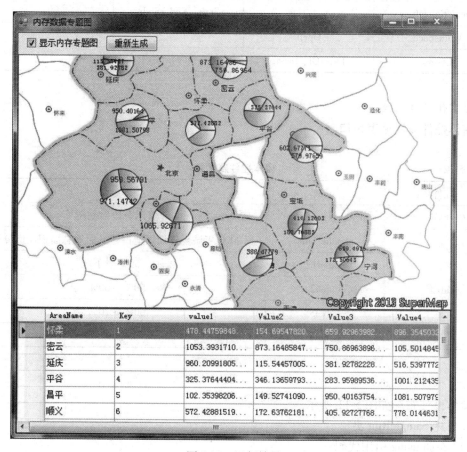

图 5-10 运行结果

5.2.3 分段专题图

1. 数据

在安装目录\SampleData\ThematicMaps\ThematicMaps.smwu 路径下打开数据文件。

2. 新建文件夹和工程

在【D:\MyProject\】文件中新建一个文件夹,命名为【第 5 章 地图制图及专题图制作】,并在其目录下建立一个文件夹,命名为【ThemeRangeDisplay】,在此文件夹里新建一个工程,将此工程命名为 ThemeRangeDisplay。

3. 关键类型/成员

关键类型/成员具体见表 5-11。

表 5-11 关键类型/成员表

控件/类	方法	属性	事件
ThemeRange	AddToHead		
ThemeRangeItem		Start、End、Style	
Workspace	Open		
Layers	IndexOf、Insert		

4. 窗体控件属性

窗体控件属性见表 5-12。

表 5-12 窗体控件属性表

控件	Name	Text
CheckBox	m_checkBoxSelect	显示分段专题图
Label	m_labelTitle	2000 年城乡建设用地占建设用地总面积比例分级
	label1	50% 以下
	Label2	50% ~60%
	Label3	60% ~70%
	Label4	70% 以上

5. 窗体设计布局

窗体设计布局如图 5-11 所示。

6. 创建 SampleRun 类

(1) 添加 using 引用代码

添加 using 引用代码具体如下:

using System. Diagnostics; //添加 System. Diagnostics 命名空间,使其对应的类可以使用

using System. Drawing; //添加 System. Drawing 命名空间,使其对应的类可以使用

using System. Windows. Forms; //添加 System. Windows. Forms 命名空间,使其对应的类可以使用

using SuperMap. Mapping; //添加 SuperMap. Mapping 命名空间,使其对应的类可以使用

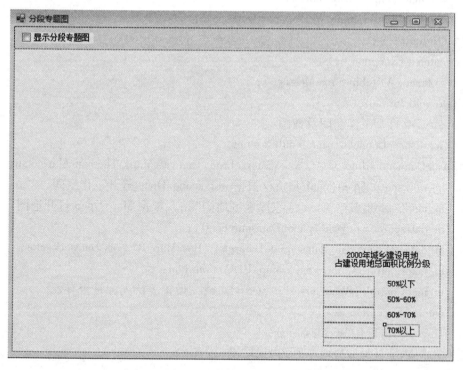

图 5-11　窗体设计布局图

using SuperMap. Data；//添加 SuperMap. Data 命名空间，使其对应的类可以使用
using SuperMap. UI；//添加 SuperMap. UI 命名空间，使其对应的类可以使用
（2）在 public class SampleRun 中添加代码
在 public class SampleRun 中添加如下代码：
//定义类对象和控件
private Workspace m_workspace；
private MapControl m_mapControl；
private Datasource m_datasource；
private DatasetVector m_datasetVector；// DatasetVector：矢量数据集类。描述矢量数据集，并提供相应的管理和操作。对矢量数据集的操作主要包括数据查询、修改、删除、建立索引等。
private Layer m_layer；// Layer：图层类
private String m_themeLayerName；
private Boolean m_isThemeLayerDisplay；
public Boolean IsThemeLayerDisplay
　　｛set｛ m_isThemeLayerDisplay = value；｝｝
/// 根据 workspace 和 mapControl 构造 SampleRun 对象
public SampleRun(Workspace workspace，MapControl mapControl)
　　｛try｛m_workspace = workspace；

```csharp
            m_mapControl = mapControl;
            m_mapControl.Map.Workspace = workspace;
            Initialize();}
        catch(Exception ex)
          {Trace.WriteLine(ex.Message);}}
    private void Initialize()
        {try{//打开工作空间及地图
        WorkspaceConnectionInfo conInfo = new
orkspaceConnectionInfo(@"..\..\SampleData\ThematicMaps\ThematicMaps.smwu");
            m_workspace.Open(conInfo);//根据 workspace.Open 打开工作空间
            m_mapControl.Map.Open("京津地区地图");//根据 Map.Open 打开地图
            m_datasource = m_workspace.Datasources[0];
            m_datasetVector = m_datasource.Datasets["BaseMap_R"] as DatasetVector;
            m_mapControl.Action = SuperMap.UI.Action.Pan;
            m_mapControl.IsWaitCursorEnabled = false;//设置等待光标是否有效}
        catch(Exception ex)
          {Trace.WriteLine(ex.Message);}    }
        /// 判断是否添加分段专题图图层
    public void SetThemeRangeDisplay()
        {try{ if(m_isThemeLayerDisplay)
            {AddThemeRangeLayer();}
            else{m_mapControl.Map.Layers.Remove(m_themeLayerName);//删除指定图层
            m_mapControl.Map.Refresh();//刷新地图}}
        catch(Exception ex)
          {Trace.WriteLine(ex.Message);}}
        /// 构造分段专题图并添加分段专题图图层
    private void AddThemeRangeLayer()
            { try{//构造分段专题图对象并设置分段字段表达式
            ThemeRange themeRange = new ThemeRange();//ThemeRange:分段专题图类。按照提供的分段方法对字段的属性值进行分段,并根据每个属性值所在的分段范围赋予相应对象的显示风格。
            themeRange.RangeExpression = "UrbanRural";
            GeoStyle style = new GeoStyle();//GeoStyle:几何风格类。用于定义点状符号、线状符号、填充符号风格及其相关属性。
            //设置风格
            style.LineColor = Color.White;
            style.LineWidth = 0.3;
            //初始化分段专题图子项并设置各自的风格
            ThemeRangeItem item0 = new ThemeRangeItem();//ThemeRangeItem:分段专题图子项
```

类。在分段专题图中,将分段字段的表达式的值按照某种分段模式分成多个范围段。每个分段都有其分段起始值、终止值、名称和风格等。每个分段所表示的范围为[Start, End)。

```
//设置起始值和终止值
item0.Start = double.MinValue;
item0.End = 50;
style.FillForeColor = Color.FromArgb(209, 182, 210);
item0.Style = style;
//通过 ThemeRangeItem 设置 item1 风格
ThemeRangeItem item1 = new ThemeRangeItem();
item1.Start = 50;
item1.End = 60;
style.FillForeColor = Color.FromArgb(205, 167, 183);
item1.Style = style;
//通过 ThemeRangeItem 设置 item2 风格
ThemeRangeItem item2 = new ThemeRangeItem();
item2.Start = 60;
item2.End = 70;
style.FillForeColor = Color.FromArgb(183, 128, 151);
item2.Style = style;
//通过 ThemeRangeItem 设置 item3 风格
ThemeRangeItem item3 = new ThemeRangeItem();
item3.Start = 70;
item3.End = 90;
style.FillForeColor = Color.FromArgb(164, 97, 136);
item3.Style = style;
//将分段专题图子项依次添加到分段专题图
themeRange.AddToHead(item0);
themeRange.AddToTail(item1);
themeRange.AddToTail(item2);
themeRange.AddToTail(item3);
//添加分段专题图层
Layers layers = m_mapControl.Map.Layers;
Int32 index = layers.IndexOf(m_datasetVector.Name+"@"+m_datasource.Alias);
m_layer = layers.Insert(index, m_datasetVector, themeRange);
m_themeLayerName = m_layer.Name;
m_mapControl.Map.Refresh();//刷新图层}
catch (Exception ex)
    {Trace.WriteLine(ex.Message);}    }
```

7. 窗体代码

（1）添加 using 引用代码

添加 using 引用代码具体如下：

using System. Diagnostics；//添加 System. Diagnostics 命名空间，使其对应的类可以使用

（2）初始化代码

初始化 SampleRun，public FormMain( ) 函数部分的代码如下：

```
//定义类对象和控件
private SampleRun m_sampleRun;
private SuperMap. Data. Workspace m_workspace;
private SuperMap. UI. MapControl m_mapControl;
public FormMain( )
    {try{ InitializeComponent( );
      m_workspace = new SuperMap. Data. Workspace( this. components );
      m_mapControl = new SuperMap. UI. MapControl( );
      m_mapControl. Dock = DockStyle. Fill;}
    catch（Exception ex）
```

（3）窗体的 Load 事件

在窗体的属性窗口中，选择 ❋ 按钮后，左键双击 Load，即可添加以下代码：

```
private void FormMain_Load( object sender, EventArgs e)
    {try{//设定图例中分段的颜色并设置图例初始状态为不显示
      m_panel0. BackColor = Color. FromArgb( 209, 182, 210 );
      m_panel1. BackColor = Color. FromArgb( 205, 167, 183 );
      m_panel2. BackColor = Color. FromArgb( 183, 128, 151 );
      m_panel3. BackColor = Color. FromArgb( 164, 97, 136 );
      m_panelDiagram. Visible = false;
      //实例化 SampleRun
      m_sampleRun = new SampleRun( m_workspace, m_mapControl );
      m_panelMap. Controls. Add( m_mapControl );
      //复选框选中
      m_checkBoxSelect. Checked = true;}
    catch（Exception ex）
    {Trace. WriteLine( ex. Message );  }}
```

（4）窗体的 FormClose 事件

在退出程序时，要断开空间的关联，关闭地图窗口、工作空间。因此，在属性窗口点击 ❋ 图标，左键双击 FormClose 事件，添加代码同前文中相应部分。

（5）添加各个控件的代码

显示分段专题图 CheckBox 控件的 CheckedChanged 事件。在 m_checkBoxSelect 的属性窗口中，选择 ❋ 按钮后，左键双击 CheckedChanged，即可添加以下代码，实现显示分段专题图：

private void m_checkBoxSelect_CheckedChanged(object sender, EventArgs e) {try{m_sampleRun.IsThemeLayerDisplay=m_checkBoxSelect.Checked；

m_panelDiagram.Visible=m_checkBoxSelect.Checked；

m_sampleRun.SetThemeRangeDisplay( )；}//通过 sampleRun.SetThemeRangeDisplay 显示分段专题图

catch（Exception ex）

{Trace.WriteLine（ex.Message）；}} //在 CheckBox 的选中事件触发时，运行 SampleRun 的功能操作

8．运行结果

窗体设计完成后点击▶按钮，运行该程序，运行后弹出窗体，如图 5-12 所示，可以通过勾选"显示分段专题图"，进行显示专题图。通过颜色的渐变显示数据的分布，窗体右下角的图例可以清晰地得出各部分的数据分布情况。

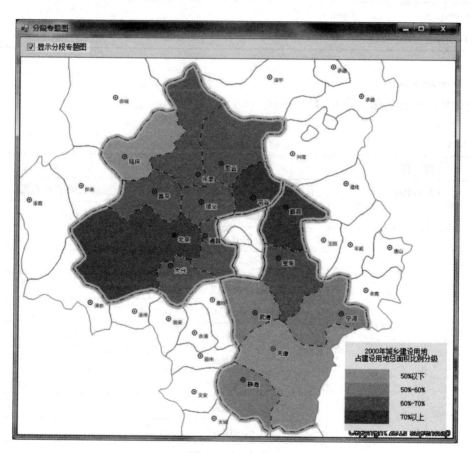

图 5-12　运行结果

5.2.4　单值专题图

1．数据

在安装目录\SampleData\ThematicMaps\ThematicMaps.smwu 路径下打开数据文件。

2. 新建文件夹和工程

在【D:\MyProject\】文件中新建一个文件夹,命名为【第 5 章 地图制图及专题图制作】,并在其目录下建立一个文件夹,命名为【单值专题图】,在此文件夹里新建一个工程,将此工程命名为 ThemeUniqueDisplay。

3. 关键类型/成员

关键类型/成员见表 5-13。

表 5-13 关键类型/成员表

控件/类	方法	属性	事件
ThemeUnique		UniqueExpression	
ThemeUniqueItem		Caption、Unique	
Workspace	Open		
GeoStyle		FillForeColor	

4. 窗体控件属性

窗体控件属性具体见表 5-14。

表 5-14 窗体控件属性表

控件	Name	Text
CheckBox	m_checkBox	显示单值专题图
Label	m_labelLegend	图例
Panel	m_panelLegend	
	m_panelCity	
	m_panelDry	
	m_panelWater	
	m_panelField	
	m_panelDesert	
	m_panelMesh	
	m_panelLake	
	m_panelBush	
	m_panelWood	
	m_panelForrest	
	m_panelGrassland	

5. 窗体设计布局

窗体设计布局如图 5-13 所示。

图 5-13　窗体设计布局图

6. 创建 SampleRun 类

(1) 添加 using 引用代码

添加 using 引用代码具体如下：

using SuperMap. Analyst. SpatialAnalyst；//添加 SuperMap. Analyst. SpatialAnalyst 命名空间，使其对应的类可以使用

using System. Diagnostics；//添加 System. Diagnostics 命名空间，使其对应的类可以使用

using System. Drawing；//添加 System. Drawing 命名空间，使其对应的类可以使用

using System. Windows. Forms；//添加 System. Windows. Forms 命名空间，使其对应的类可以使用

using SuperMap. Data；//添加 SuperMap. Data 命名空间，使其对应的类可以使用

using SuperMap. Mapping；//添加 SuperMap. Mapping 命名空间，使其对应的类可以使用

using SuperMap. UI；//添加 SuperMap. UI 命名空间，使其对应的类可以使用

(2) 在 public class SampleRun 中添加代码

在 public class SampleRun 中添加如下代码：

//定义类对象和控件

private Workspace m_workspace；

private MapControl m_mapControl；

private String m_themeLayerName；

private List<LabelItem> m_items；

public List<LabelItem> Items

{get { return m_items；} }

/// 根据 workspace 和 mapControl 构造 SampleRun 对象

public SampleRun(Workspace workspace，MapControl mapControl)

　　{try{ m_workspace = workspace；

```
            m_mapControl = mapControl;
            m_mapControl.Map.Workspace = m_workspace;
            m_items = new List<LabelItem>();
            Initialize();}
            catch(Exception ex)
            {Trace.WriteLine(ex.Message);}}
        private void Initialize()
            {try{//打开工作空间及地图
WorkspaceConnectionInfo workspaceConnectionInfo = new
WorkspaceConnectionInfo(@"..\..\SampleData\ThematicMaps\ThematicMaps.smwu");
workspaceConnectionInfo.Type = WorkspaceType.SMWU;//获取工作空间类型
m_workspace.Open(workspaceConnectionInfo);//通过 workspace.Open 打开工作空间
Dataset dataset = m_workspace.Datasources[0].Datasets["Landuse_R"];
m_mapControl.Map.Layers.Add(dataset, true);
// 平移地图
m_mapControl.Action = SuperMap.UI.Action.Pan;
// 底图不可选，防止专题图图层被覆盖
for(Int32 i = 0; i < m_mapControl.Map.Layers.Count; i++)
    {Layer m_layer = m_mapControl.Map.Layers[i];
     m_layer.IsSelectable = false;}//设置图层中对象是否被选择
    m_mapControl.Map.Refresh();}//刷新地图
     catch(Exception ex)
        {Trace.WriteLine(ex.Message);}}
    /// 显示单值专题图
public void themeuniqueDisplay(Boolean isDisplayThemeLayer)
      {try{// 先移除图层
    m_mapControl.Map.Layers.Remove(m_themeLayerName);
       if(isDisplayThemeLayer)
      {// 当只有普通图层时添加专题图层
DatasetVector datasetVector = m_workspace.Datasources[0].Datasets["Landuse_R"] as
DatasetVector;
       // 设置单值专题图
       ThemeUnique themeunique = new ThemeUnique();//ThemeUnique：单值专题图类。将
字段或表达式的值相同的要素采用相同的风格来显示，从而用来区分不同的类别。
       themeunique.UniqueExpression = "LandType";
       // 实例化若干单值专题图子项分别设置每个子项的属性
       Color itemColor = Color.FromArgb(206, 101, 156);
       String itemName = "城市";
       GeoStyle geostyle1 = new GeoStyle();//GeoStyle：几何风格类。用于定义点状符号、线状
```

符号、填充符号风格及其相关属性。

```
//设置风格
geostyle1.FillForeColor = itemColor;
geostyle1.FillOpaqueRate = 80;
geostyle1.LineSymbolID = 5;
//通过 ThemeUniqueItem 设置 item1 风格
ThemeUniqueItem item1 = new ThemeUniqueItem();
item1.Caption = itemName;
item1.IsVisible = true;
item1.Style = geostyle1;
item1.Unique = itemName;
LabelItem item = new LabelItem(itemColor, itemName);
m_items.Add(item);
itemColor = Color.FromArgb(181, 178, 181);
GeoStyle geostyle2 = new GeoStyle();
geostyle2.LineSymbolID = 5;
geostyle2.FillForeColor = itemColor;
geostyle2.FillOpaqueRate = 80;
//通过 ThemeUniqueItem 设置 item2 风格
ThemeUniqueItem item2 = new ThemeUniqueItem();
itemName = "旱地";
item2.Caption = itemName;
item2.IsVisible = true;
item2.Style = geostyle2;
item2.Unique = itemName;
item = new LabelItem(itemColor, itemName);
m_items.Add(item);
itemColor = Color.FromArgb(255, 255, 115);
GeoStyle geostyle3 = new GeoStyle();
geostyle3.LineSymbolID = 5;
geostyle3.FillForeColor = itemColor;
geostyle3.FillOpaqueRate = 45;
//通过 ThemeUniqueItem 设置 item3 风格
ThemeUniqueItem item3 = new ThemeUniqueItem();
itemName = "水浇地";
item3.Caption = itemName;
item3.IsVisible = true;
item3.Style = geostyle3;
item3.Unique = itemName;
```

```
item = newLabelItem(itemColor, itemName);
m_items.Add(item);
itemColor = Color.FromArgb(254, 175, 136);
GeoStyle geostyle4 = new GeoStyle();
geostyle4.LineSymbolID = 5;
geostyle4.FillForeColor = itemColor;
geostyle4.FillOpaqueRate = 100;
//通过 ThemeUniqueItem 设置 item4 风格
ThemeUniqueItem item4 = new ThemeUniqueItem();
itemName = "水田";
item4.Caption = itemName;
item4.IsVisible = true;
item4.Style = geostyle4;
item4.Unique = itemName;
item = newLabelItem(itemColor, itemName);
m_items.Add(item);
itemColor = Color.FromArgb(115, 77, 0);
GeoStyle geostyle5 = new GeoStyle();
geostyle5.LineSymbolID = 5;
geostyle5.FillForeColor = itemColor;
geostyle5.FillOpaqueRate = 100;
//通过 ThemeUniqueItem 设置 item5 风格
ThemeUniqueItem item5 = new ThemeUniqueItem();
itemName = "沙漠";
item5.Caption = itemName;
item5.IsVisible = true;
item5.Style = geostyle5;
item5.Unique = itemName;
item = newLabelItem(itemColor, itemName);
m_items.Add(item);
itemColor = Color.FromArgb(173, 170, 0);
GeoStyle geostyle6 = new GeoStyle();
geostyle6.LineSymbolID = 5;
geostyle6.FillForeColor = Color.FromArgb(173, 170, 0);
geostyle6.FillOpaqueRate = 100;
//通过 ThemeUniqueItem 设置 item6 风格
ThemeUniqueItem item6 = new ThemeUniqueItem();
itemName = "沼泽";
item6.Caption = itemName;
```

```
item6.IsVisible = true;
item6.Style = geostyle6;
item6.Unique = itemName;
item = new LabelItem(itemColor, itemName);
m_items.Add(item);
itemColor = Color.FromArgb(151,219,242);
GeoStyle geostyle7 = new GeoStyle();
geostyle7.LineSymbolID = 5;
geostyle7.FillForeColor = itemColor;
geostyle7.FillOpaqueRate = 100;
//通过 ThemeUniqueItem 设置 item7 风格
ThemeUniqueItem item7 = new ThemeUniqueItem();
itemName = "湖泊水库";
item7.Caption = itemName;
item7.IsVisible = true;
item7.Style = geostyle7;
item7.Unique = itemName;
item = new LabelItem(itemColor, itemName);
m_items.Add(item);
itemColor = Color.FromArgb(90,138,66);
GeoStyle geostyle8 = new GeoStyle();
geostyle8.LineSymbolID = 5;
geostyle8.FillForeColor = itemColor;
geostyle8.FillOpaqueRate = 50;
//通过 ThemeUniqueItem 设置 item8 风格
ThemeUniqueItem item8 = new ThemeUniqueItem();
itemName = "灌丛";
item8.Caption = itemName;
item8.IsVisible = true;
item8.Style = geostyle8;
item8.Unique = itemName;
item = new LabelItem(itemColor, itemName);
m_items.Add(item);
itemColor = Color.FromArgb(0,113,74);
GeoStyle geostyle9 = new GeoStyle();
geostyle9.LineSymbolID = 5;
geostyle9.FillForeColor = itemColor;
geostyle9.FillOpaqueRate = 60;
//通过 ThemeUniqueItem 设置 item9 风格
```

```
ThemeUniqueItem item9 = new ThemeUniqueItem();
itemName = "用材林";
item9.Caption = itemName;
item9.IsVisible = true;
item9.Style = geostyle9;
item9.Unique = itemName;
item = new LabelItem(itemColor, itemName);
m_items.Add(item);
itemColor = Color.FromArgb(0, 170, 132);
GeoStyle geostyle10 = new GeoStyle();
geostyle10.LineSymbolID = 5;
geostyle10.FillForeColor = itemColor;
geostyle10.FillOpaqueRate = 80;
//通过 ThemeUniqueItem 设置 item10 风格
ThemeUniqueItem item10 = new ThemeUniqueItem();
itemName = "经济林";
item10.Caption = itemName;
item10.IsVisible = true;
item10.Style = geostyle10;
item10.Unique = itemName;
item = new LabelItem(itemColor, itemName);
m_items.Add(item);
itemColor = Color.FromArgb(90, 179, 40);
GeoStyle geostyle11 = new GeoStyle();
geostyle11.LineSymbolID = 5;
geostyle11.FillForeColor = itemColor;
geostyle11.FillOpaqueRate = 30;
//通过 ThemeUniqueItem 设置 item11 风格
ThemeUniqueItem item11 = new ThemeUniqueItem();
itemName = "草地";
item11.Caption = itemName;
item11.IsVisible = true;
item11.Style = geostyle11;
item11.Unique = itemName;
item = new LabelItem(itemColor, itemName);
m_items.Add(item);
//向 themeunique 对象中逐个添加单值专题图子项
themeunique.Add(item1);
themeunique.Add(item2);
```

themeunique. Add(item3);
themeunique. Add(item4);
themeunique. Add(item5);
themeunique. Add(item6);
themeunique. Add(item7);
themeunique. Add(item8);
themeunique. Add(item9);
themeunique. Add(item10);
themeunique. Add(item11);
//将制作好的专题图添加到地图中显示
Layerlayer=this. m_mapControl. Map. Layers. Add(datasetVector, themeunique, true);//添加图层
　　m_themeLayerName=layer. Name;}
　　this. m_mapControl. Map. Refresh();//刷新}
　　catch(Exception ex)
　　　　{Trace. WriteLine(ex. Message); } }
　　public class LabelItem
　　　　{Color m_color;
　　　String m_caption;
　　public LabelItem(Color color, String caption)
　　　　{m_color=color;
　　　m_caption=caption; }
　　public Color Color
　　　　{get { return m_color; } }
　　public String Caption
　　　　{get { return m_caption; } } }

7. 窗体代码

(1)添加 using 引用代码

添加 using 引用具体代码如下:

using System. Diagnostics;//添加 System. Diagnostics 命名空间,使其对应的类可以使用
using SuperMap. Data;//添加 SuperMap. Data 命名空间,使其对应的类可以使用
using SuperMap. Mapping;//添加 SuperMap. Mapping 命名空间,使其对应的类可以使用
using SuperMap. UI;//添加 SuperMap. UI 命名空间,使其对应的类可以使用

(2)初始化代码

初始化 SampleRun,以及 public FormMain()函数部分的代码如下:
//定义类对象和控件
　private SampleRun m_sampleRun;

```
private SuperMap. Data. Workspace m_workspace；
private SuperMap. UI. MapControl m_mapControl；
public FormMain( )
    {try{ InitializeComponent( )；
  this. m_workspace=new SuperMap. Data. Workspace(this. components)；//创建新对象
  this. m_mapControl=new SuperMap. UI. MapControl( )；
  m_mapControl. Dock=DockStyle. Fill；}
  catch (Exception ex)
  {Trace. WriteLine( ex. Message)；} }
```

(3) 窗体的 Load 事件

在窗体的属性窗口中，选择 按钮后，左键双击 Load，即可添加以下代码：

```
private void FormMain_Load( object sender, EventArgs e)
    { try {//实例化 SampleRun
    m_sampleRun=new SampleRun(m_workspace, m_mapControl)；
    base. Controls. Add(m_mapControl)；
    m_checkBox. Checked=true；}
    catch (Exception ex)
    {Trace. WriteLine( ex. Message)；} }
```

(4) 窗体的 FormClose 事件

窗体的 FormClose 事件，在退出程序时，要断开空间的关联，关闭地图窗口、工作空间。因此，在属性窗口点击 图标，左键双击 FormClose 事件，添加代码同前文中相应部分。

(5) 添加各个控件的代码

① 显示单值专题图 CheckBox 控件的 CheckedChanged 事件。在 m_checkBox 的属性窗口中，选择 按钮后，左键双击 CheckedChanged，即可添加以下代码，实现显示单值专题图：

```
private void m_checkBox_CheckedChanged( object sender, EventArgs e)
     { try { m _ sampleRun. themeuniqueDisplay ( m _ checkBox. Checked)；//通过
sampleRun. themeuniqueDisplay 显示单值专题图
    m_panelLegend. Visible=m_checkBox. Checked；//设置控件的可见性
    catch (Exception ex)
    {Trace. WriteLine(ex. Message)；} }//在 CheckBox 的选中事件触发时，运行 SampleRun
的功能操作
```

② 控件 m_panelLegend 的 Paint 事件。在 m_panelLegend 的属性窗口中，选择 按钮后，左键双击 Paint，即可添加以下代码：

```
//设置单值专题图图例中不同项目的颜色
private void m_panelLegend_Paint( object sender, PaintEventArgs e)
```

```
{try{ if (m_checkBox.Checked && m_sampleRun.Items.Count > 0)
{m_labelBush.Text = m_sampleRun.Items[0].Caption;
m_panelBush.BackColor = m_sampleRun.Items[0].Color;
m_panelCity.BackColor = m_sampleRun.Items[1].Color;
m_labelCity.Text = m_sampleRun.Items[1].Caption;
m_panelDesert.BackColor = m_sampleRun.Items[2].Color;
m_labelDesert.Text = m_sampleRun.Items[2].Caption;
m_panelDry.BackColor = m_sampleRun.Items[3].Color;
m_labelDry.Text = m_sampleRun.Items[3].Caption;
m_panelField.BackColor = m_sampleRun.Items[4].Color;
m_labelField.Text = m_sampleRun.Items[4].Caption;
m_panelForrest.BackColor = m_sampleRun.Items[5].Color;
m_labelForrest.Text = m_sampleRun.Items[5].Caption;
m_panelGrassland.BackColor = m_sampleRun.Items[6].Color;
m_labelGrassland.Text = m_sampleRun.Items[6].Caption;
m_panelLake.BackColor = m_sampleRun.Items[7].Color;
m_labelLake.Text = m_sampleRun.Items[7].Caption;

m_panelMesh.BackColor = m_sampleRun.Items[8].Color;
m_labelMesh.Text = m_sampleRun.Items[8].Caption;
m_panelWater.BackColor = m_sampleRun.Items[9].Color;
m_labelWater.Text = m_sampleRun.Items[9].Caption;
m_panelWood.BackColor = m_sampleRun.Items[10].Color;
m_labelWood.Text = m_sampleRun.Items[10].Caption;}}
catch (Exception ex)
{Trace.WriteLine(ex.Message);}}
```

8. 运行结果

窗体设计完成后点击▶按钮,运行该程序,运行后弹出窗体如图 5-14 所示,可以通过勾选"显示单值专题图"进行显示专题图。通过颜色的渲染进行专题图的显示。

5.2.5 统一风格标签专题图

1. 数据

在安装目录\SampleData\ThematicMaps\ThematicMaps.smwu 路径下打开数据文件。

2. 新建文件夹和工程

在【D:\MyProject\】文件中新建一个文件夹,命名为【第 5 章 地图制图及专题图制作】,并在其目录下建立一个文件夹,命名为【统一风格标签专题图】,在此文件夹里新建一个工程,将此工程命名为 UniformStyleThemeLabel。

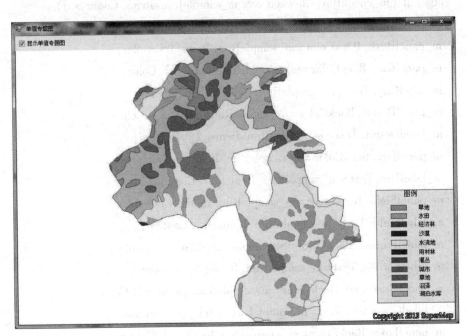

图 5-14 运行结果

3. 关键类型/成员

关键类型/成员见表 5-15。

表 5-15 关键类型/成员表

控件/类	方法	属性	事件
Layers	Add、Remove	UniqueExpression	
ThemeLabel		LabelExpression	
Workspace	Open		

4. 窗体控件属性

窗体控件属性见表 5-16。

表 5-16 窗体控件属性表

控件	Name	Text
CheckBox	m_checkBox	显示统一风格标签专题图

5. 窗体设计布局

窗体设计布局如图 5-15 所示。

6. 创建 SampleRun 类

(1) 添加 using 引用代码

图 5-15　窗体设计布局图

添加 using 引用代码具体如下：

using System. Diagnostics; //添加 System. Diagnostics 命名空间，使其对应的类可以使用
using System. Drawing; //添加 System. Drawing 命名空间，使其对应的类可以使用
using System. Windows. Forms; //添加 System. Windows. Forms 命名空间，使其对应的类可以使用
using SuperMap. Data; //添加 SuperMap. Data 命名空间，使其对应的类可以使用
using SuperMap. UI; //添加 SuperMap. UI 命名空间，使其对应的类可以使用
using SuperMap. Mapping; //添加 SuperMap. Mapping 命名空间，使其对应的类可以使用

（2）在 public class SampleRun 中添加代码

在 public class SampleRun 中添加如下代码：
//定义类对象和控件
private Workspace m_workspace;
private MapControl m_mapControl;
private DatasetVector m_dataset;
private String m_themeLayerName;
/// 根据 workspace 和 mapControl 构造 SampleRun 对象
public SampleRun(Workspace workspace, MapControl mapControl)
　　{ try { m_workspace = workspace;
　　　　m_mapControl = mapControl;
　　　　m_mapControl. Map. Workspace = workspace;
　　　　Initialize( ); }
　　　catch (Exception ex)
　　　{ Trace. WriteLine(ex. Message) ; } }

```csharp
/// 打开需要的工作空间文件及地图
private void Initialize()
{try{ WorkspaceConnectionInfo conInfo = new
WorkspaceConnectionInfo(@"..\..\SampleData\ThematicMaps\ThematicMaps.smwu");
m_workspace.Open(conInfo);//通过 workspace.Open 打开工作空间
m_dataset = m_workspace.Datasources[0].Datasets["BaseMap_P"] as DatasetVector;
Dataset datasetBase = m_workspace.Datasources[0].Datasets["BaseMap_R"];
this.m_mapControl.Map.Layers.Add(datasetBase, true);//添加图层
this.m_mapControl.Map.Layers.Add(m_dataset, true);
m_mapControl.Map.Refresh();//刷新地图
// 调整 mapControl 的状态
m_mapControl.Action = SuperMap.UI.Action.Pan;}
catch(Exception ex)
{Trace.WriteLine(ex.Message);}}
/// 通过属性 m_isThemeLayerDisplay 来控制是否显示统一风格标签专题图
public void CheckUniformStyleThemeLabelDisplay(Boolean m_isThemeLayerDisplay)
{try{if(m_isThemeLayerDisplay)
{AddUniformStyleThemeLabelLayer();}
else
{m_mapControl.Map.Layers.Remove(m_themeLayerName);//移除指定对象
m_mapControl.Map.Refresh();}}
catch(Exception ex)
{Trace.WriteLine(ex.Message);}}
/// 设置专题图的属性，添加专题图图层到地图
private void AddUniformStyleThemeLabelLayer()
{try{ThemeLabel themeLabel = new ThemeLabel();//ThemeLabel：标签专题图类。
用文本的形式在图层上直接显示属性表中的数据，其实质就是对图层的标注。
    TextStyle textStyle = new TextStyle();//TextStyle：文本风格类
    //设置文本风格
    textStyle.ForeColor = Color.FromArgb(115, 0, 74);
    textStyle.BackColor = Color.FromArgb(231, 227, 231);
    textStyle.Bold = true;
    textStyle.Outline = true;
    textStyle.FontHeight = 8;
    themeLabel.LabelExpression = "Name";
    //设置统一风格
    themeLabel.UniformStyle = textStyle;
    //将制作好的专题图添加到地图中显示
    Layer themeLayer = m_mapControl.Map.Layers.Add(m_dataset, themeLabel, true);
```

m_themeLayerName=themeLayer.Name;
m_mapControl.Map.Refresh();}
catch(Exception ex)
{Trace.WriteLine(ex.Message);}}

7. 窗体代码

(1)添加 using 引用代码

添加 using 引用代码具体如下：

using System.Diagnostics;//添加 System.Diagnostics 命名空间,使其对应的类可以使用

(2)初始化代码

初始化 SampleRun,以及 public FormMain()函数部分的代码如下：

//定义类风格
private SampleRun m_sampleRun;
private SuperMap.Data.Workspace m_workspace;
private SuperMap.UI.MapControl m_mapControl;
public FormMain()
    {try{InitializeComponent();
    this.m_workspace=new SuperMap.Data.Workspace(this.components);//创建新对象
    this.m_mapControl=new SuperMap.UI.MapControl();
    //初始化控件
    m_mapControl.Dock=DockStyle.Fill;
    base.Controls.Add(m_mapControl);}
    catch(Exception ex)
    {Trace.WriteLine(ex.Message);}}

(3)窗体的 Load 事件

在窗体的属性窗口中,选择 ≶ 按钮后,左键双击 Load,即可添加以下代码：

private void FormMain_Load(object sender, EventArgs e)
    {try{//实例化 SampleRun
        m_sampleRun=new SampleRun(m_workspace, m_mapControl);
        //复选框选中
        m_checkBox.Checked=true;}
    catch(Exception ex)
        {Trace.WriteLine(ex.Message);}}

(4)窗体的 FormClose 事件

在退出程序时,要断开空间的关联,关闭地图窗口、工作空间。因此,在属性窗口点击 ≶ 图标,左键双击 FormClose 事件,添加代码同前文的相应部分。

(5)添加各个控件的代码

显示统一风格标签专题图 CheckBox 控件的 CheckedChanged 事件。在 m_checkBox 的属性窗口中,选择 ≶ 按钮后,左键双击 CheckedChanged,即可添加以下代码,实现显示统一风格标签专题图：

```
private void m_checkBox_CheckedChanged(object sender, EventArgs e)
    {try{ m_sampleRun.CheckUniformStyleThemeLabelDisplay(m_checkBox.Checked);
}//通过sampleRun.CheckUniformStyleThemeLabelDisplay显示统一风格专题图
    catch(Exception ex)
    {Trace.WriteLine(ex.Message);}}
    //在CheckBox的选中事件触发时，运行SampleRun的功能操作
```

8. 运行结果

窗体设计完成后点击▶按钮，运行该程序，运行后弹出窗体，如图5-16所示，可以通过勾选"显示统一风格标签专题图"进行专题图显示，对地物进行风格统一的标注。

图 5-16 运行结果

5.2.6 混合风格标签专题图

1. 数据

在安装目录\SampleData\ThematicMaps\ThematicMaps.smwu路径下打开数据文件。

2. 新建文件夹和工程

在【D:\MyProject\】文件中新建一个文件夹，命名为【第 5 章　地图制图及专题图制作】，并在其目录下建立一个文件夹，命名为【MixedTextStyleThemeLabel】，在此文件夹里新建一个工程，将此工程命名为 MixedTextStyleThemeLabel。

3. 关键类型/成员

关键类型/成员见表 5-17。

表 5-17　　　　　　　　　　　　　关键类型/成员表

控件/类	方法	属性	事件
Workspace	Open		
Layers	Add		
ThemeLabel		UniformMixedStyle	
MixedTextStyle		MixedTextStyle	

4. 窗体控件属性

窗体控件属性见表 5-18。

表 5-18　　　　　　　　　　　　　窗体控件属性表

控件	Name	Text
CheckBox	m_checkBoxDisplay	显示混合风格标签专题图

5. 窗体设计布局

窗体设计布局如图 5-17 所示。

6. 创建 SampleRun 类

（1）添加 using 引用代码

添加 using 引用代码具体如下：

using SuperMap.Data; //添加 SuperMap.Data 命名空间，使其对应的类可以使用
using SuperMap.Mapping; //添加 SuperMap.Mapping 命名空间，使其对应的类可以使用
using SuperMap.UI; //添加 SuperMap.UI 命名空间，使其对应的类可以使用

（2）在 public class SampleRun 中添加代码

在 public class SampleRun 中添加以下代码：

//定义类对象及控件
private Workspace m_workspace;
private MapControl m_mapControl;
private DatasetVector m_datasetVector;
private Layer m_layerThemeLabel;
private ThemeLabel m_themeLabel;

图 5-17　窗体设计布局图

private MixedTextStyle m_mixedTextStyle;
/// 根据 workspace 和 map 构造 SampleRun 对象
public SampleRun(Workspace workspace, MapControl mapControl)
    {m_workspace=workspace;
    m_mapControl=mapControl;
    m_mapControl.Map.Workspace=workspace;
    Initialize();}
private void Initialize()
    {try{//打开工作空间及地图
    WorkspaceConnectionInfo conInfo = new WorkspaceConnectionInfo(@"../../SampleData/ThematicMaps/ThematicMaps.smwu");
    m_workspace.Open(conInfo);//通过 workspace.Open 打开工作空间
    m_mapControl.Map.Open("京津地区交通干线图");//通过 Map.Open 打开地图
    // 调整 mapControl 的状态
    m_mapControl.Action=SuperMap.UI.Action.Pan;
    //要表达的数据
    m_datasetVector = m_workspace.Datasources[0].Datasets["Road_L"] as DatasetVector;
    //设置反走样
    m_mapControl.Map.IsAntialias=true;
    //设置重叠显示
    m_mapControl.Map.IsOverlapDisplayed=false;
    //默认开启功能
    SetThemeVisible(true);}

```
                catch (Exception ex)
                    {Trace.WriteLine(ex.Message);}}
/// 设置是否显示混合标签专题图
public void SetThemeVisible(Boolean value)//定义 SetThemeVisible 初始化函数
            {try{if (m_layerThemeLabel == null)
            {//构造标签专题图
            ThemeLabel themeLabel = new ThemeLabel();
            themeLabel.LabelExpression = "Name";
            //添加矩阵标签专题图
            m_layerThemeLabel = m_mapControl.Map.Layers.Add(m_datasetVector, themeLabel, true);
            m_themeLabel = m_layerThemeLabel.Theme as ThemeLabel;}
            if (value)
            {DisplayMixedTextStyleTheme();}
            else
            {DisplayNormalTheme();}//显示混合风格专题图
            m_mapControl.Map.Refresh();}
            catch (Exception ex)
            {Trace.WriteLine(ex.Message);}}
/// 显示复合标签专题图
private void DisplayMixedTextStyleTheme()//定义 DisplayMixedTextStyleTheme 初始化函数
                { try{if (m_mixedTextStyle == null)//设置标签专题图统一的文本复合风格
            {m_mixedTextStyle = new MixedTextStyle();
            m_mixedTextStyle.SplitIndexes = new Int32[]{1};
            TextStyle[] textStyles = new TextStyle[2];
            textStyles[0] = new TextStyle();
            textStyles[0].ForeColor = Color.Red;
            textStyles[0].FontHeight = 8;
            textStyles[0].Bold = true;
            textStyles[1] = new TextStyle();
            textStyles[1].ForeColor = Color.Blue;
            textStyles[1].FontHeight = 8;
            textStyles[1].Bold = true;
            m_mixedTextStyle.Styles = textStyles;}
            m_themeLabel.UniformMixedStyle = m_mixedTextStyle;}
                catch (Exception ex)
                    { Trace.WriteLine(ex.Message);}}
                /// 显示普通专题图
private void DisplayNormalTheme()
```

```
            try{ m_themeLabel.UniformMixedStyle=null; }
            catch(Exception ex)
              { Trace.WriteLine(ex.Message); } }
```

7. 窗体代码

(1) 添加 using 引用代码

添加 using 引用代码具体如下:

using System.Diagnostics; //添加 System.Diagnostics 命名空间,使其对应的类可以使用

(2) 实例化代码

实例化 SampleRun 以及 public FormMain()函数部分的代码如下:

```
//定义类对象
private SampleRun m_sampleRun;
private SuperMap.Data.Workspace m_workspace;
private SuperMap.UI.MapControl m_mapControl;
public FormMain()
    {try{ InitializeComponent();
    this.Load+=new EventHandler(FormMain_Load);
    this.m_workspace=new SuperMap.Data.Workspace(this.components); //创建新对象
    this.m_mapControl=new SuperMap.UI.MapControl();
    //初始化控件
    m_mapControl.Dock=DockStyle.Fill;
    m_mapControl.IsWaitCursorEnabled=false;
    //实例化 SampleRun
    m_sampleRun=new SampleRun(m_workspace, m_mapControl);
    //添加指定控件
    base.Controls.Add(m_mapControl);
    base.Controls.SetChildIndex(m_mapControl, 0); }
    catch(Exception ex)
    {Trace.WriteLine(ex.Message); } }
```

(3) 窗体的 Load 事件

在窗体的属性窗口中,选择 ✁ 按钮后,左键双击 Load,即可添加以下代码:

```
private void FormMain_Load(object sender, EventArgs e)
    { m_mapControl.Map.ViewEntire(); //地图全幅显示}
```

(4) 窗体的 FormClose 事件

在退出程序时,要断开空间的关联,关闭地图窗口、工作空间。因此,在属性窗口点击 ✁ 图标,左键双击 FormClose 事件,添加代码同前文的相应部分。

(5) 添加各个控件的代码

m_checkBoxDisplay 控件的 CheckedChanged 事件,在控件的属性窗口点击 ✁ 图标,左键双击 CheckedChanged,添加如下代码,实现显示混合风格专题图:

private void m_checkBoxDisplay_CheckedChanged(object sender, EventArgs e)
　　{ try { m _ sampleRun. SetThemeVisible (m _ checkBoxDisplay. Checked);//通过 sampleRun. SetThemeVisible 显示混合风格专题图}
catch（Exception ex）
{Trace. WriteLine(ex. Message）; } }

8. 运行结果

窗体设计完成后单击▶按钮，运行该程序，运行后弹出窗体，如图 5-18 所示。在图中通过勾选"显示混合风格标签专题图"来显示专题图，可以实现以不同风格进行标注的操作。

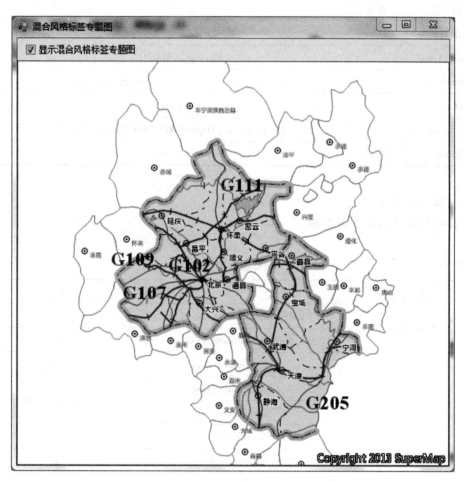

图 5-18　运行结果

5.2.7　布局打印

1. 数据

在安装目录\SampleData\ThematicMaps\Thematicmaps. smwu 路径下打开数据文件。

2. 新建文件夹和工程

在【D:\MyProject\】文件中新建一个文件夹，命名为【第 5 章　地图制图及专题图制作】，并在其目录下建立一个文件夹，命名为【Layout】，在此文件夹里新建一个工程，将此工程命名为 Layout。

3. 关键类型/成员

关键类型/成员见表 5-19。

表 5-19　　　　　　　　　　　关键类型/成员表

控件/类	方法	属　性	事件
LayoutElements	AddNew		
GeoNorthArrow		BindingGeoMapID	
GeoMapScale		LeftDivisionCount、ScaleUnit、SegmentCount	
MapLayout		Elements	
MapLayoutControl		LayoutAction、MapAction、ActiveGeoMapID	
Printer	Print		

4. 窗体控件属性

窗体控件属性具体见表 5-20。

表 5-20　　　　　　　　　　　窗体控件属性表

控　件	Name	Text
ComboBox	toolStripComboBoxActionMode	
	toolStripComboBoxPrinter	
绘图 Button	toolStripButtonPoint	●
	toolStripButtonLine	
	toolStripButtonPolyLine	
	toolStripButtonFreeLine	
	toolStripButtonCurve	
	toolStripButtonParallel	
	toolStripButtonPolygon	
	toolStripButtonRectangle	
	toolStripButtonRoundRectangle	
	toolStripButtonCircle3P	
	toolStripButtonParallelogram	

续表

控 件	Name	Text
命令 Button	toolStripButtonLockMap	锁定地图
	toolStripButtonZoomIn	放大
	toolStripButtonZoomOut	缩小
	toolStripButtonPan	平移
	toolStripButtonPrint	打印
Label	toolStripLabel1	打印服务

5. 窗体设计布局

窗体设计布局如图 5-19 所示。

图 5-19　窗体设计布局图

6. 创建 SampleRun 类

（1）添加 using 引用代码

添加 using 引用代码具体如下：

using System. Diagnostics；//添加 System. Diagnostics 命名空间，使其对应的类可以使用
using System. Drawing；//添加 System. Drawing 命名空间，使其对应的类可以使用
using System. Windows. Forms；//添加 System. Windows. Forms 命名空间，使其对应的类可以使用
using SuperMap. Data；//添加 SuperMap. Data 命名空间，使其对应的类可以使用
using SuperMap. UI；//添加 SuperMap. UI 命名空间，使其对应的类可以使用
using SuperMap. Mapping；//添加 SuperMap. Mapping 命名空间，使其对应的类可以使用
using SuperMap. Layout；//添加 SuperMap. Layout 命名空间，使其对应的类可以使用

(2)在 public class SampleRun 中添加代码

在 public class SampleRun 中添加如下代码:

```
//定义类对象和控件
private Workspace m_workspace;
private MapLayoutControl m_mapLayoutControl;
private Int32 m_mapID;
/// 根据 workspace 和 map 构造 SampleRun 对象
public SampleRun(Workspace workspace, MapLayoutControl mapLayoutControl)
    {try{m_workspace=workspace;
        m_mapLayoutControl=mapLayoutControl;
        Initialize();}
    catch(Exception ex)
        {Trace.WriteLine(ex.Message);}}
private void Initialize()
    {try{//打开工作空间及地图
        WorkspaceConnectionInfo conInfo=new WorkspaceConnectionInfo
        (@"../../SampleData/ThematicMaps/Thematicmaps.smwu");
        m_workspace.Open(conInfo);//通过 workspace.Open 打开工作空间
        InitializeLayout();//通过
        m_mapLayoutControl.ElementAdded+=new ElementEventHandler
        (m_mapLayoutControl_ElementAdded);//添加对象出发事件
        m_mapLayoutControl.TrackMode=TrackMode.Edit;// TrackMode:该枚举定义了
```

绘制方式类型常量。用来定义地图控件中绘制对象时,是在图层中创建一个新对象还是在内存中创建一个新对象,或者是在 CAD 图层中绘制地图几何对象 Edit:在图层中创建新对象。

```
        m_mapLayoutControl.MapLayout.Zoom(4);
        SetLayoutAction(SuperMap.UI.Action.CreatePoint);}
    catch(Exception ex)
        {Trace.WriteLine(ex.Message);}}
private void InitializeLayout()
    {try {LayoutElements elements=m_mapLayoutControl.MapLayout.Elements;//
LayoutElements:布局元素集合类。
//构造 GeoMap
GeoMap geoMap=new GeoMap();//GeoMap:地图几何对象类。该类用于布局中添加地图。
geoMap.MapName="京津地区交通干线图";
//设置 GeoMap 对象的外切矩形
Rectangle2D rect=new Rectangle2D(new Point2D(850,1300),new Size2D(1500,1500));
```

// Rectangle2D：矩形类。用来表示坐标值为双精度的矩形对象，即其左边界坐标值、下边界坐标值、右边界坐标值、上边界坐标值均为双精度类型，其中左边界坐标值小于等于右边界坐标值，下边界坐标值小于等于上边界坐标值。

GeoRectangle geoRect = new GeoRectangle(rect, 0);// GeoRectangle：二维矩形几何对象类。该类主要用于 CAD 图层，是 Geometry 对象的子对象。

```
geoMap.Shape = geoRect;
elements.AddNew(geoMap);
m_mapID = elements.GetID();
//构造指北针
GeoNorthArrow northArrow = new GeoNorthArrow(NorthArrowStyleType.EightDirection,
        new Rectangle2D(new Point2D(1400, 2250), new Size2D(350, 350)),
0);// 
```

GeoNorthArrow：指北针几何对象类。该类的对象是地图布局中的指北针对象。

```
        northArrow.BindingGeoMapID = m_mapID;
        elements.AddNew(northArrow);
        //构造比例尺
    GeoMapScale scale = new GeoMapScale(m_mapID, new Point2D(125, 400),
50, 50);// 
```

GeoMapScale：地图比例尺几何类。该类用于向地图布局中添加比例尺对象。

```
        scale.LeftDivisionCount = 2;
        scale.ScaleUnit = Unit.Kilometer;
        scale.SegmentCount = 4;
        elements.AddNew(scale); }
        catch (Exception ex)
        { Trace.WriteLine(ex.Message); } }
/// 改变 LayoutMapControl 的 LayoutMapAction
public void SetLayoutAction(SuperMap.UI.Action action)
    { try {m_mapLayoutControl.LayoutAction = action; }
    catch (Exception ex)
    { Trace.WriteLine(ex.Message); } }
/// 改变 LayoutMapControl 的 MapAction
public void SetMapAction(SuperMap.UI.Action action)
    {try{ m_mapLayoutControl.MapAction = action; }
    catch (Exception ex)
    { Trace.WriteLine(ex.Message); }}
///锁定地图
public void LockMap(Boolean isLocked)
    { try{ Int32 mapID = -1;
        if (isLocked)
        {mapID = m_mapID; }
        m_mapLayoutControl.ActiveGeoMapID = mapID; }
```

```csharp
            catch (Exception ex)
                { Trace.WriteLine(ex.Message); } }
/// 打印布局
public void PrintLayout(String printerName)
    { try { m_mapLayoutControl.MapLayout.Printer.PrinterName = printerName;
        m_mapLayoutControl.MapLayout.Printer.Print(); }
      catch (Exception ex)
        { Trace.WriteLine(ex.Message); } }
/// 对象添加事件
public void m_mapLayoutControl_ElementAdded(object sender, ElementEventArgs e)
      { try { LayoutElements elements = m_mapLayoutControl.MapLayout.Elements;//
LayoutElements:布局元素集合类。
        if(elements.SeekID(e.ID))
          { Geometry gemetry = elements.GetGeometry();
            if (gemetry! = null)
              { GeoNorthArrow northArrow = gemetry as GeoNorthArrow;
                if (northArrow! = null)
                  { northArrow.BindingGeoMapID = m_mapID; }
                GeoMapScale mapScale = gemetry as GeoMapScale;
                    if (mapScale! = null)
                      { mapScale.BindingGeoMapID = m_mapID; }
                      elements.SetGeometry(gemetry);
                      elements.Refresh();
                      m_mapLayoutControl.MapLayout.Refresh(); } } }
        catch (Exception ex)
          { Trace.WriteLine(ex.Message); } }
```

7. 窗体的代码

(1) 添加 using 引用

添加 using 引用代码具体如下：

using System.Diagnostics; //添加 System.Diagnostics 命名空间，使其对应的类可以使用
using SuperMap.Layout; //添加 SuperMap.Layout 命名空间，使其对应的类可以使用
using SuperMap.UI; //添加 SuperMap.UI 命名空间，使其对应的类可以使用
using SuperMap.Data; //添加 SuperMap.Data 命名空间，使其对应的类可以使用
using System.Threading; //添加 System.Threading 命名空间，使其对应的类可以使用
using System.Reflection; //添加 System.Reflection 命名空间，使其对应的类可以使用

(2) 实例化代码

实例化 SampleRun、Workspace、MapLayoutControl 以及 public FormMain() 函数部分的代码如下：

```csharp
//定义类对象和控件
private SampleRun m_sampleRun;
private Workspace m_workspace;
private MapLayoutControl m_mapLayoutControl;
public FormMain()
{   try
    {   InitializeComponent();
        m_workspace = new SuperMap.Data.Workspace(this.components);
        m_mapLayoutControl = new MapLayoutControl();//创建新对象
        base.Controls.Add(m_mapLayoutControl);
        base.Controls.SetChildIndex(m_mapLayoutControl, 0);
        //初始化控件
        m_mapLayoutControl.Dock = DockStyle.Fill;
        m_mapLayoutControl.MapLayout.Workspace = m_workspace;
        m_mapLayoutControl.IsHorizontalScrollbarVisible = false;
        m_mapLayoutControl.IsVerticalScrollbarVisible = false;
        //实例化 SampleRun
        m_sampleRun = new SampleRun(m_workspace, m_mapLayoutControl);
        // 填充控件
        toolStripComboBoxActionMode.Items.Clear();
        toolStripComboBoxActionMode.Items.Add("绘制点");
        toolStripComboBoxActionMode.Items.Add("绘制线");
        toolStripComboBoxActionMode.Items.Add("绘制面");
        toolStripComboBoxActionMode.Items.Add("绘制指北针");
        toolStripComboBoxActionMode.Items.Add("绘制比例尺");
        toolStripComboBoxActionMode.SelectedIndex = 0;
        foreach (String printer in System.Drawing.Printing.PrinterSettings.InstalledPrinters)
        { toolStripComboBoxPrinter.Items.Add(printer); }
        toolStripComboBoxPrinter.SelectedIndex = 0;}
    catch (Exception ex)
    { Trace.WriteLine(ex.Message); } }
```

(3) 窗体的 FormClose 事件

在退出程序时,要断开空间的关联,关闭地图窗口、工作空间。因此,在属性窗口中双击 FormClose 事件,添加代码同前文中的相应部分。

(4) 添加各个控件的代码

① toolStripComboBoxActionMode 控件的 SelectedIndexChanged 事件。在 toolStrip

ComboBoxActionMode 的属性窗口中,选择 按钮后,左键双击 SelectedIndexChanged,即可添加以下代码:

```
//选择绘制对象的类型
private void toolStripComboBoxActionMode_SelectedIndexChanged(object sender, EventArgs e)
    try{ switch (toolStripComboBoxActionMode.SelectedIndex)
        {// 绘制点
        case 0:
        {//设置控件是否可见
            toolStripButtonPoint.Enabled = true;
            toolStripButtonLine.Enabled = false;
            toolStripButtonPolyLine.Enabled = false;
            toolStripButtonFreeLine.Enabled = false;
            toolStripButtonCurve.Enabled = false;
            toolStripButtonParallel.Enabled = false;
            toolStripButtonPolygon.Enabled = false;
            toolStripButtonRectangle.Enabled = false;
            toolStripButtonRoundRectangle.Enabled = false;
            toolStripButtonCircle3P.Enabled = false;
            toolStripButtonParallelogram.Enabled = false;
            m_sampleRun.SetLayoutAction(SuperMap.UI.Action.CreatePoint);}
            break;
        case 1:
        { toolStripButtonPoint.Enabled = false;
            toolStripButtonLine.Enabled = true;
            toolStripButtonPolyLine.Enabled = true;
            toolStripButtonFreeLine.Enabled = true;
            toolStripButtonCurve.Enabled = true;
            toolStripButtonParallel.Enabled = true;
            toolStripButtonPolygon.Enabled = false;
            toolStripButtonRectangle.Enabled = false;
            toolStripButtonRoundRectangle.Enabled = false;
            toolStripButtonCircle3P.Enabled = false;
            toolStripButtonParallelogram.Enabled = false;
            m_sampleRun.SetLayoutAction(SuperMap.UI.Action.CreateLine);}
            break;
        case 2:
        { toolStripButtonPoint.Enabled = false;
```

```
toolStripButtonLine.Enabled = false;
toolStripButtonPolyLine.Enabled = false;
toolStripButtonFreeLine.Enabled = false;
toolStripButtonCurve.Enabled = false;
toolStripButtonParallel.Enabled = false;
toolStripButtonPolygon.Enabled = true;
toolStripButtonRectangle.Enabled = true;
toolStripButtonRoundRectangle.Enabled = true;
toolStripButtonCircle3P.Enabled = true;
 toolStripButtonParallelogram.Enabled = true;
 m_sampleRun.SetLayoutAction(SuperMap.UI.Action.CreatePolygon);}
break;
case 3:
{ toolStripButtonPoint.Enabled = false;
 toolStripButtonLine.Enabled = false;
 toolStripButtonPolyLine.Enabled = false;
 toolStripButtonFreeLine.Enabled = false;
 toolStripButtonCurve.Enabled = false;
 toolStripButtonParallel.Enabled = false;
 toolStripButtonPolygon.Enabled = false;
 toolStripButtonRectangle.Enabled = false;
 toolStripButtonRoundRectangle.Enabled = false;
 toolStripButtonCircle3P.Enabled = false;
 toolStripButtonParallelogram.Enabled = false;
m_sampleRun.SetLayoutAction(SuperMap.UI.Action.CreateNorthArrow);}
 break;
 case 4:
{ toolStripButtonPoint.Enabled = false;
 toolStripButtonLine.Enabled = false;
 toolStripButtonPolyLine.Enabled = false;
 toolStripButtonFreeLine.Enabled = false;
 toolStripButtonCurve.Enabled = false;
 toolStripButtonParallel.Enabled = false;
 toolStripButtonPolygon.Enabled = false;
 toolStripButtonRectangle.Enabled = false;
 toolStripButtonRoundRectangle.Enabled = false;
 toolStripButtonCircle3P.Enabled = false;
```

```
              toolStripButtonParallelogram. Enabled = false;
              m_sampleRun. SetLayoutAction(SuperMap. UI. Action. CreateMapScale);}
              break;
            default:
              break;}}
          catch(Exception ex)
          { Trace. WriteLine(ex. Message);}}
```

② buttonPoint 控件的 Click 事件。在 buttonPoint 的属性窗口中，选择 按钮后，左键双击 Click，即可添加以下代码，实现绘制点：

```
private void buttonPoint_Click(object sender, EventArgs e)
    {m_sampleRun. SetLayoutAction(SuperMap. UI. Action. CreatePoint);}
```

// 通过 sampleRun. SetLayoutAction:绘制点

③ buttonLine 控件的 Click 事件。在 buttonPoint 的属性窗口中，选择 按钮后，左键双击 Click，即可添加代码:m_sampleRun. SetLayoutAction(SuperMap. UI. Action. CreatePolyline，实现绘制线。

④ buttonFreeLine 控件的 Click 事件。在 buttonPoint 的属性窗口中，选择 按钮后，左键双击 Click，即可添加代码：m_sampleRun. SetLayoutAction（SuperMap. UI. Action. CreateFreePolyline），实现自由画线。

⑤ buttonCurve 控件的 Click 事件 在 buttonPoint 的属性窗口中，选择 按钮后，左键双击 Click，即可添加代码:m_sampleRun. SetLayoutAction(SuperMap. UI. Action. CreateCurve)，实现绘制贝兹线。

⑥ buttonParallel 控件的 Click 事件。在 buttonPoint 的属性窗口中，选择 按钮后，左键双击 Click，即可添加代码：m_sampleRun. SetLayoutAction（SuperMap. UI. Action. CreateParallel），实现绘制平行线。

⑦ buttonPolygon 控件的 Click 事件。在 buttonPoint 的属性窗口中，选择 按钮后，左键双击 Click，即可添加代码：m_sampleRun. SetLayoutAction（SuperMap. UI. Action. CreatePolygon），实现绘制多边形。

⑧ buttonRectangle 控件的 Click 事件。在 buttonPoint 的属性窗口中，选择 按钮后，左键双击 Click，即可添加代码：m_sampleRun. SetLayoutAction（SuperMap. UI. Action. CreateRectangle），实现绘制矩形。

⑨ buttonRoundRectangle 控件的 Click 事件。在 buttonPoint 的属性窗口中，选择 按钮后，左键双击 Click，即可添加代码:m_sampleRun. SetLayoutAction(SuperMap. UI. Action. CreateRoundRectangle)，实现绘制圆角矩形。

⑩ button3pCircle 控件的 Click 事件。在 buttonPoint 的属性窗口中，选择 按钮后，左键双击 Click，即可添加代码：m_sampleRun. SetLayoutAction（SuperMap. UI. Action. CreateCircle3P），实现绘制三点圆。

⑪ buttonParallelogram 控件的 Click 事件。在 buttonPoint 的属性窗口中，选择 按钮后，

左键双击 Click,即可添加代码:m_sampleRun. SetLayoutAction(SuperMap. UI. Action. CreateParallelogram),实现绘制平行四边形。

⑫ toolStripButtonZoomIn 控件的 Click 事件。在 buttonPoint 的属性窗口中,选择 按钮后,左键双击 Click,即可添加以下代码,实现放大:
　　private void toolStripButtonZoomIn_Click(object sender, EventArgs e)
　　　　{ if (toolStripButtonLockMap. CheckState==CheckState. Checked)
　　　　{m_sampleRun. SetMapAction(SuperMap. UI. Action. ZoomIn); }
　　　　// 通过 sampleRun. SetLayoutAction:放大地图
　　　　 Else
　　　　　　{ m_sampleRun. SetLayoutAction(SuperMap. UI. Action. ZoomIn); } }

⑬ toolStripButtonZoomOut 控件的 Click 事件。在 buttonPoint 的属性窗口中,选择 按钮后,左键双击 Click,即可添加以下代码,实现缩小:
　　private void toolStripButtonZoomOut_Click(object sender, EventArgs e)
　　　　{ if (toolStripButtonLockMap. CheckState==CheckState. Checked)
　　　　　　{m_sampleRun. SetMapAction(SuperMap. UI. Action. ZoomOut); }
　　　　// 通过 sampleRun. SetLayoutAction:缩小地图
　　　　else
　　　　　　{ m_sampleRun. SetLayoutAction(SuperMap. UI. Action. ZoomOut); } }

⑭ toolStripButtonPan 控件的 Click 事件。在 buttonPoint 的属性窗口中,选择 按钮后,左键双击 Click,即可添加以下代码,实现平移:
　　private void toolStripButtonPan_Click(object sender, EventArgs e)
　　　　{ if (toolStripButtonLockMap. CheckState==CheckState. Checked)
　　　　{m_sampleRun. SetMapAction(SuperMap. UI. Action. Pan); }
　　　　// 通过 sampleRun. SetLayoutAction:平移地图
　　　　else{m_sampleRun. SetLayoutAction(SuperMap. UI. Action. Pan); }}

⑮ toolStripButtonLockMap 控件的 Click 事件。在 buttonPoint 的属性窗口中,选择 按钮后,左键双击 Click,即可添加以下代码,实现锁定:
　　private void toolStripButtonLockMap_Click(object sender, EventArgs e)
　　　　{ try{ Boolean isLock=toolStripButtonLockMap. CheckState==CheckState. Checked;
　　　　　　m_sampleRun. LockMap(isLock); // 通过 sampleRun. SetLayoutAction:锁定地图
　　　　　　this. toolStripComboBoxActionMode. SelectedIndex=0;
　　　　　　this. toolStripButtonPoint. Enabled=! isLock;
　　　　　　this. toolStripComboBoxActionMode. Enabled=! isLock; }
　　　　catch (Exception ex)
　　　　{Trace. WriteLine(ex. Message); } }

⑯ toolStripButtonPrint 控件的 Click 事件。在 buttonPoint 的属性窗口中,选择 按钮后,左键双击 Click,即可添加以下代码,实现打印:

```
private void toolStripButtonPrint_Click(object sender, EventArgs e)
    {try{m_sampleRun.PrintLayout(toolStripComboBoxPrinter.SelectedItem as String);}
  // 通过 sampleRun.SetLayoutAction:打印地图
    catch(Exception ex)
    {Trace.WriteLine(ex.Message);}}
```

8. 运行结果

窗体设计完成后点击▶按钮，运行该程序，运行后弹出窗体，如图 5-20 所示，可以对地图进行编辑，添加点、线、面，并对编辑好的地图进行打印。

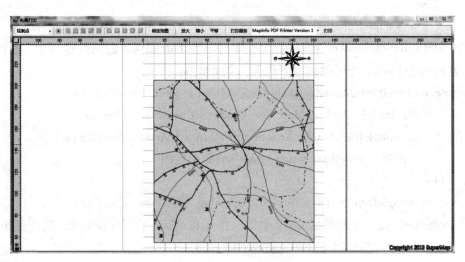

图 5-20　运行结果

第6章 查询功能

6.1 图查属性与属性查图

1. 数据

在安装目录\SampleData\World\world.smwu 路径下打开数据文件。

2. 新建文件夹和工程

在【D:\MyProject\】文件中新建一个文件夹，命名为【第6章 查询功能】，并在其目录下建立一个文件夹，命名为【SQLQuery】，在此文件夹里新建一个工程，将此工程命名为 SQLQuery。

3. 关键类型/成员

关键类型/成员见表6-1。

表6-1　　　　　　　　　　　关键类型/成员表

控件类	方法	属性	事件
Workspace	Open		
Selection	ToRecordsetFromRecordset		Click
QueryParameter	Dispose	AttributeFilter CursorType	
Recordset	Dispose		

4. 窗体控件属性

控件属性具体见表6-2。

表6-2　　　　　　　　　　　控件属性表

控件	Name	Text
toolStrip1	toolStrip1	toolStrip1
Button	toolStripOpen	打开工作空间
	toolStripPan	漫游
	toolStripZoomIn	放大
	toolStripZoomOut	缩小

续表

控件	Name	Text
toolStrip1	toolStrip1	toolStrip1
Button	toolStripZoomFree	自由缩放
	toolStripViewEntire	全幅显示
	toolStripSelect	选择
	toolStripQueryProperty	图查属性
	toolStripSQLQuery	属性查图
mapControl	mapControl1	mapControl1
openFileDialog	openFileDialog1	openFileDialog1
workspace	workspace1	workspace1
dataGridView	dataGridView1	dataGridView1
Label	toolStripLabel1	输入查询条件：
toolStripTextBox	toolStripTextBox1	toolStripTextBox1

5. 窗体设计布局

窗体设计布局如图 6-1 所示。

图 6-1　窗体设计布局图

6. 窗体代码

（1）添加 using 引用代码

添加 using 引用代码具体如下：

using SuperMap.Data; //添加 SuperMap.Data 命名空间，使其对应的类可以使用

using SuperMap.Mapping；//添加 SuperMap.Mapping 命名空间，使其对应的类可以使用

using SuperMap.UI；//添加 SuperMap.UI 命名空间，使其对应的类可以使用

(2) 窗体的 FormClose 事件

在退出程序时，要断开空间的关联，关闭地图窗口、工作空间。因此，在属性窗口点击 图标，左键双击 FormClose 事件，添加以下代码：

```
private void FormDemo_FormClosing(object sender, FormClosingEventArgs e)
    { mapControl1.Dispose();
    workspace1.Close();//关闭工作空间
    workspace1.Dispose();}
```

(3) 添加控件代码

① toolStripOpen 控件的 Click 事件。双击 toolStripOpen 控件，添加如下代码：

```
private void toolStripOpen_Click(object sender, EventArgs e)
    {//设置公用打开对话框
    openFileDialog1.Filter = "SuperMap 工作空间文件(*.smwu)|*.smwu";
    //判断打开的结果，如果打开就执行下列操作
    if(openFileDialog1.ShowDialog() == DialogResult.OK)
      {//避免连续打开工作空间导致异常信息
       mapControl1.Map.Close();
       workspace1.Close();
       mapControl1.Map.Refresh();
       //定义打开工作空间文件名
       String fileName = openFileDialog1.FileName;
       //打开工作空间文件
       WorkspaceConnectionInfo connectionInfo = new WorkspaceConnectionInfo(fileName);
       //打开工作空间
       workspace1.Open(connectionInfo);
       //建立 MapControl 与 Workspace 的连接
       mapControl1.Map.Workspace = workspace1;
       //判断工作空间中是否有地图
       if(workspace1.Maps.Count == 0)
           { MessageBox.Show("当前工作空间中不存在地图!");
            return; }
       //通过名称打开工作空间中的地图
       mapControl1.Map.Open("世界地图_Day");
       //刷新地图窗口
       mapControl1.Map.Refresh();}}
```

② toolStripSelect 控件的 Click 事件。双击 toolStripSelect 控件，添加代码：mapControl1.Action = SuperMap.UI.Action.Select2，实现选择对象。

③ toolStripPan 控件的 Click 事件。双击 toolStripPan 控件,添加代码:mapControl1. Action = SuperMap. UI. Action. Pan,实现漫游。

④ toolStripZoomIn 控件的 Click 事件。双击 toolStripZoomIn 控件,添加代码:mapControl1. Action = SuperMap. UI. Action. ZoomIn,实现放大地图。

⑤ toolStripZoomOut 控件的 Click 事件。双击 toolStripZoomOut 控件,添加代码:mapControl1. Action = SuperMap. UI. Action. ZoomOut,实现缩小地图。

⑥ toolStripZoomFree 控件的 Click 事件。双击 toolStripZoomFree 控件,添加代码:mapControl1. Action = SuperMap. UI. Action. ZoomFree,实现自由缩放。

⑦ toolStripViewEntire 控件的 Click 事件。双击 oolStripViewEntire 控件,添加代码:mapControl1. Map. ViewEntire( ),实现全屏显示。

⑧ toolStripQueryProperty 控件的 Click 事件。双击 toolStripQueryProperty 控件,添加如下代码:

```
private void toolStripQueryProperty_Click(object sender, EventArgs e)
    {//获取选择集
    Selection[ ] selection = mapControl1. Map. FindSelection(true);
    //判断选择集是否为空
    if (selection == null || selection. Length == 0)
    {MessageBox. Show("请选择要查询属性的空间对象");
    return;}
    //将选择集转换为记录
    Recordset recordset = selection[0]. ToRecordset( );
    this. dataGridView1. Columns. Clear( );
    this. dataGridView1. Rows. Clear( );
    for (int i = 0; i < recordset. FieldCount; i++)
    {//定义并获得字段名称
      String fieldName = recordset. GetFieldInfos( )[i]. Name;
      //将得到的字段名称添加到 dataGridView 列中
      this. dataGridView1. Columns. Add(fieldName, fieldName); }
    //初始化 row
    DataGridViewRow row = null;
    //根据选中的记录,将选中的对象的信息添加到 dataGridView 中显示
    while (! recordset. IsEOF)
    { row = new DataGridViewRow( );
     for (int i = 0; i < recordset. FieldCount; i++)
     {//定义并获得字段值
       Object fieldValue = recordset. GetFieldValue(i);
       //将字段值添加到 dataGridView 中对应的位置
       DataGridViewTextBoxCell cell = new DataGridViewTextBoxCell( );
       if (fieldValue ! = null)
```

```
            {cell.ValueType=fieldValue.GetType();
             cell.Value=fieldValue;}
             row.Cells.Add(cell);    }
             this.dataGridView1.Rows.Add(row);
             recordset.MoveNext();}
             this.dataGridView1.Update();
             recordset.Dispose();}
```

⑨ toolStripSQLQuery 控件的 Click 事件，双击 toolStripSQLQuery 控件，添加如下代码：

```
//查询
private void toolStripSQLQuery_Click(object sender, EventArgs e)
    {//判断 toolStripTextBox1 的输入内容是否为空
     if (toolStripTextBox1.Text.Length==0)
     { MessageBox.Show("查询信息不能为空");
       return;}
//定义图层个数
Int32 layerCount=mapControl1.Map.Layers.Count;
//判断当前地图窗口中是否有打开的图层
 if (layerCount==0)
 {MessageBox.Show("请先打开一个矢量数据集!");
   return;}
//定义查询条件信息
QueryParameter queryParameter=new QueryParameter();
queryParameter.AttributeFilter=toolStripTextBox1.Text;
queryParameter.CursorType=CursorType.Static;
Boolean hasGeometry=false;
//遍历每一个图层，实现多图层查询
foreach (Layer layer in mapControl1.Map.Layers)
 {//得到矢量数据集并强制转换为矢量数据集类型
   DatasetVector dataset=layer.Dataset as DatasetVector;
   if (dataset==null)
   {continue;}
   //通过查询条件对矢量数据集进行查询，从数据集中查询出属性数据
   Recordset recordset=dataset.Query(queryParameter);
   //判断是否有查询结果
   if (recordset.RecordCount > 0)
   { hasGeometry=true;}
    //把查询得到的数据加入到选择集中(使其高亮显示)
    Selection selection=layer.Selection;
    selection.FromRecordset(recordset);
```

```
        recordset.Dispose();}
    //没有查询结果,弹出提示
    if(! hasGeometry)
    {MessageBox.Show("没有符合查询条件的结果或查询条件有误,请重新确认后查
询!");}
        //当可创建对象使用完毕后,使用Dispose方法来释放所占用的内部资源
        queryParameter.Dispose();
        //属新地图窗口显示
        mapControl1.Refresh();
        hasGeometry=false;}
```

7. 运行结果

窗体设计完成后点击 ▶ 按钮,运行该程序,运行后弹出窗体,如图6-2和图6-3所示。

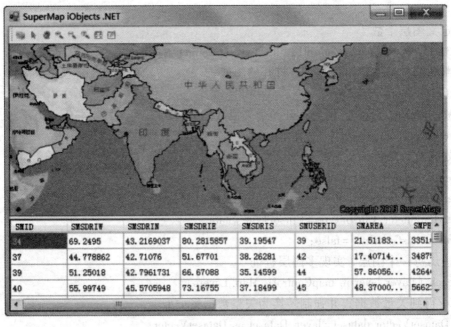

图6-2 图查属性

如图6-2,点击左上角的小图标即可打开地图,点击"选择"、" 放大"、" 缩小"、" 漫游"、" 全幅显示"、"自由缩放"按钮,分别对地图进行选择、放大、缩小、漫游、全幅显示、自由缩放的操作。点击"选择",选中地图,然后点击"图查属性",在图下就可以显示出当前选择对象的属性。

如图6-3,点击左上角的小图标即可打开地图,点击"选择"、" 放大"、" 缩小"、" 漫游"、" 全幅显示"、" 自由缩放"按钮,分别对地图进行选择、放大、缩小、漫游、全幅显示、自由缩放的操作。在工具栏中输入查询条件,例如,SmID<25,点击"属性查图"按钮,即可在图上高亮显示出查询结果。

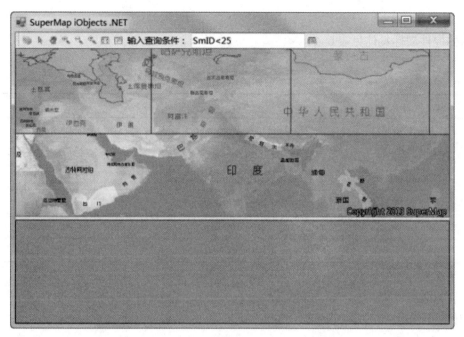

图 6-3　属性查图

6.2　空间查询

1. 数据

在安装目录\SampleData\World\world.smwu 路径下打开数据文件。

2. 新建文件夹和文件

在【D:\MyProject\】文件中新建一个文件夹，命名为【第 6 章　查询功能】，并在其目录下建立一个文件夹，命名为【SpatialQuery】，在此文件夹里新建一个工程，将此工程命名为 SpatialQuery。

3. 关键类型/成员

关键类型/成员见表 6-3。

表 6-3　　　　　　　　　　　　　　关键类型/成员表

控件/类	方法	属性	事件
Workspace	Open		
QueryParameter		SpatialQueryObject	
SpatialQueryMode		Contain、Intersect、Disjoint	
DatasetVector	Query		
Layer		Selection	

4. 窗体控件属性

窗体控件属性具体见表 6-4。

表 6-4　　　　　　　　　　　　窗体控件属性表

控件	Name	Text
toolStrip	m_toolStripButtons	m_toolStripButtons
Button	m_toolStripButtonContain	包含查询
Button	m_toolStripButtonIntersect	相交查询
Button	m_toolStripButtonDisjoint	分离查询
statusStrip	m_statusStrip	m_statusStrip
Label	m_toolStripStatusLabel	toolStripStatusLabel1

5. 窗体设计布局

窗体设计布局如图 6-4 所示。

图 6-4　窗体设计布局图

6. 创建 SampleRun 类

(1) 添加 using 引用代码

添加 using 引用代码具体如下：

using System.Diagnostics; //添加 System.Diagnostics 命名空间，使其对应的类可以使用

using System.Drawing; //添加 System.Drawing 命名空间，使其对应的类可以使用

using System.Windows.Forms; //添加 System.Windows.Forms 命名空间，使其对应的类可以使用

using SuperMap.Data; //添加 SuperMap.Data 命名空间，使其对应的类可以使用

using SuperMap.UI; //添加 SuperMap.UI 命名空间，使其对应的类可以使用
using SuperMap.Mapping; //添加 SuperMap.Mapping 命名空间，使其对应的类可以使用

（2）在 public class SampleRun 中添加代码

在 public class SampleRun 中添加如下代码：

```csharp
//定义类对象
private Workspace m_workspace;
private MapControl m_mapControl;
    //查询用到的数据
private readonly String m_mapName = "世界地图";
private readonly String m_queryObjectLayerName = "Ocean@world";
private readonly String m_queriedLayerName = "单值专题图";
    /// 根据 workspace 构造 SampleRun 对象
public SampleRun(Workspace workspace, MapControl mapControl)
    {m_workspace = workspace;
    m_mapControl = mapControl;
    m_mapControl.Map.Workspace = workspace;
    Initialize();}
private void Initialize()
    {try{//打开工作空间及地图
        WorkspaceConnectionInfo conInfo = new WorkspaceConnectionInfo(@"..\..\SampleData\World\World.smwu");
        m_workspace.Open(conInfo);//通过 workspace.Open 打开工作空间
        this.m_mapControl.Map.Open(m_mapName);
        // 调整 mapControl 的状态
        for (int i=0; i < m_mapControl.Map.Layers.Count; i++)
            {Layer layer = m_mapControl.Map.Layers[i];
            if (layer.Caption! = m_queryObjectLayerName)
            { layer.IsSelectable = false; }
            else{ layer.IsSelectable = true; } }
            Layer layer1 = m_mapControl.Map.Layers[m_queriedLayerName];
            layer1.Selection.Style.LineColor = Color.Green; }
        catch (Exception ex)
            {Trace.WriteLine(ex.Message); } }
    /// 按照各种算子进行查询
public void Query(SpatialQueryMode mode)
    {try{ //获取地图中的选择集，并转换为记录集
        Selection[] selections = m_mapControl.Map.FindSelection(true);
        Selection selection = selections[0];
```

```csharp
            Recordset recordset = selection.ToRecordset();
            //设置查询参数
            QueryParameter parameter = new QueryParameter();
            parameter.SpatialQueryObject = recordset;
            parameter.SpatialQueryMode = mode;
            //对指定查询的图层进行查询
            Layer layer = this.GetLayerByCaption(m_queriedLayerName);
            DatasetVector dataset = layer.Dataset as DatasetVector;
            Recordset recordset2 = dataset.Query(parameter);
            layer.Selection.FromRecordset(recordset2);
            layer.Selection.Style.LineColor = Color.Red;
            recordset2.Dispose();
            //刷新地图
            m_mapControl.Map.Refresh();
            recordset.Dispose();}
        catch(Exception ex)
        { Trace.WriteLine(ex.Message); } }
private Layer GetLayerByCaption(String layerCaption)
    { Layers layers = m_mapControl.Map.Layers;
     Layer result = null;
     foreach(Layer layer in layers)
         { if(String.Compare(layer.Caption, layerCaption, true) == 0)
             { result = layer;
               break; } }
             return result; }
```

7. 窗体代码

（1）添加 using 引用代码

添加 using 引用代码具体如下：

using System.Diagnostics; //添加 System.Diagnostics 命名空间，使其对应的类可以使用

（2）实例化代码

实例化 SampleRun 以及 public FormMain() 函数部分的代码如下：

```csharp
//定义类对象
private SampleRun m_sampleRun;
private SuperMap.Data.Workspace m_workspace;
private SuperMap.UI.MapControl m_mapControl;
private readonly String m_statusTextSelect = "请在地图中选择查询的对象";
private readonly String m_statusTextQuery = "请点击相应的按钮进行查询";
public FormMain()
```

```
{ try{ InitializeComponent();
    this.Load+=new EventHandler(FormMain_Load);
    this.m_workspace=new
    SuperMap.Data.Workspace();//新建对象
    this.m_mapControl=new SuperMap.UI.MapControl();//新建对象
//初始化控件
    m_mapControl.Dock=DockStyle.Fill;
    m_mapControl.GeometrySelected+=new SuperMap.UI.GeometrySelectedEventHandler(m_mapControl_GeometrySelected);
    m_toolStripStatusLabel.Text=m_statusTextSelect;
      base.Controls.Add(m_mapControl);
      base.Controls.SetChildIndex(m_mapControl,0);
//实例化 SampleRun
    m_sampleRun=new SampleRun(m_workspace,m_mapControl);}
      catch(Exception ex)
      {Trace.WriteLine(ex.Message);}}
```

(3) 窗体的 Load 事件

在窗体的属性窗口中,选择 按钮后,左键双击 Load,即可添加以下代码:

```
private void FormMain_Load(object sender, EventArgs e)
{ try{ m_mapControl.Map.ViewEntire();//全幅显示
      m_mapControl.Map.Refresh();}
      catch(Exception ex)
      {Trace.WriteLine(ex.Message);} }
```

(4) 添加各个控件的代码

① m_mapControl 控件的 GeometrySelected 事件。在控件的属性窗口单击 图标,左键双击 GeometrySelected,添加如下代码:

```
private void m_mapControl_GeometrySelected(object sender, SuperMap.UI.GeometrySelectedEventArgs e)
{try{ if(e.Count!=0)
      {//设置控件的有效性
        m_toolStripButtonContain.Enabled=true;
        m_toolStripButtonIntersect.Enabled=true;
        m_toolStripButtonDisjoint.Enabled=true;
        m_toolStripStatusLabel.Text=m_statusTextQuery;}}
      catch(Exception ex)
      {Trace.WriteLine(ex.Message);} }
```

② m_toolStripButtonContain 控件的 Click 事件。在控件的属性窗口单击 图标,左键双

击 Click，添加如下代码：

//包含查询

```
private void m_toolStripButtonContain_Click(object sender, EventArgs e)
    {try{m_sampleRun.Query(SuperMap.Data.SpatialQueryMode.Contain);//通过 sampleRun.Query 进行包含查询
        SetDisable();}
    catch(Exception ex)
        {Trace.WriteLine(ex.Message);}}
```

③ m_toolStripButtonIntersect 控件的 Click 事件。在控件的属性窗口单击 ⚡ 图标，左键双击 Click，添加如下代码：

//相交查询

```
private void m_toolStripButtonIntersect_Click(object sender, EventArgs e)
    {try{m_sampleRun.Query(SuperMap.Data.SpatialQueryMode.Intersect);
//通过 sampleRun.Query 进行相交查询
        SetDisable();}
catch(Exception ex)
{Trace.WriteLine(ex.Message);}}
```

④ m_toolStripButtonDisjoint 控件的 Click 事件。在控件的属性窗口单击 ⚡ 图标，左键双击 Click，添加如下代码：

//分离查询

```
private void m_toolStripButtonDisjoint_Click(object sender, EventArgs e)
    {try{m_sampleRun.Query(SuperMap.Data.SpatialQueryMode.Disjoint);//通过 sampleRun.Query 进行分离查询
        SetDisable();}
    catch(Exception ex)
        {Trace.WriteLine(ex.Message);}}
```

⑤ 定义 SetDisable 初始化函数，需添加以下代码：

```
private void SetDisable()
    {try{//设置控件有效性
        m_toolStripButtonContain.Enabled=false;
        m_toolStripButtonIntersect.Enabled=false;
        m_toolStripButtonDisjoint.Enabled=false;
        m_toolStripStatusLabel.Text=m_statusTextSelect;}
    catch(Exception ex)
        {Trace.WriteLine(ex.Message);}}
```

(5) 窗体的 FormClose 事件

在退出程序时，要断开空间的关联，关闭地图窗口、工作空间。因此，在属性窗口点

击 图标，左键双击 FormClose 事件，添加代码同前文中的相应部分。

8. 运行结果

窗体设计完成后点击 按钮，运行该程序，运行后弹出窗体，如图 6-5 所示。如图根据鼠标点击的正方形框中，查询出包含、相交、分离的地块。

图 6-5　空间查询

第7章 空间分析

7.1 网络分析

7.1.1 最近设施查找分析

1. 数据

在安装目录\SampleData\City\Changchun.udb 路径下打开数据文件。

2. 新建文件夹和工程

在【D:\MyProject\】文件中新建一个文件夹,命名为【第7章 空间分析】,并在其目录下建立一个文件夹,命名为【网络分析】,在此文件夹里新建文件夹,命名为 FindClosestFacility,在此文件夹里新建一个工程,将此工程命名为 FindClosestFacility。

3. 关键类型/成员

关键类型/成员具体见表7-1。

表7-1 关键类型/成员表

控件/类	方法	属性	事件
MapControl			MouseDown、MouseMove
TransportationAnalyst-Setting		NetworkDataset、EdgeIDField、NodeIDField、Tolerance、WeightFieldInfo、FNodeIDField、TNodeIDField	
TransportationAnalyst	Load、FindClosestFacility	AnalystSetting	
TransportationAnalyst-Parameter		Points、BarrierEdges、BarrierNodes、WeightName、NodesReturn、EdgesReturn、PathGuidesReturn、StopIndexesReturn、RoutesReturn	
TransportationAnalyst-Result		Routes、PathGuides	

4. 窗体控件属性

窗体控件属性具体见表7-2。

表 7-2　　　　　　　　　　　　　窗体控件属性表

控　件	Name	Text
toolStrip	toolStrip1	toolStrip1
Button	toolStripButton_Pan	漫游
	toolStripButton_EntireView	全幅显示
	toolStripButtonAnalyst	分析
	toolStripButtonClear	清空
Radio Button	radioButtonFacilityNodes	选取设施点
	radioButtonBarrierNodes	选取障碍点
	radioButtonEventNode	选取事件点
dataGridView	dataGridView	

5. 窗体设计布局

窗体设计布局如图 7-1 所示。

图 7-1　窗体设计布局图

6. 创建 SampleRun 类

（1）添加 using 引用代码

添加 using 引用代码具体如下：

using System. Diagnostics；//添加 System. Diagnostics 命名空间，使其对应的类可以使用

using System. Drawing；//添加 System. Drawing 命名空间，使其对应的类可以使用

using System. Windows. Forms；//添加 System. Windows. Forms 命名空间，使其对应的类可以使用

using SuperMap. Data；//添加 SuperMap. Data 命名空间，使其对应的类可以使用

using SuperMap.UI; //添加SuperMap.UI命名空间，使其对应的类可以使用

using SuperMap.Mapping; //添加SuperMap.Mapping命名空间，使其对应的类可以使用

using SuperMap.Analyst.NetworkAnalyst; //添加SuperMap.Analyst.NetworkAnalyst命名空间，使其对应的类可以使用

（2）在public class SampleRun中添加代码

在public class SampleRun中添加如下代码：

//定义类对象及控件

private Workspace m_workspace;

private MapControl m_mapControl;

private DataGridView m_dataGridView;

private SelectMode m_selectMode;// SelectMode:选择枚举，用于选择点

private Boolean m_selectEventNode;

private Boolean m_selectBarrier;

private Boolean m_selectFacilityNode;

private List<Int32> m_barrierEdges;

private List<Int32> m_barrierNodes;

private List<Int32> m_nodesList;

private DatasetVector m_datasetLine;// DatasetVector:矢量数据集类。描述矢量数据集，并提供相应的管理和操作。对矢量数据集的操作主要包括数据查询、修改、删除、建立索引等。

private Recordset m_recordset;// Recordset:记录集类。通过此类，可以实现对矢量数据集中的数据进行操作。

private DatasetVector m_datasetPoint;

private Point2Ds m_Points;// Point2Ds:点集合对象。此类管理线对象或线对象的子对象上的所有节点。由于线对象或线对象的子对象都是有向的，所以其点集合对象为有序的点的集合。

private TrackingLayer m_trackingLayer;// TrackingLayer:跟踪图层类。在SuperMap中，每个地图窗口都有一个跟踪图层，确切地说，每个地图显示时都有一个跟踪图层。跟踪图层是一个空白的透明图层，总是在地图各图层的最上层，主要用于在一个处理或分析过程中，临时存放一些图形对象，以及一些文本等。只要地图显示，跟踪图层就会存在，用户不可以删除跟踪图层，也不可以改变其位置。

private Layer m_layerLine;

private Layer m_layerPoint;

private GeoPoint m_geoPoint;// GeoPoint:二维点几何对象类，派生于Geometry类。

private static String m_datasetName = "RoadNet";

private static String m_nodeID = "SmNodeID";

private static String m_edgeID = "SmEdgeID";

private Int32 m_eventNode;

```csharp
private Point m_mousePoint;
private TransportationAnalyst m_analyst;// TransportationAnalyst:交通网络分析类。该类用于提供路径分析、旅行商分析、服务区分析、多旅行商(物流配送)分析、最近设施查找和选址分区分析等交通网络分析的功能。
private TransportationAnalystResult m_analystResult;// TransportationAnalystResult:交通网络分析结果类。该类用于获取分析结果的路由集合、分析途径的节点集合以及弧段集合、行驶导引集合、站点集合和权值集合以及各站点的花费。通过该类的设置,可以灵活地得到最佳路径分析、旅行商分析、物流配送和最近设施查找等分析的结果。
public enum SelectMode
{SELECTPOINT, SELECTBARRIER, SELECTEVENT, SELECTPAN, NONE}
/// 根据 workspace 和 mapControl 构造 SampleRun 对象
public SampleRun(Workspace workspace, MapControl mapControl, DataGridView dataGridView)
{try{ m_workspace = workspace;
m_mapControl = mapControl;
m_dataGridView = dataGridView;
m_mapControl.Map.Workspace = workspace;
Initialize();}
catch(Exception ex)
{Trace.WriteLine(ex.Message);}}
/// 初始化控件及数据
private void Initialize()
{try{ DatasourceConnectionInfo connectionInfo = new DatasourceConnectionInfo(
@"..\..\SampleData\City\Changchun.udb", "findClosestFacility",
"");
connectionInfo.EngineType = EngineType.UDB;// EngineType:该枚举定义了空间数据库引擎类型常量
m_workspace.Datasources.Open(connectionInfo);//根据 Datasources.Open 打开工作空间
m_datasetLine = (DatasetVector)m_workspace.Datasources[0]
.Datasets[m_datasetName] as DatasetVector;
m_datasetPoint = m_datasetLine.ChildDataset;
m_selectFacilityNode = true;
m_selectBarrier = false;
m_selectEventNode = false;
m_Points = new Point2Ds();
m_barrierEdges = new List<Int32>();
m_barrierNodes = new List<Int32>();
m_selectMode = SelectMode.SELECTPOINT;
```

```
m_nodesList = new List<Int32>( );
m_trackingLayer = m_mapControl. Map. TrackingLayer;
//加载点数据集及线数据集并设置各自的风格
m_layerLine = m_mapControl. Map. Layers. Add( m_datasetLine, true );
LayerSettingVector lineSetting = ( LayerSettingVector) m_layerLine
                . AdditionalSetting;
GeoStyle lineStyle = new GeoStyle( );//通过 GeoStyle 设置风格
lineStyle. LineColor = Color. LightGray;
lineStyle. LineWidth = 0.1;
lineSetting. Style = lineStyle;
m_layerPoint = m_mapControl. Map. Layers. Add(
m_datasetPoint, true );//添加图层
LayerSettingVector pointSetting = ( LayerSettingVector) m_layerPoint
                . AdditionalSetting;
GeoStyle pointStyle = new GeoStyle( );
pointStyle. LineColor = Color. DarkGray;
pointStyle. MarkerSize = new Size2D( 2.5, 2.5 );//设置标记点大小
pointSetting. Style = pointStyle;
// 调整 mapControl 的状态
m_mapControl. Action = SuperMap. UI. Action. Select;
m_mapControl. IsWaitCursorEnabled = false;
m_mapControl. Map. Refresh( );
m_mapControl. MouseDown += new MouseEventHandler( m_mapControl_MouseDown );
m_mapControl. MouseMove += new mouseEventHandler( m_mapControl_MouseMove );
//加载模型
 Load( );}
catch ( System. Exception ex )
{Trace. WriteLine( ex. Message ); } }
/// 加载图层
private void Load( )
    {try{// 设置网络分析基本环境,这一步骤需要设置分析权重、节点、弧段标识字段、容限
    TransportationAnalystSetting  setting  =  new  TransportationAnalystSetting ( );//
TransportationAnalystSetting:交通网络分析环境设置类。该类用于提供交通网络分析时所需要的所有参数信息。交通网络分析环境设置类的各个参数的设置直接影响分析的结果。
    //设置弧段的 ID、名字、字段
    setting. NetworkDataset = m_datasetLine;
    setting. EdgeIDField = m_edgeID;
    setting. EdgeNameField = "roadName";
```

```
        setting.NodeIDField = m_nodeID;
        setting.Tolerance = 0.01559;//设置容限
```
WeightFieldInfos weightFieldInfos = new WeightFieldInfos();//WeightFieldInfos:权值字段信息集合类。该类是权值字段信息对象(WeightFieldInfo)的集合，用于对权值字段信息对象进行管理，如添加、删除、获取指定名称或索引的权值字段信息对象等。

WeightFieldInfo weightFieldInfo = new WeightFieldInfo();//WeightFieldInfo:权值字段信息类型，存储了网络分析中权值字段的相关信息，包括正向权值字段与反向权值字段。

```
        weightFieldInfo.FTWeightField = "smLength";//设置正向阻力值
        weightFieldInfo.TFWeightField = "smLength";//设置反向阻力值
        weightFieldInfo.Name = "length";
        weightFieldInfos.Add(weightFieldInfo);
        setting.WeightFieldInfos = weightFieldInfos;
        setting.FNodeIDField = "SmFNode";//设置标志弧段起始节点ID字段
        setting.TNodeIDField = "SmTNode";//设置标志弧段终止节点ID字段
        //构造交通网络分析对象，加载环境设置对象
        m_analyst = new TransportationAnalyst();
        m_analyst.AnalystSetting = setting;
        m_analyst.Load();}
        catch(Exception ex)
        {Trace.WriteLine(ex.Message);}}
    /// 清除跟踪层
    public void ClearAll()
        {try{ //清除控件
            m_dataGridView.Rows.Clear();
            m_mapControl.Map.Layers[0].Selection.Clear();
            m_mapControl.Map.Layers[1].Selection.Clear();
            m_mapControl.Map.TrackingLayer.Clear();
            m_nodesList.RemoveRange(0,m_nodesList.Count);
            m_barrierEdges.RemoveRange(0,m_barrierEdges.Count);
            m_barrierNodes.RemoveRange(0,m_barrierNodes.Count);
            m_eventNode = -1;
            m_mapControl.Map.Refresh();}
    catch(Exception ex)
        {Trace.WriteLine(ex.Message);}}
    /// 更换鼠标状态
    public void ChangeAction(SelectMode mode)
        {try{ if(mode == SelectMode.SELECTPAN)
            {m_mapControl.Map.Layers[0].IsEditable = false;}//设置对象是否处于可编
```

辑状态

```
            m_selectMode=mode;
            m_mapControl. Map. Refresh( );}
        catch ( Exception ex)
            {Trace. WriteLine( ex. Message); }}
///开始分析
public void StartAnalyst( )
    {try{ int index=m_trackingLayer. IndexOf( "route");
        if (index ! =-1)
            { m_trackingLayer. Remove(index); }//根据对象索引值判断是否移除图层
        m_mapControl. Map. Refresh( );
        TransportationAnalystParameter parameter=new TransportationAnalystParameter( );
        //设置设施点
        int[ ] facilityNodes=new int[ m_nodesList. Count];
        for ( int i=0; i < facilityNodes. Length; i++)
            {facilityNodes[ i]=m_nodesList[ i]; }
        //设置属性
        parameter. Nodes=facilityNodes;
        parameter. WeightName="length";
        parameter. IsEdgesReturn=true;
        parameter. IsNodesReturn=true;
        parameter. IsRoutesReturn=true;
        parameter. IsPathGuidesReturn=true;
        //设置障碍点及障碍边
        int[ ] barrierEdges=new int[ m_barrierEdges. Count];
        for ( int i=0; i < barrierEdges. Length; i++)
            { barrierEdges[ i]=m_barrierEdges[ i]; }
        parameter. BarrierEdges=barrierEdges;
        int[ ] barrierNodes=new int[ m_barrierNodes. Count];
        for ( int i=0; i < barrierNodes. Length; i++)
            { barrierNodes[ i]=m_barrierNodes[ i]; }
        parameter. BarrierNodes=barrierNodes;
        //进行分析,这里设置查找设施点的数量
        m_analystResult=m_analyst. FindClosestFacility( parameter,
        m_eventNode, 1, true, 0);
        if ( m_analystResult==null)
            {MessageBox. Show("分析失败");
            return; }
        else{ ShowResult( );}//显示结果
```

FillResultTable(0);//填充结果导引}}
catch(Exception ex)
{Trace.WriteLine(ex.Message);}}
/// 显示结果
public void ShowResult()
{try{GeoLineM[] geoLineMs = m_analystResult.Routes;// GeoLineM:路由对象,是一组具有X,Y坐标与线性度量值(M值)的点组成的线性地物对象。例如高速公路上的里程碑,交通管制部门经常使用高速公路上的里程碑来标注并管理高速公路的路况、车辆的行驶限速和高速事故点等。
//设置属性
GeoLineM geoLineM = geoLineMs[0];
GeoStyle style = new GeoStyle();
style.LineColor = Color.Blue;
style.LineWidth = 1;
geoLineM.Style = style;
for(Int32 i = 0; i < m_trackingLayer.Count; i++)
{if(m_trackingLayer.Get(i).Type == GeometryType.GeoLineM)//清除上次结果 GeometryType:该枚举定义了一系列几何对象类型常量。
{m_trackingLayer.Remove(i);}}
m_trackingLayer.Add(geoLineM, "route");
m_mapControl.Map.Refresh();}
catch(System.Exception ex)
{Trace.WriteLine(ex.Message);}}
/// 显示结果,制定显示路线
public void FillResultTable(Int32 pathNum)
{try{//清除原数据,添加初始点信息
m_dataGridView.Rows.Clear();
Object[] objs = new Object[4];
objs[0] = m_dataGridView.RowCount;
objs[1] = "从起始点出发";
objs[2] = "--";
objs[3] = "--";
m_dataGridView.Rows.Add(objs);
//得到形式导引对象,根据导引子项类型的不同进行不同的填充
PathGuide[] pathGuides = m_analystResult.PathGuides;
PathGuide pathGuide = pathGuides[pathNum];
for(int j = 1; j < pathGuide.Count; j++)
{PathGuideItem item = pathGuide[j];
objs[0] = m_dataGridView.RowCount;

```
//导引子项为站点的添加方式
if ( item. IsStop )
{String side = "无";
if ( item. SideType == SideType. Left )
side = "左侧";
if ( item. SideType == SideType. Right )
side = "右侧";
if ( item. SideType == SideType. Middle )
side = "上";
String dis = item. Distance. ToString( );
if ( j ! = pathGuide. Count - 1 )
{objs[1] = "到达[" +item. Index+ "号路由点], 在道路" +side+dis; }
else{ objs[1] = "到达终点, 在道路" +side+dis; }
objs[2] = "";
objs[3] = "";
m_dataGridView. Rows. Add( objs ); }
//导引子项为弧段的添加方式
if ( item. IsEdge )
{String direct = "直行";
if ( item. DirectionType == DirectionType. East )
direct = "东";
if ( item. DirectionType == DirectionType. West )
direct = "西";
if ( item. DirectionType == DirectionType. South )
direct = "南";
if ( item. DirectionType == DirectionType. North )
direct = "北";
String weight = item. Weight. ToString( );
String roadName = item. Name;
String roadString = roadName. Equals( "" ) ? "匿名路段" : roadName;
objs[1] = "沿着[" +roadString+ "], 朝" +direct+ "行走" +weight;
objs[2] = weight;
objs[3] = item. Length;
m_dataGridView. Rows. Add( objs ); }}}
catch ( Exception e )
{ Trace. WriteLine( e. Message ); }}
/// 选择设施点
public void FindFacilityNodes( )
{try{ Recordset recordset = m_layerPoint. Selection. ToRecordset( );
```

```
            Geometry geometry = recordset. GetGeometry( );
            //设置属性
            m_selectFacilityNode = true;
            m_selectBarrier = false;
            m_selectEventNode = false;
            int id = m_recordset. GetID( );
            if ( geometry. Type = = GeometryType. GeoPoint)
            {m_nodesList. Add( m_recordset. GetID( ));//添加对象
            m_trackingLayer. Add( m_geoPoint, "FacilityNode" +id);}}//添加对象到跟踪
图层
        catch (Exception ex)
            {Trace. WriteLine( ex. Message);}}
/// 选择障碍点
public void FindBarrierNodes( )
    {try{Recordset recordset = m_layerPoint. Selection. ToRecordset( );
        Geometry geometry = recordset. GetGeometry( );
        m_selectFacilityNode = false;
        m_selectBarrier = true;
        m_selectEventNode = false;
        Int32 id = m_recordset. GetID( );//获取当前记录对应的几何对象的 ID 值
        if ( geometry. Type = = GeometryType. GeoPoint)
        {m_nodesList. Add( m_recordset. GetID( ));//添加对象
        m_trackingLayer. Add( m_geoPoint, "FacilityNode" +id);}}//添加对象到跟踪
图层
    catch (Exception ex)
        {Trace. WriteLine( ex. Message);}}
/// 选择事件点
public void FindEventNode( )
    {try{Recordset recordset = m_layerPoint. Selection. ToRecordset( );
        Geometry geometry = recordset. GetGeometry( );
        m_selectFacilityNode = false;
        m_selectBarrier = false;
        m_selectEventNode = true;
        m_eventNode = m_recordset. GetID( );
        if ( geometry. Type = = GeometryType. GeoPoint)
            {int index = m_trackingLayer. IndexOf( "EventNode");
            if (index ! = -1)//根据索引值判断是否删除跟踪层的指定对象
            m_trackingLayer. Remove( index );
            m_trackingLayer. Add( m_geoPoint, "EventNode");}}
```

```csharp
catch (Exception ex)
    {Trace.WriteLine(ex.Message);}}
/// MapControl MouseMove 事件
private void m_mapControl_MouseMove(object sender, MouseEventArgs e)
    {try{m_mapControl.DoMouseMove(e);
        if(m_mapControl.Action==SuperMap.UI.Action.Select
                    || m_mapControl.Action==SuperMap.UI.Action.Select2)
            {//获取鼠标点对应的地图点
            m_mousePoint=new Point(e.X,e.Y);
            Point2D point2D=m_mapControl.Map.PixelToMap(m_mousePoint);
            //根据当前比例尺设置捕捉框的大小
            double scale=(3 * 10E-4)/m_mapControl.Map.Scale;
            Selection selection=m_layerPoint.HitTest(point2D,4/3 * scale);
            int index=m_trackingLayer.IndexOf("geoLine");//根据索引值判断是否删除
跟踪层的指定对象
            if(index!=-1)
            m_trackingLayer.Remove(index);
            if(selection!=null && selection.Count>0)
            {Recordset recordset=selection.ToRecordset();
            GeoPoint geoPoint=(GeoPoint)recordset.GetGeometry();
            recordset.Dispose();
            double pointX=geoPoint.X;
            double pointY=geoPoint.Y;
            //构造捕捉框
            Point2Ds point2Ds=new Point2Ds();
            point2Ds.Add(new Point2D(pointX-scale,pointY-scale));
            point2Ds.Add(new Point2D(pointX+scale,pointY-scale));
            point2Ds.Add(new Point2D(pointX+scale,pointY+scale));
            point2Ds.Add(new Point2D(pointX-scale,pointY+scale));
            point2Ds.Add(new Point2D(pointX-scale,pointY-scale));
            GeoLine geoLine=new GeoLine(point2Ds);
            //刷新地图
            m_mapControl.SelectionTolerance=2;
            m_trackingLayer.Add(geoLine,"geoLine");
            m_mapControl.Map.Refresh();}}}
        catch(System.Exception ex)
        {Trace.WriteLine(ex.Message);}}
/// MapControl MouseDown 事件
private void m_mapControl_MouseDown(object sender, MouseEventArgs e)
```

```
{try{if(e.Button==MouseButtons.Left)
    {Selection selection=m_layerPoint.Selection;
     if(selection==null || selection.Count==0)
     {selection=m_layerLine.Selection;}//根据选择集中选择对象个数获取对象个数
     if(m_mapControl.Action==SuperMap.UI.Action.Select && e.Clicks==1
        && (m_selectMode==SelectMode.SELECTPOINT || m_selectMode==SelectMode.SELECTBARRIER || m_selectMode==SelectMode.SELECTEVENT))
     {if(selection.Count<=0)
      {MessageBox.Show("坐标点超出选择容限,不能作为分析点");}
      else{//根据选择的不同,构造点对象
       m_recordset=selection.ToRecordset();
       Geometry geometry=m_recordset.GetGeometry();
       AddPoint(geometry);
       m_recordset.Dispose();}}}}
catch(System.Exception ex)
    {Trace.WriteLine(ex.Message);}}
/// 添加分析经过点
public void AddPoint(Geometry geometry)
    {try{//在跟踪图层上添加点
     GeoPoint geoPoint=new GeoPoint();
     if(geometry.Type==GeometryType.GeoPoint)// GeometryType:该枚举定义了一系列几何对象类型常量。
     {geoPoint=(GeoPoint)geometry;}
     else{geoPoint=new GeoPoint(((GeoLine)geometry).InnerPoint);}
     GeoStyle style=new GeoStyle();
     if(m_selectFacilityNode)
     {style.MarkerSize=new Size2D(10,10);
      style.LineColor=Color.LightGreen;
      geoPoint.Style=style;
      int id=m_recordset.GetID();
      if(geometry.Type==GeometryType.GeoPoint)
      {m_nodesList.Add(m_recordset.GetID());}//根据对象的类型判断是否添加图层
      m_trackingLayer.Add(geoPoint,"FacilityNode"+id);}}
     else if(m_selectBarrier)
     {style.LineColor=Color.Red;
      style.MarkerSymbolID=8622;
      style.MarkerSize=new Size2D(10,10);
```

```
                geoPoint.Style = style;
                //构造障碍点
                Int32 id = m_recordset.GetID();
                if (geometry.Type == GeometryType.GeoPoint)// GeometryType:该枚举定义了
一系列几何对象类型常量。
                    {m_barrierNodes.Add(m_recordset.GetID());//根据对象的类型判断是否添
加图层
                    m_trackingLayer.Add(geoPoint, "barrierNode" +id);}
                else{m_barrierEdges.Add(m_recordset.GetID());
                    m_trackingLayer.Add(geoPoint, "barrierEdge" +id);}
                m_mapControl.Map.Refresh();}
            else if (m_selectEventNode)
                {//设置属性
                style.MarkerSize = new Size2D(10, 10);
                style.LineColor = Color.Orange;
                geoPoint.Style = style;
                m_eventNode = m_recordset.GetID();
                if (geometry.Type == GeometryType.GeoPoint)
                {int index = m_trackingLayer.IndexOf("EventNode");
                if (index! = -1)
                {m_trackingLayer.Remove(index);//根据索引值判断是否移除指定图层
                m_mapControl.Map.Refresh();}
                m_trackingLayer.Add(geoPoint, "EventNode");}//添加图层
                else {int index = m_trackingLayer.IndexOf("EventNode");
                if (index! = -1)
                { m_trackingLayer.Remove(index);
                m_mapControl.Map.Refresh();}
                m_eventNode = -1;
                MessageBox.Show("选择失败,请重新选择");}}
                geoPoint.Style = style;
                m_mapControl.Map.Refresh();}
        catch (Exception ex)
            { Trace.WriteLine(ex.Message);}}
```

7. 窗体代码

(1) 添加 using 引用代码

添加 using 引用代码具体如下:

using System.Diagnostics;//添加 System.Diagnostics 命名空间,使其对应的类可以使用
using SuperMap.Data;//添加 SuperMap.Data 命名空间,使其对应的类可以使用
using SuperMap.UI;//添加 SuperMap.UI 命名空间,使其对应的类可以使用

using SuperMap.Mapping;//添加 SuperMap.Mapping 命名空间,使其对应的类可以使用

(2)在 public FormMain()函数前添加代码
在 public FormMain()函数前添加如下代码:
//定义类对象
private SampleRun m_sampleRun;
private SuperMap.Data.Workspace m_workspace;
private SuperMap.UI.MapControl m_mapControl;
private Boolean m_hasAnalysted;

(3)窗体的 Load 事件
在窗体的属性窗口中,选择 按钮后,左键双击 Load,即可添加以下代码:
private void FormMain_Load(object sender, EventArgs e)
    {try{//创建新对象和初始化控件
        this.m_workspace = new SuperMap.Data.Workspace();
        this.m_mapControl = new SuperMap.UI.MapControl();
        dataGridView.AutoResizeColumns();
        dataGridView.AutoSizeColumnsMode = DataGridViewAutoSizeColumnsMode.Fill;
        dataGridView.Columns.Clear();
        dataGridView.Columns.Add("序号","序号");
        dataGridView.Columns.Add("导引","导引");
        dataGridView.Columns.Add("耗费","耗费");
        dataGridView.Columns.Add("距离","距离");
        //实例化 SampleRun
        m_sampleRun = new SampleRun(m_workspace, m_mapControl, dataGridView);
        for (int i=0; i < dataGridView.Columns.Count; i++)
        {dataGridView.Columns[i].ReadOnly = true;}
        //初始化控件
        m_mapControl.Dock = DockStyle.Fill;
        this.splitContainer1.Panel1.Controls.Add(m_mapControl);
        this.Controls.SetChildIndex(m_mapControl, 0);
        m_mapControl.Paint += new PaintEventHandler(m_mapControl_Paint);}
    catch (Exception ex)
        {Trace.WriteLine(ex.Message);}}

(4)表格清空
没有选择对象的时候表格清空,需添加如下代码:
private void m_mapControl_Paint(object sender, PaintEventArgs e)
    {try{if (m_mapControl.Map.Layers[0].Selection.Count < 1)
        {dataGridView.Columns.Clear();
         dataGridView.Rows.Clear();}}

catch(Exception ex)
{Trace.WriteLine(ex.Message);}}

(5)添加各个控件的代码

① 漫游 Button 控件的 Click 事件。左键双击漫游 Button 控件,即可添加如下该按钮的 Click 事件代码,实现漫游:

private void toolStripButton_Pan_Click(object sender,EventArgs e)
{try{m_mapControl.Action=SuperMap.UI.Action.Pan;//通过 Action.Pan 实现漫游
　　m_mapControl.Map.Refresh();}
catch(Exception ex)
{Trace.WriteLine(ex.Message);}}

② 全屏 Button 控件的 Click 事件。左键双击全屏 Button 控件,即可添加如下该按钮的 Click 事件代码:

private void toolStripButtonEntireView_Click(object sender,EventArgs e)
{m_mapControl.Map.ViewEntire();//全幅显示
m_mapControl.Map.Refresh();}

③ 分析 Button 控件的 Click 事件。左键双击分析 Button 控件,即可添加如下该按钮的 Click 事件代码:

private void toolStripButtonAnalyst_Click(object sender,EventArgs e)
{m_sampleRun.StartAnalyst();}//通过 sampleRun.StartAnalyst 分析

④ 清除 Button 控件的 Click 事件。左键双击清除 Button 控件,即可添加如下该按钮的 Click 事件代码:

private void toolStripButtonClear_Click(object sender,EventArgs e)
{m_sampleRun.ClearAll();//通过 sampleRun.ClearAll 清除
if(m_hasAnalysted)
{m_hasAnalysted=false;}}

⑤ 选取设施点 RadioButton 控件的 CheckedChanged 事件。在 radioButtonFacilityNodes 的属性窗口中,选择 ⚡ 按钮后,左键双击 CheckedChanged,即可添加以下代码,实现选取设施点:

private void radioButtonFacilityNodes_CheckedChanged(object sender,EventArgs e)
{m_sampleRun.FindFacilityNodes();//通过 sampleRun.FindFacilityNodes 选取设施点
m_sampleRun.ChangeAction(SampleRun.SelectMode.SELECTPOINT);}

⑥ 选取障碍点 RadioButton 控件的 CheckedChanged 事件。在 radioButtonBarrierNodes 的属性窗口中,选择 ⚡ 按钮后,左键双击 CheckedChanged,即可添加以下代码,实现选取障碍点:

private void radioButtonBarrierNodes_CheckedChanged(object sender,EventArgs e)
{m_sampleRun.FindBarrierNodes();//通过 sampleRun.FindBarrierNodes 选取障碍点
m_sampleRun.ChangeAction(SampleRun.SelectMode.SELECTBARRIER);}

⑦ 选取事件点 RadioButton 控件的 CheckedChanged 事件。在 radioButtonEventNode 的属性窗口中，选择 按钮后，左键双击 CheckedChanged，即可添加以下代码，实现选取事件点：

private void radioButtonEventNode_CheckedChanged(object sender, EventArgs e)
　　{m_sampleRun.FindEventNode();//通过 sampleRun.FindEventNode 选取事件点
　　m_sampleRun.ChangeAction(SampleRun.SelectMode.SELECTEVENT);}

（6）窗体的 FormClose 事件

在退出程序时，要断开空间的关联，关闭地图窗口、工作空间。因此在属性窗口点击 图标，左键双击 FormClose 事件，添加代码同前文相应部分。

8. 运行结果

窗体设计完成后点击 按钮，运行该程序，运行后弹出窗体，如图7-2所示，图中绿色圆点为设施点，黄色圆点为事件地点，红色三角为障碍地点，点击"分析"按钮，即可分析出事件发生地点最近的设施地点。

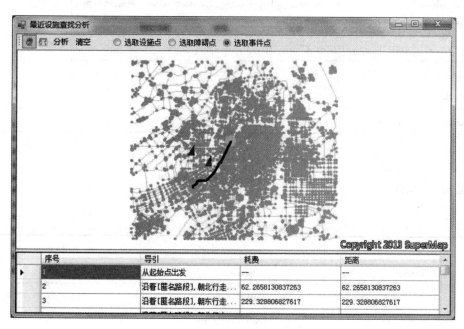

图 7-2　最近设施查找分析图

7.1.2　选址分区分析

1. 数据

在安装目录\SampleData\City\Changchun.udb 路径下打开数据文件。

2. 新建文件夹和工程

在【D:\MyProject\】文件中新建一个文件夹，命名为【第7章　空间分析】，并在其目录下建立一个文件夹命名为【网络分析】，在此文件夹里新建一个文件夹命名为 FindLocation，在此文件夹新建一个工程，将此工程命名为 FindLocation。

3. 关键类型/成员

关键类型/成员见表7-3。

表7-3　　　　　　　　　　　　　　　关键类型/成员表

控件/类	方法	属性	事件
MapControl			GeometrySelected、MouseDown、MouseMove
TransportationAnalystSetting		NetworkDataset、EdgeIDField、NodeIDField、Tolerance、WeightFieldInfos、FNodeIDField、TNodeIDFieldTNodeIDField	
TransportationAnalyst	Load	AnalystSetting、FindLocation	
LocationAnalystParameter		FromCenter、ExpectedSupplyCenterCount、WeightName、SupplyCenters	
SupplyCenters	Add、Remove	Item、Count、ID、MaxWeight、Type	
LocationAnalystResult		DemandResults、SupplyResults	
Colors	MakeRandom		

4. 窗体控件属性

窗体控件属性具体见表7-4。

表7-4　　　　　　　　　　　　　　　窗体控件属性表

控件	Name	Text
TextBox	TextBox1	
Button	buttonDelete	删除
	buttonAnalyst	分析
	buttonClear	清除结果
Label	Label1	中心点列表
	Label2	中心点点数
dataGridView	dataGridView	

5. 窗体设计布局

窗体设计布局如图7-3所示。

6. 创建 SampleRun 类

（1）添加 using 引用代码

7.1 网络分析

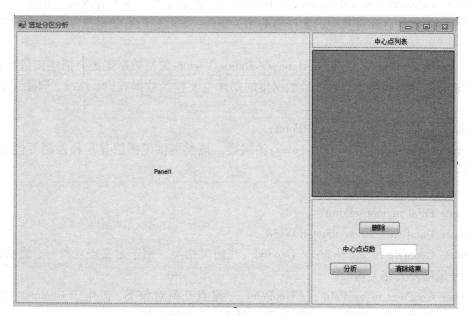

图 7-3 窗体设计布局图

添加 using 引用代码具体如下：

using System. Diagnostics；//添加 System. Diagnostics 命名空间，使其对应的类可以使用

using System. Drawing；//添加 System. Drawing 命名空间，使其对应的类可以使用

using System. Windows. Forms；//添加 System. Windows. Forms 命名空间，使其对应的类可以使用

using SuperMap. Data；//添加 SuperMap. Data 命名空间，使其对应的类可以使用

using SuperMap. UI；//添加 SuperMap. UI 命名空间，使其对应的类可以使用

using SuperMap. Mapping；//添加 SuperMap. Mapping 命名空间，使其对应的类可以使用

using SuperMap. Analyst. NetworkAnalyst；//添加 SuperMap. Analyst. NetworkAnalyst 命名空间，使其对应的类可以使用

（2）在 public class SampleRun 中添加代码

在 public class SampleRun 中添加如下代码：

//设置选址分析中用到的数据集及字段 定义类对象

private static String m_datasetName = "RoadNet"；

private static String m_nodeID = "SmNodeID"；

private static String m_edgeID = "SmEdgeID"；

private Workspace m_workspace；// Workspace：工作空间类。工作空间是用户的工作环境，主要完成数据的组织和管理，包括打开、关闭、创建、保存工作空间文件。

private MapControl m_mapControl；// MapControl：地图控件类。该类是用于提供地图的显示界面，同时为地图与数据的互操作提供途径。

private Datasource m_datasource;// Datasource:数据源类。该类管理投影信息、数据源与数据库的连接信息和对其中的数据集的相关操作,如通过已有数据集复制生成新的数据集等。

private DatasetVector m_datasetLine;// DatasetVector:矢量数据集类。描述矢量数据集,并提供相应的管理和操作。对矢量数据集的操作主要包括数据查询、修改、删除、建立索引等。

private DatasetVector m_datasetPoint;

private Layer m_layerPoint;// Layer:图层类。该类提供了图层显示和控制等便于地图管理的一系列属性。

private Layer m_layerLine;

private Point m_mousePoint;

private DataGridView m_dataGridView;

private GeoStyle m_geoStyle;// GeoStyle:几何风格类。用于定义点状符号、线状符号、填充符号风格及其相关属性。

private GeoPoint m_geoPoint;// GeoPoint:二维点几何对象类,派生于 Geometry 类。该类一般用于描述点状地理实体,例如,气象站点、公交站点等。

private TrackingLayer m_trackingLayer;// TrackingLayer:跟踪图层类。在 SuperMap 中,每个地图窗口都有一个跟踪图层,确切地说,每个地图显示时都有一个跟踪图层。跟踪图层是一个空白的透明图层,总是在地图各图层的最上层,主要用于在一个处理或分析过程中,临时存放一些图形对象,以及一些文本等。只要地图显示,跟踪图层就会存在,跟踪图层不能被删除,也不能被改变位置。

//分析需要的交通网络分析对象

private TransportationAnalyst m_analyst;// TransportationAnalyst:交通网络分析类。该类用于提供路径分析、旅行商分析、服务区分析、多旅行商(物流配送)分析、最近设施查找和选址分区分析等交通网络分析的功能。

private SupplyCenters m_supplyCenters;// SupplyCenters:资源供给中心集合类。该类是资源供给中心对象(SupplyCenter)的集合,用于管理资源供给中心对象,如添加、删除资源供给中心,获取资源供给中心对象的个数,获取指定序号的资源供给中心对象,以及将资源供给中心集合对象转换为资源供给中心对象数组等。

private DemandResult[ ] m_demandResult;// DemandResult:需求结果类。该类用于返回需求结果的相关信息,包括需求节点的 ID 和资源供给中心 ID。

private SupplyResult[ ] m_supplyResult; // SupplyResult:资源供给结果类。该类提供了资源供给的结果,包括资源供给中心的类型、ID、最大阻值、需求点的数量、平均耗费和总耗费等。

/// 根据 workspace 和 mapControl 构造 SampleRun 对象

public SampleRun (Workspace workspace, MapControl mapControl, DataGridView dataGridView)

{ try { m_workspace = workspace;

m_mapControl = mapControl;

```
            m_dataGridView = dataGridView;
            m_mapControl.Map.Workspace = workspace;
            Initialize();}
            catch(Exception ex)
            {Trace.WriteLine(ex.Message);}}
            /// 初始化控件及数据
    private void Initialize()
        {try{ DatasourceConnectionInfo connectionInfo = new DatasourceConnectionInfo(
            @"..\..\SampleData\City\Changchun.udb"," findLocation","");//
DatasourceConnectionInfo:数据源连接信息类,包括进行数据源连接的所有信息,如所要连接
的服务器名称、数据库名称、用户名、密码等。当保存工作空间时,工作空间中的数据源
的连接信息都将存储到工作空间文件中。
            connectionInfo.EngineType = EngineType.UDB;// EngineType:该枚举定义了空间
数据库引擎类型常量。
            m_datasource = m_workspace.Datasources.Open(connectionInfo);//打开工作空间
            m_datasetLine = (DatasetVector)m_workspace.Datasources[0]
                    Datasets[m_datasetName] as DatasetVector;
            m_datasetPoint = m_datasetLine.ChildDataset;
            m_supplyCenters = new SupplyCenters();
            m_trackingLayer = m_mapControl.Map.TrackingLayer;
            m_trackingLayer.IsAntialias = true;
            m_geoStyle = new GeoStyle();
            m_geoStyle.MarkerSize = new Size2D(10,10);
            //加载点数据集及线数据集并设置各自的风格
            m_layerLine = m_mapControl.Map.Layers.Add(m_datasetLine,true);
            LayerSettingVector lineSetting = (LayerSettingVector)m_layerLine
            AdditionalSetting;// LayerSettingVector:矢量图层设置类。该类主要用来设置矢
量图层的显示风格。矢量图层用单一的符号或风格绘制所有的要素。当用户只想可视化地
显示空间数据,只关心空间数据中各要素在什么位置,而不关心各要素在数量或性质上的
不同时,可以用普通图层来显示要素数据。
            GeoStyle lineStyle = new GeoStyle();
            lineStyle.LineColor = Color.LightGray;//设置颜色
            lineStyle.LineWidth = 0.1;//设置线宽
            lineSetting.Style = lineStyle;//设置风格
            m_layerPoint = m_mapControl.Map.Layers.Add(m_datasetPoint,true);//添加图层
            LayerSettingVector pointSetting = (LayerSettingVector)m_layerPoint.AdditionalSetting;
            GeoStyle pointStyle = new GeoStyle();
            pointStyle.LineColor = Color.DarkGray;
            pointStyle.MarkerSize = new Size2D(2.5,2.5);//设置点标志大小
```

```csharp
            pointSetting.Style = pointStyle;
            // 调整 mapControl 的状态
            m_mapControl.IsWaitCursorEnabled = false;
            m_mapControl.Map.Refresh();
            //注册事件
            m_mapControl.MouseMove += new MouseEventHandler(m_mapControl_MouseMove);
            m_mapControl.GeometrySelected += new GeometrySelectedEventHandler(m_mapControl_GeometrySelected);
            Load();}
        catch(System.Exception ex)
        {Trace.WriteLine(ex.Message);}}
    /// 加载图层
    private void Load()
        {try{ // 设置网络分析基本环境，这一步骤需要设置分析权重、节点、弧段标识字段、容限
            TransportationAnalystSetting setting = new TransportationAnalystSetting();//TransportationAnalystSetting:交通网络分析环境设置类。该类用于提供交通网络分析时所需要的所有参数信息。交通网络分析环境设置类的各个参数的设置直接影响分析的结果。
            setting.NetworkDataset = m_datasetLine;
            setting.EdgeIDField = m_edgeID;//设置标识弧段 ID 的字段
            setting.NodeIDField = m_nodeID;  //设置标识网络节点 ID 的字段
            setting.Tolerance = 0.01559;
            WeightFieldInfos weightFieldInfos = new WeightFieldInfos();//WeightFieldInfos:权值字段信息集合类。该类是权值字段信息对象(WeightFieldInfo)的集合,用于管理权值字段信息对象,如添加、删除、获取指定名称或索引的权值字段信息对象等。
            WeightFieldInfo weightFieldInfo = new WeightFieldInfo();//WeightFieldInfo:权值字段信息类型,存储了网络分析中权值字段的相关信息,包括正向权值字段与反向权值字段。
            weightFieldInfo.FTWeightField = "smLength";//设置正向阻力字段
            weightFieldInfo.TFWeightField = "smLength";//设置反向阻力字段
            weightFieldInfo.Name    = "length";//设置权值字段名称
            weightFieldInfos.Add(weightFieldInfo);
            setting.WeightFieldInfos = weightFieldInfos;
            setting.FNodeIDField = "SmFNode";
            setting.TNodeIDField = "SmTNode";
            //构造网络分析对象,加载环境设置对象
            m_analyst = new TransportationAnalyst();
            m_analyst.AnalystSetting = setting;
```

```csharp
            m_analyst.Load();}
        catch(Exception ex)
        {Trace.WriteLine(ex.Message);}}
/// 选址分析，得出结果
public void StartAnalyst(Int32 supplyCenterCount)
    {try{// 设置选址分区分析参数
        LocationAnalystParameter parameter = new LocationAnalystParameter();//
LocationAnalystParameter:选址分区分析参数类。为选址分区分析提供必要的参数信息，包括是否从资源中心分配，资源供给中心集合，权值字段信息的名称，转向权值字段，以及期望的供给中心数量等。
        parameter.IsFromCenter = true;//设置是否从资源供给中心开始分配资源
        parameter.ExpectedSupplyCenterCount = 0;//设置期望用于最终设施选址的资源供给中心的数量
        parameter.WeightName = "length";//设置权值字段信息的名称
        parameter.SupplyCenters = m_supplyCenters;//设置资源供给中心集合
        Int32 length = m_supplyCenters.Count;
        if (length < supplyCenterCount)
        {supplyCenterCount = length;}
        parameter.ExpectedSupplyCenterCount = supplyCenterCount;
        //根据给定的参数进行选址分析
        LocationAnalystResult result = m_analyst.FindLocation(parameter);// LocationAnalystResult:选址分区分析结果类。
        if(null != result)
        {m_demandResult = result.DemandResults;
        m_supplyResult = result.SupplyResults;
        ShowResult();}
        else{MessageBox.Show("分析失败，请检查中心点点数");}}
    catch(Exception ex)
        {Trace.WriteLine(ex.Message);}}
/// 对结果进行渲染显示
private void ShowResult()
    {try{ //分析前删除已存在的结果
        Int32 count = m_trackingLayer.Count;
        for (Int32 i = 0; i < count; i++)
        {Int32 index = m_trackingLayer.IndexOf("result");
        if (index != -1)
        m_trackingLayer.Remove(index);}
        m_mapControl.Map.Refresh();
        //对不同的中心点对应的结果设置不同的风格
```

```
GeoStyle geoStyle = new GeoStyle();
geoStyle.MarkerSize = new Size2D(6,6);
Colors colors = Colors.MakeRandom(m_supplyResult.Length);
for(Int32 i=0;i < m_supplyResult.Length;i++)
{geoStyle.LineColor = colors[i];
//将同一个中心点对应的结果保存至数组中
List<Int32> arrayList = new List<Int32>();
for(Int32 j=0;j < m_demandResult.Length;j++)
{if(m_supplyResult[i].ID == m_demandResult[j].SupplyCenterID)
{arrayList.Add(m_demandResult[j].ID);}}
Int32[] ids = new Int32[arrayList.Count];
for(Int32 j=0;j < ids.Length;j++)
{ids[j] = arrayList[j];}
//由数组得到相应对象设置风格后添加到跟踪图层中
DatasetVector datasetVector =(DatasetVector)m_layerPoint.Dataset;
Recordset recordset = datasetVector.Query(ids,CursorType.Static);//Recordset:
```
记录集类。通过此类，可以实现对矢量数据集中的数据进行操作。
```
recordset.MoveFirst();
while(! recordset.IsEOF)
{Geometry geometry = recordset.GetGeometry();//Geometry:所有几何类
```
型(GeoPoint，GeoLine，GeoRegion等)的基类。该类是一个抽象类。提供一些基本的几何类型的属性与方法。该类用来描述地理实体的空间特征，并提供相关的处理方法。根据地理空间实体特征的不同，分别用点(GeoPoint)、线(GeoLine)、面(GeoRegion)等类型进行描述。
```
geometry.Style = geoStyle;
m_trackingLayer.Add(geometry,"result");
recordset.MoveNext();}
m_mapControl.Map.Refresh();
recordset.Dispose();}
//重新绘制中心点和节点，使其不被分析结果遮挡
for(Int32 i=0;i<m_dataGridView.Rows.Count;i++)
{Int32 nodeId = Int32.Parse(m_dataGridView.Rows[i].Cells[1].Value.ToString());
Int32 indexOfCenter = m_trackingLayer.IndexOf("centerPoint"+nodeId);
Geometry centerPoint = m_trackingLayer.Get(indexOfCenter);
int indexOfCenterID = m_trackingLayer.IndexOf(nodeId.ToString());
Geometry centerID = m_trackingLayer.Get(indexOfCenterID);//获取此跟踪图层中
```
指定索引的几何对象
```
m_trackingLayer.Add(centerPoint,"centerPoint"+nodeId);//向跟踪图层中添加
```

几何对象

```
            m_trackingLayer.Add(centerID,nodeId.ToString());
            m_mapControl.Map.RefreshTrackingLayer();}}//刷新地图
        catch(System.Exception ex)
            {Trace.WriteLine(ex.Message);}}
        ///对象选择事件
    public void m_mapControl_GeometrySelected(object sender,GeometrySelectedEventArgs e)
        {//构造点对象
        Recordset recordset=m_layerPoint.Selection.ToRecordset();
            try{Geometry geometry=recordset.GetGeometry();
            m_geoPoint=(GeoPoint)geometry;
            //添加中心点
            Int32 centerID=Int32.Parse(recordset.GetFieldValue(m_nodeID).ToString());
            AddCenter(centerID);}
        catch(Exception ex)
            {Trace.WriteLine(ex.Message);}
        finally{ recordset.Dispose();}}
    ///判断对象是否已经被选择
    private bool IsHasSelected(Int32 selectedId,GeometryType type)
        {try{//判断是否在中心点列表中
        for(Int32 i=1; i < m_dataGridView.RowCount; i++)
            {Int32 id=Int32.Parse(m_dataGridView.Rows[i - 1].Cells[1].Value.ToString());
            if(id==selectedId)
            {return true;}}}
        catch(System.Exception ex)
            {Trace.WriteLine(ex.Message);}
        return false;}
        ///根据对象添加中心点
    private void AddCenter(Int32 centerID)
        {try{if(! IsHasSelected(centerID,GeometryType.GeoPoint))// GeometryType:该枚举定义了一系列几何对象类型常量。
            {m_geoStyle.LineColor=Color.Green;
            m_geoPoint.Style=m_geoStyle;
            //构造中心点列表对应的模型
            Object[] values=new Object[4];
            values[0]=m_dataGridView.RowCount;
            values[1]=centerID;
```

```
            values[2] = 700;
            values[3] = "可选中心点?";
            m_dataGridView.Rows.Add(values);
            //设置中心点初始相应属性,并添加到中心点集合中
            SupplyCenter supplyCenter = new SupplyCenter();//SupplyCenter:资源供给
中心类,存储了选址分区分析(FindLocation 方法)中资源供给中心的信息,包括资源供给
中心的 ID、最大耗费和类型。
            supplyCenter.ID = centerID;
            supplyCenter.MaxWeight = Convert.ToInt32(values[2]);//设置资源供给中
心的最大耗费
            supplyCenter.Type = SupplyCenterType.OptionalCenter;// SupplyCenterType:
该枚举定义了资源供给中心类型常量。网络分析中资源中心点的类型,主要用于选址分
区。
            m_supplyCenters.Add(supplyCenter);
            TextPart textPart = new TextPart(centerID.ToString(), new Point2D(m_geoPoint.
X, m_geoPoint.Y));
            GeoText geoText = new GeoText(textPart);
            TextStyle textStyle = new TextStyle();
            textStyle.FontHeight = 5;//设置文本高度
            textStyle.ForeColor = Color.Black;//设置文本前景色
            textStyle.Bold = true;//设置文本是否为粗体
            geoText.TextStyle = textStyle;//设置文本风格
            m_trackingLayer.Add(m_geoPoint, "centerPoint" + centerID);
            m_trackingLayer.Add(geoText, centerID.ToString());
            m_mapControl.Map.RefreshTrackingLayer();}}
        catch(System.Exception ex)
        {Trace.WriteLine(ex.Message);}}
    /// 点捕捉方法
    private void SnapPoint()
        {try{Point2D point2D = m_mapControl.Map.PixelToMap(m_mousePoint);
            //根据当前比例尺设置捕捉框的大小
            Double scale = (3 * 10E-4) / m_mapControl.Map.Scale;
            Selection selection = m_layerPoint.HitTest(point2D, 4 / 3 * scale);// Selection:
选择集类。该类用于处理地图上被选中的对象。
            Int32 index = m_trackingLayer.IndexOf("geoLine");
            if(index! = -1)
            m_trackingLayer.Remove(index);
            if(selection! = null && selection.Count > 0)
            {Recordset recordset = selection.ToRecordset();//Recordset:记录集类。通过此
```

类，可以实现对矢量数据集中的数据进行操作。

```
            GeoPoint geoPoint = (GeoPoint) recordset.GetGeometry();
            recordset.Dispose();
            Double pointX = geoPoint.X;
            Double pointY = geoPoint.Y;
            //构造捕捉框
            Point2Ds point2Ds = new Point2Ds();
            point2Ds.Add(new Point2D(pointX - scale, pointY - scale));
            point2Ds.Add(new Point2D(pointX+scale, pointY - scale));
            point2Ds.Add(new Point2D(pointX+scale, pointY+scale));
            point2Ds.Add(new Point2D(pointX - scale, pointY+scale));
            point2Ds.Add(new Point2D(pointX - scale, pointY - scale));
            GeoLine geoLine = new GeoLine(point2Ds);
            //刷新地图
              m_mapControl.SelectionTolerance = 2;
              m_trackingLayer.Add(geoLine, "geoLine");
              m_mapControl.Map.Refresh();}}
            catch (Exception ex)
            {Trace.WriteLine(ex.Message);}}
            /// 鼠标移动中实现对点的捕捉
    public void m_mapControl_MouseMove(object sender, MouseEventArgs e)
        {try{m_mapControl.DoMouseMove(e);
            if (m_mapControl.Action == SuperMap.UI.Action.Select
                        || m_mapControl.Action == SuperMap.UI.Action.Select2)
            {//获取鼠标点和对应的地图点
            m_mousePoint = new Point(e.X, e.Y);
            SnapPoint();}}
            catch (System.Exception ex)
            {Trace.WriteLine(ex.Message);}}
            /// 删除中心点
    public void DeleteCenters()
        {try{ for (Int32 i = m_dataGridView.SelectedRows.Count - 1; i >= 0; i--)
          {Int32 selectedRowIndex = m_dataGridView.SelectedRows[i].Index;
          m_supplyCenters.Remove(selectedRowIndex);
          //删除列表中的记录，并刷新中心点列表
          Int32 nodeId = Int32.Parse(m_dataGridView.Rows
                    [selectedRowIndex].Cells[1].Value.ToString());
          m_dataGridView.Rows.RemoveAt(selectedRowIndex);
          for (Int32 j = 0; j < m_dataGridView.RowCount - 1; j++)
```

```
{m_dataGridView.Rows[j].Cells[0].Value=j+1;}
    //删除跟踪图层上对应的点和节点
    Int32 count=m_trackingLayer.Count;
    for(Int32 k=0; k < count; k++)
    {Int32 indexOfCenter=m_trackingLayer.IndexOf("centerPoint"+nodeId);
     if(indexOfCenter! =-1)
     {m_trackingLayer.Remove(indexOfCenter);}
     Int32 indexOfCenterID=m_trackingLayer.IndexOf(nodeId.ToString());
     if(indexOfCenterID! =-1)
     {m_trackingLayer.Remove(indexOfCenterID);}
     m_mapControl.Map.RefreshTrackingLayer();}}}
catch(Exception ex)
{Trace.WriteLine(ex.Message);}}
    /// 修改中心点类型
public void ModifyType(String centerType, Int32 rowIndex)
    {try{Int32 nodeId=Int32.Parse(m_dataGridView.Rows[rowIndex].Cells[1].Value.ToString());
     Int32 index=IndexOf(nodeId);
     //重新构造中心点并替换原中心点
     SupplyCenter supplyCenter=m_supplyCenters[index];
     SupplyCenterType type = centerType = = "可选中心点?" SupplyCenterType.OptionalCenter
     SupplyCenterType.FixedCenter;
     supplyCenter.Type=type;
     m_supplyCenters[index]=supplyCenter;}
     catch(Exception ex)
     {Trace.WriteLine(ex.Message);}}
    /// 修改中心最大阻力值
public void ModifyMaxWeight(Int32 rowIndex)
    {try{Int32 nodeId=Int32.Parse(m_dataGridView.Rows[rowIndex].Cells[1].Value.ToString());
     Int32 index=IndexOf(nodeId);
     //重新构造中心点,并替换原中心点
     SupplyCenter supplyCenter=m_supplyCenters[index];
     supplyCenter.MaxWeight=Double.Parse(m_dataGridView.Rows[rowIndex].Cells[2].Value.ToString());
     m_supplyCenters[index]=supplyCenter;}
     catch(Exception ex)
     {Trace.WriteLine(ex.Message);}}
```

/// 得到 ID 为 anodeID 的中心点索引值
```
private Int32 IndexOf(Int32 nodeID)
    {try{Int32 count=m_supplyCenters.Count;
        for(Int32 i=0; i < count; i++)
            {if(m_supplyCenters[i].ID==nodeID)
                {return i;}}}
    catch(System.Exception ex)
    {Trace.WriteLine(ex.Message);}
    return -1;}
```
/// 清除结果
```
public void ClearResult()
    {try{Int32 count=m_trackingLayer.Count;
      for(Int32 i=0; i < count; i++)
          {Int32 index=m_trackingLayer.IndexOf("result");
          if(index! =-1) m_trackingLayer.Remove(index);}
          m_mapControl.Map.Refresh();}
      catch(System.Exception ex)
          {Trace.WriteLine(ex.Message);}}
```

7. 窗体代码

(1) 添加 using 引用代码

添加 using 引用代码具体如下：

using System.Diagnostics; //添加 System.Diagnostics 命名空间，使其对应的类可以使用

using SuperMap.Data; //添加 SuperMap.Data 命名空间，使其对应的类可以使用

using SuperMap.UI; //添加 SuperMap.UI 命名空间，使其对应的类可以使用

using SuperMap.Mapping; //添加 SuperMap.Mapping 命名空间，使其对应的类可以使用

(2) 在 public FormMain() 函数前添加代码

public FormMain() 函数前添加如下代码：

//定义类对象

private SampleRun m_sampleRun;

private DataGridView dataGridView;

private SuperMap.Data.Workspace m_workspace;

private SuperMap.UI.MapControl m_mapControl;

private Boolean m_hasAnalysted;

(3) 窗体的 Load 事件

窗体的 Load 事件，在主窗体加载时，初始化 SampleRun，DataGridView。在窗体的属性窗口中，选择 按钮后，左键双击 Load，即可添加以下代码：

```
private void FormMain_Load(object sender, EventArgs e)
    {try{this.m_workspace=new SuperMap.Data.Workspace();
```

```
this.m_mapControl = new SuperMap.UI.MapControl();
this.WindowState = FormWindowState.Maximized;
dataGridView.AutoResizeColumns();
dataGridView.AutoSizeColumnsMode = DataGridViewAutoSizeColumnsMode.Fill;
dataGridView.Columns.Clear();
dataGridView.Columns.Add("序号","序号");
dataGridView.Columns.Add("节点标识","节点标识");
dataGridView.Columns.Add("最大阻力值","最大阻力值");
DataGridViewComboBoxColumn centerTypeColumn = new DataGridViewComboBoxColumn();
centerTypeColumn.Name = "中心点类型";//设置列名称
centerTypeColumn.HeaderText = "中心点类型;
centerTypeColumn.Items.Add("可选中心点");
centerTypeColumn.Items.Add("固定中心点");
dataGridView.Columns.Add(centerTypeColumn);
for (int i = 0; i < dataGridView.Columns.Count; i++)
{if (i < 2)
{dataGridView.Columns[i].ReadOnly = true;}
else{dataGridView.Columns[i].ReadOnly = false;}}
dataGridView.Columns[0].Width = 60;
dataGridView.Columns[3].Width = 120;
dataGridView.RowHeadersWidth = 30;
textBoxCenterPointNum.Text = "2";
//实例化 SampleRun
m_sampleRun = new SampleRun(m_workspace, m_mapControl, dataGridView);
m_mapControl.Dock = DockStyle.Fill;
this.splitContainer1.Panel1.Controls.Add(m_mapControl);
this.Controls.SetChildIndex(m_mapControl, 0);}
catch (Exception ex)
{Trace.WriteLine(ex.Message);}}
```

(4) 添加各个控件的代码

① 删除 Button 控件的 Click 事件。左键双击漫游 Button 控件，即可添加该按钮的 Click 事件代码，实现删除功能：

```
private void buttonDelete_Click(object sender, EventArgs e)
    {
          m_sampleRun.DeleteCenters();//通过 sampleRun.DeleteCenters 实现删除功能
    }
```

② 分析 Button 控件的 Click 事件。左键双击全屏 Button 控件，即可添加该按钮的 Click

事件代码,实现分析功能:
```
private void buttonAnalyst_Click(object sender, EventArgs e)
    {try{if(textBoxCenterPointNum.Text ! ="")
    {Int32 supplyCenterCount = Int32.Parse(textBoxCenterPointNum.Text);
    m_sampleRun.StartAnalyst(supplyCenterCount);}//通过 sampleRun.StartAnalyst 实现分析功能
    else{MessageBox.Show("请输入分析中心点个数");}}
    catch(Exception ex)
    {Trace.WriteLine(ex.Message);}}
```

③ 清除结果 Button 控件的 Click 事件。左键双击分析 Button 控件,即可添加该按钮的 Click 事件代码,实现清除功能:
```
private void buttonClear_Click(object sender, EventArgs e)
    {try{m_sampleRun.ClearResult();//通过 sampleRun.ClearResult 实现清除结果功能
        if(m_hasAnalysted)
        {m_hasAnalysted=false;}}
    catch(Exception ex)
    {Trace.WriteLine(ex.Message);}}
```

④ 改变属性表 dataGridView 控件的 CellValueChanged 事件。在 dataGridView 的属性窗口中,选择 ![] 按钮后,左键双击 CellValueChanged,即可添加以下代码:
```
private void dataGridView_CellValueChanged(object sender, DataGridViewCellEventArgs e)
    {try{if(e.ColumnIndex==dataGridView.Columns["最大阻力值"].Index)
        {m_sampleRun.ModifyMaxWeight(e.RowIndex);}//通过 sampleRun.ModifyMaxWeight 实现此功能
        if(e.ColumnIndex==dataGridView.Columns["中心点类型"].Index)
        {m_sampleRun.ModifyType(dataGridView.Rows[e.RowIndex].Cells[3].Value.ToString(), e.RowIndex);}}
    catch(Exception ex)
    {Trace.WriteLine(ex.Message);}}
```

(5)窗体的 FormClose 事件

在退出程序时,要断开空间的关联,关闭地图窗口、工作空间。因此,在属性窗口点击 ![] 图标,左键双击 FormClose 事件,添加代码同前文中的相应部分。

8. 运行结果

窗体设计完成后点击 ![] 按钮,运行该程序,运行后弹出窗体,如图 7-4 所示。图中绿色圆点为所选点标号为 4019,在右侧的中心点列表中可以改变其属性,包括最大阻力值及中心点类型,在属性表的下方可以设置期望用于最终设施选址的资源供给中心数量,点击"分析"按钮便可出现分析结果。

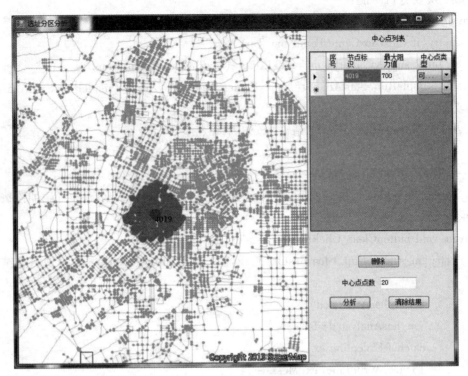

图 7-4　选址分区分析图

7.1.3　最佳路径分析

1. 数据

在安装目录\SampleData\City\Changchun.udb 路径下打开数据文件。

2. 新建文件夹和工程

在【D:\MyProject\】文件中新建一个文件夹，命名为【第 7 章　空间分析】，并在其目录下建立一个文件夹，命名为【网络分析】。在此文件夹里新建一个文件夹，命名为 FindPath，在此文件夹里新建一个工程，将此工程命名为 FindPath。

3. 关键类型/成员

关键类型/成员见表 7-5。

表 7-5　　　　　　　　　　　关键类型/成员表

控件/类	方法	属　性	事件
MapControl			GeometrySelected、MouseDown
Transportation-AnalystSetting		NetworkDataset、EdgeIDField、NodeIDField、Tolerance、WeightFieldInfo、FNodeIDField、TNodeIDField	

续表

控件/类	方法	属性	事件
Transportation-Analyst	Load、FindPath	AnalystSetting	
Transportation-AnalystParameter		Points、BarrierEdges、BarrierNodes、WeightName、NodesReturn、EdgesReturn、PathGuidesReturn、StopIndexesReturn、RoutesReturn	
Transportation-AnalystResult	Add、Remove	Routes、PathGuides	

4. 窗体控件属性

窗体控件属性见表7-6。

表7-6　　　　　　　　　　窗体控件属性表

控件	Name	Text
Toolstrip	Toolstrip	Toolstrip
Radiobutton	radioButtonPoint	路径经过点
	radioButtonBarrier	障碍
dataGridView	dataGridView	dataGridView
Button	toolStripButtonAnalyst	分析
	toolStripButtonPlay	导引
	toolStripButtonStop	停止
	toolStripButtonClear	清除
timer	timer	

5. 窗体设计布局

窗体设计布局如图7-5所示。

6. 创建SampleRun类

(1) 添加using引用代码

添加using引用代码具体如下：

using System.Diagnostics；//添加System.Diagnostics命名空间，使其对应的类可以使用

using System.Drawing；//添加System.Drawing命名空间，使其对应的类可以使用

using System.Windows.Forms；//添加System.Windows.Forms命名空间，使其对应的类可以使用

using SuperMap.Data；//添加SuperMap.Data命名空间，使其对应的类可以使用

using SuperMap.UI；//添加SuperMap.UI命名空间，使其对应的类可以使用

图 7-5　窗体设计布局图

　　using SuperMap. Mapping；//添加 SuperMap. Mapping 命名空间，使其对应的类可以使用

　　using SuperMap. Analyst. NetworkAnalyst；//添加 SuperMap. Analyst. NetworkAnalyst 命名空间，使其对应的类可以使用

　　(2)在 public class SampleRun 中添加代码

　　在 public class SampleRun 中添加如下代码：

　　//定义类对象

　　private static String m_datasetName = "RoadNet"；

　　private static String m_nodeID = "SmNodeID"；

　　private static String m_edgeID = "SmEdgeID"；

　　private SelectMode m_selectMode；// SelectMode：选择模式枚举。

　　private MapControl m_mapControl；// MapControl：地图控件类。该类用于为地图的显示提供界面，同时为地图与数据的互操作提供了途径。

　　private Workspace m_workspace；// Workspace：工作空间是用户的工作环境，主要完成数据的组织和管理，包括打开、关闭、创建、保存工作空间文件。

　　private DatasetVector m_datasetLine；// DatasetVector：矢量数据集类。描述矢量数据集，并提供相应的管理和操作。对矢量数据集的操作主要包括数据查询、修改、删除、建立索引等。

　　private DatasetVector m_datasetPoint；

　　private Layer m_layerLine；// Layer：图层类。该类提供了图层显示和控制等便于地图管理的一系列属性。

　　private Layer m_layerPoint；

　　private TrackingLayer m_trackingLayer；// TrackingLayer：跟踪图层类。

```csharp
private Point2Ds m_Points;
private GeoStyle m_style;// GeoStyle:几何风格类。用于定义点状符号、线状符号、填充符号风格及其相关属性。
private List<Int32> m_barrierNodes;
private List<Int32> m_barrierEdges;
private TransportationAnalyst m_analyst;// TransportationAnalyst:交通网络分析类。该类用于提供路径分析、旅行商分析、服务区分析、多旅行商(物流配送)分析、最近设施查找和选址分区分析等交通网络分析的功能。
private TransportationAnalystResult m_result;// TransportationAnalystResult:交通网络分析结果类。该类用于获取分析结果的路由集合、分析途经的节点集合以及弧段集合、行驶导引集合、站点集合和权值集合以及各站点的花费。通过该类的设置,可以灵活地得到最佳路径分析、旅行商分析、物流配送和最近设施查找等分析的结果。
private DataGridView m_dataGridView;//:
private Timer m_timer;
private int m_count;
private int m_flag;
/// 选择模式枚举
public enum SelectMode
    {SelectPoint, SelectBarrier, None}
/// 根据 workspace、mapControl 及 DataGridView 构造 SampleRun 对象
public SampleRun(Workspace workspace, MapControl mapControl,
    DataGridView dataGridView)
    {m_workspace = workspace;
    m_mapControl = mapControl;
    m_dataGridView = dataGridView;
    m_mapControl.Map.Workspace = workspace;
    Initialize();}
/// 打开网络数据集并初始化相应变量
private void Initialize()
    {try{//打开数据源得到点线数据集
        DatasourceConnectionInfo connectionInfo = new DatasourceConnectionInfo(
            @"..\..\SampleData\City\Changchun.udb", "FindPath2D", "");//
DatasourceConnectionInfo:数据源连接信息类。包括进行数据源连接的所有信息,如所要连接的服务器名称,数据库名称、用户名、密码等。当保存工作空间时,工作空间中的数据源的连接信息都将存储到工作空间文件中。
        connectionInfo.EngineType = EngineType.UDB;// EngineType:该枚举定义了空间数据库引擎类型常量。
        m_workspace.Datasources.Open(connectionInfo);//打开工作空间
        m_datasetLine = (DatasetVector)m_workspace.Datasources[0].Datasets[m_
```

datasetName];
m_datasetPoint = m_datasetLine.ChildDataset;
m_trackingLayer = m_mapControl.Map.TrackingLayer;
m_trackingLayer.IsAntialias = true;
//初始化各变量
m_flag = 1;
m_Points = new Point2Ds();
m_style = new GeoStyle();
m_barrierNodes = new List<Int32>();
m_barrierEdges = new List<Int32>();
m_selectMode = SelectMode.SelectPoint;
m_timer = new Timer();
m_timer.Interval = 200;
m_timer.Enabled = false;
//加载点数据集及线数据集并设置各自风格
m_layerLine = m_mapControl.Map.Layers.Add(m_datasetLine, true);
m_layerLine.IsSelectable = false;
LayerSettingVector lineSetting = (LayerSettingVector) m_layerLine.AdditionalSetting;// LayerSettingVector:矢量图层设置类。该类主要用来设置矢量图层的显示风格。矢量图层用单一的符号或风格绘制所有的要素。当用户只想可视化地显示空间数据，只关心空间数据中各要素在什么位置，而不关心各要素在数量或性质上的不同时，可以用普通图层来显示要素数据。
GeoStyle lineStyle = new GeoStyle();
lineStyle.LineColor = Color.LightGray;//设置线状符号的颜色
lineStyle.LineWidth = 0.1;//设置线状符号的宽度
lineSetting.Style = lineStyle;//设置风格
m_layerPoint = m_mapControl.Map.Layers.Add(m_datasetPoint, true);
LayerSettingVector pointSetting = (LayerSettingVector) m_layerPoint.AdditionalSetting;
GeoStyle pointStyle = new GeoStyle();
pointStyle.LineColor = Color.DarkGray;
pointStyle.MarkerSize = new Size2D(2.5, 2.5);//设置点标志风格
pointSetting.Style = pointStyle;//设置图层风格
//调整 mapControl 的状态
m_mapControl.Action = SuperMap.UI.Action.Select;
m_mapControl.Map.IsAntialias = true;
m_mapControl.IsWaitCursorEnabled = false;
m_mapControl.Map.Refresh();
m_mapControl.MouseDown += new MouseEventHandler(m_mapControl_

MouseDown);
　　　　　　m_mapControl.GeometrySelected+=new GeometrySelectedEventHandler(m_mapControl_GeometrySelected);
　　　　　　m_timer.Tick+=new EventHandler(m_timer_Tick);
　　　　　　Load();}
　　　　catch(Exception e)
　　　　{Trace.WriteLine(e.Message);}}
　　　/// MapControl MouseDown 事件
　　private void m_mapControl_MouseDown(object sender, MouseEventArgs e)
　　　{try{if(e.Button==MouseButtons.Left)
　　　　{Point point=new Point(e.X, e.Y);
　　　　Point2D mapPoint=m_mapControl.Map.PixelToMap(point);
　　　　if(m_mapControl.Map.Bounds.Contains(mapPoint))
　　　　{if((m_mapControl.Action==SuperMap.UI.Action.Select
　　　　　 || m_mapControl.Action==SuperMap.UI.Action.Select2)
　　　　　 && m_selectMode==SelectMode.SelectPoint)
　　　　　{AddPoint(mapPoint);}}}}
　　catch(Exception ex)
　　　{Trace.WriteLine(ex.Message);}}
/// 加载环境设置对象
public void Load()
　　　{try{// 设置网络分析基本环境，这一步骤需要设置分析权重、节点、弧段标识字段、容限
TransportationAnalystSetting setting=new TransportationAnalystSetting();
//TransportationAnalystSetting：交通网络分析环境设置类。该类用于提供交通网络分析时所需要的所有参数信息。交通网络分析环境设置类的各个参数的设置直接影响分析的结果。
　　　setting.NetworkDataset=m_datasetLine;//设置进行分析的网络数据集
　　　setting.EdgeIDField=m_edgeID;//设置标识网络中弧段 ID 的字段
　　　setting.NodeIDField=m_nodeID;//设置标识网络节点 ID 的字段
　　　setting.EdgeNameField="roadName";//设置存储弧段名称的字段的字段名
　　　setting.Tolerance=89;//设置节点到弧段的距离容限
　　　WeightFieldInfos weightFieldInfos=new WeightFieldInfos();//WeightFieldInfos：权值字段信息集合类。该类是权值字段信息对象(WeightFieldInfo)的集合，用于对权值字段信息对象进行管理，如添加、删除、获取指定名称或索引的权值字段信息对象等。
　　　WeightFieldInfo weightFieldInfo=new WeightFieldInfo();//WeightFieldInfo：权值字段信息类型，存储了网络分析中权值字段的相关信息，包括正向权值字段与反向权值字段。
　　　weightFieldInfo.FTWeightField="smLength";//设置正向权值字段

```
weightFieldInfo.TFWeightField = "smLength";//设置反向权值字段
weightFieldInfo.Name = "length";//设置权值字段信息的名称
weightFieldInfos.Add(weightFieldInfo);
setting.WeightFieldInfos = weightFieldInfos;
setting.FNodeIDField = "SmFNode";
setting.TNodeIDField = "SmTNode";
//构造交通网络分析对象,加载环境设置对象
m_analyst = new TransportationAnalyst();
m_analyst.AnalystSetting = setting;//设置交通网络分析环境设置对象
m_analyst.Load();}
catch(Exception e)
{Trace.WriteLine(e.Message);}}
/// 进行最短路径分析
public bool Analyst()
{try{ m_count=0;
    TransportationAnalystParameter parameter = new TransportationAnalystParameter();//TransportationAnalystParameter:交通网络分析参数设置类。该类主要用来对交通网络分析的参数进行设置。
    //设置障碍点及障碍边
    int[] barrierEdges = new int[m_barrierEdges.Count];
    for(int i=0; i < barrierEdges.Length; i++)
    {barrierEdges[i] = m_barrierEdges[i];}
    parameter.BarrierEdges = barrierEdges;
    int[] barrierNodes = new int[m_barrierNodes.Count];
    for(int i=0; i < barrierNodes.Length; i++)
    {barrierNodes[i] = m_barrierNodes[i];}
    parameter.BarrierNodes = barrierNodes;
    parameter.WeightName = "length";
    //设置最佳路径分析的还回对象
    parameter.Points = m_Points;
    parameter.IsNodesReturn = true;//设置分析结果中是否包含分析途径的节点集合
    parameter.IsEdgesReturn = true;//设置分析结果中是否包含经过弧段集合
    parameter.IsPathGuidesReturn = true;//设置分析结果中是否包含行驶导引集合
    parameter.IsRoutesReturn = true;
    //进行分析并显示结果
    m_result = m_analyst.FindPath(parameter, false);
    if(m_result == null)
    {MessageBox.Show("分析失败");
```

```
            return false;}
            ShowResult();
            FillDataGridView(0);
            m_selectMode=SelectMode.None;
            return true;}
        catch(Exception e)
        {Trace.WriteLine(e.Message);
            return false;}}
/// 显示结果
public void ShowResult()
    {try{//删除原有结果
        int count=m_trackingLayer.Count;
        for(int i=0;i < count;i++)
        {int index=m_trackingLayer.IndexOf("result");
        if(index！=-1)
        {m_trackingLayer.Remove(index);}}
        FillDataGridView(0);
    for(int i=0;i < m_result.Routes.Length;i++)
        {GeoLineM geoLineM=m_result.Routes[0];// GeoLineM:路由对象，是一组具有
```

X，Y 坐标与线性度量值(M 值)的点组成的线性地物对象。比如高速公路上的里程碑，交通管制部门经常使用高速公路上的里程碑来标注并管理高速公路的路况、车辆的行驶限速和高速事故点等。

```
            m_style.LineColor=Color.Blue;//设置线颜色
            m_style.LineWidth=1;//设置线宽
            geoLineM.Style=m_style;//设置风格
            m_trackingLayer.Add(geoLineM,"result");}
            m_mapControl.Map.RefreshTrackingLayer();}
        catch(Exception e)
        {Trace.WriteLine(e.Message);}}
/// 填充 DataGridView
public void FillDataGridView(Int32 pathNum)
    {try{//清除原数据，添加初始点信息
        m_dataGridView.Rows.Clear();
        Object[] objs=new Object[4];
        objs[0]=m_dataGridView.RowCount;
        objs[1]="从起始点出发;
        objs[2]="--";
        objs[3]="--";
        m_dataGridView.Rows.Add(objs);
```

```
//得到行驶导引对象，根据导引子项类型的不同进行不同的填充
PathGuide[ ] pathGuides=m_result.PathGuides;
PathGuide pathGuide=pathGuides[pathNum];
for (int j=1; j < pathGuide.Count; j++)
{PathGuideItem item=pathGuide[j];
objs[0]=m_dataGridView.RowCount;
//导引子项为站点的添加方式
if (item.IsStop)
{String side="无";
if (item.SideType==SideType.Left)
side="左侧;
if (item.SideType==SideType.Right)
side="右侧;
if (item.SideType==SideType.Middle)
side="道路";
String dis=item.Distance.ToString( );
if (item.Index==-1 && item.ID==-1)
{continue; }
if (j! =pathGuide.Count - 1)
{objs[1]="到达 ["+item.Index+"号路由点], 在道路"+side+dis; }
else{objs[1]="到达终点, 在道路"+side+dis; }
objs[2]="";
objs[3]="";
m_dataGridView.Rows.Add(objs); }
//导引子项为弧段的添加方式
if (item.IsEdge)
{String direct="直行";
if (item.DirectionType==DirectionType.East)
direct="东";
if (item.DirectionType==DirectionType.West)
direct="西";
if (item.DirectionType==DirectionType.South)
direct="南";
if (item.DirectionType==DirectionType.North)
direct="北";
String weight=item.Weight.ToString( );
String roadName=item.Name;
if (weight.Equals("0") && roadName.Equals(""))
{objs[1]="朝"+direct+"行走"+item.Length;
```

```
            objs[2] = weight;
            objs[3] = item.Length;
            m_dataGridView.Rows.Add(objs);}
        else{ String roadString = roadName.Equals("") ? "匿名路段" : roadName;
            objs[1] = "沿着["+roadString+"],朝"+direct+"行走"+item.Length;
            objs[2] = weight;
            objs[3] = item.Length;
            m_dataGridView.Rows.Add(objs);}}}}
    catch(Exception e)
    {Trace.WriteLine(e.Message);}}
/// 添加分析经过点
public void AddPoint(Point2D mapPoint)
    {try{//在跟踪图层上添加点
        m_Points.Add(mapPoint);//添加点对象
        GeoPoint geoPoint = new GeoPoint(mapPoint);// GeoPoint:二维点几何对象类,
派生于Geometry类。该类一般用于描述点状地理实体,比如气象站点,公交站点等。
        m_style.LineColor = Color.Green;//设置线颜色
        m_style.MarkerSize = new Size2D(8,8);
        geoPoint.Style = m_style;//设置风格
        m_trackingLayer.Add(geoPoint, "Point");
        //在跟踪图层上添加文本对象
        TextPart part = new TextPart();//TextPart:文本子对象类。用于表示文本对象
的子对象,主要存储子对象的文本、旋转角度、锚点等信息,并提供对子对象进行处理的
相关方法。
        part.X = geoPoint.X;
        part.Y = geoPoint.Y;
        part.Text = m_flag.ToString();
        m_flag++;
        GeoText geoText = new GeoText(part);// GeoText:文本类,派生于Geometry
类。该类主要用于对地物要素进行标识和必要的注记说明。
        m_trackingLayer.Add(geoText, "Point");
        m_mapControl.Map.RefreshTrackingLayer();}
    catch(Exception ex)
    {Trace.WriteLine(ex.Message);}}
/// 对象选择事件
public void m_mapControl_GeometrySelected(Object sender, SuperMap.UI.
GeometrySelectedEventArgs e)
```

```csharp
{if(m_selectMode! =SelectMode.SelectBarrier)
return;
Selection selection = m_layerPoint.Selection;// Selection:选择集类。该类用于处理地图上被选中的对象。
if(selection.Count<=0)
{selection = m_layerLine.Selection;}
m_style.LineColor = Color.Red;//设置线的颜色
m_style.MarkerSize = new Size2D(6,6);
m_style.LineWidth = 0.5;//设置线宽
Recordset recordset = selection.ToRecordset();//Recordset:记录集类。通过此类，可以实现对矢量数据集中的数据进行操作。
try{ Geometry geometry = recordset.GetGeometry();
//捕捉到点时，将点对象添加到障碍点列表中
if(geometry.Type == GeometryType.GeoPoint)
{GeoPoint geoPoint = (GeoPoint)geometry;// GeoPoint:二维点几何对象类，派生于Geometry类。该类一般用于描述点状地理实体，比如气象站点，公交站点等。
int id = recordset.GetID();
m_barrierNodes.Add(id);
geoPoint.Style = m_style;//设置几何对象风格
m_trackingLayer.Add(geoPoint,"barrierNode");}//添加图层
//捕捉到线时，将线对象添加到障碍线列表中
if(geometry.Type == GeometryType.GeoLine)// GeometryType:该枚举定义了一系列几何对象类型常量。
{GeoLine geoLine = (GeoLine)geometry;
int id = recordset.GetID();
m_barrierEdges.Add(id);
geoLine.Style = m_style;
m_trackingLayer.Add(geoLine,"barrierEdge");}//添加图层
m_mapControl.Map.Refresh();}//刷新地图
catch(Exception ex)
{Trace.WriteLine(ex.Message);}
finally{recordset.Dispose();}}
/// 开始导引
public void Play()
{m_timer.Start();}
/// 停止导引
public void Stop()
```

```
        {m_timer.Stop();}
    /// 清除分析结果
    public void Clear()
        {try{if(m_timer! =null)
            m_timer.Stop();
            m_dataGridView.Rows.Clear();
            m_flag=1;
            m_Points=new Point2Ds();
            m_barrierNodes=new List<Int32>();
            m_barrierEdges=new List<Int32>();
            m_mapControl.Map.Layers[0].Selection.Clear();//获取当前地图所包含的图
层集合对象
            mapControl.Map.Layers[1].Selection.Clear();
            m_mapControl.Map.TrackingLayer.Clear();//获取跟踪图层
            m_mapControl.Map.Refresh();}
        catch(Exception ex)
            {Trace.WriteLine(ex.Message);}}
    /// 设置选择模式
    public void SetSelectMode(SelectMode mode, Boolean canSelectLine)
        {m_selectMode=mode;
        m_layerLine.IsSelectable=canSelectLine;}//设置图层中对象是否可以选择
    /// 进行行驶导引
    private void m_timer_Tick(object sender, EventArgs e)
        {try{int index=m_trackingLayer.IndexOf("playPoint");
        if(index! =-1)
            {m_trackingLayer.Remove(index);}
            GeoLineM lineM=m_result.Routes[0];// GeoLineM:路由对象,是一组具有
X、Y坐标与线性度量值(M值)的点组成的线性地物对象。比如高速公路上的里程碑,交
通管制部门经常使用高速公路上的里程碑来标注并管理高速公路的路况、车辆的行驶限速
和高速事故点等。
            PointM pointM=lineM.GetPart(0)[m_count];// PointM:精度为Double的路由
点类。路由点是指具有线性度量值的点。M代表路由点的度量值(Measure value)。
            //构造模拟对象
            GeoPoint point=new GeoPoint(pointM.X, pointM.Y);// GeoPoint:二维点几何
对象类,派生于Geometry类。该类一般用于描述点状地理实体,比如气象站点、公交站点
等。
            GeoStyle style=new GeoStyle();//GeoStyle:几何风格类。用于定义点状符号、
线状符号、填充符号风格及其相关属性。
```

```
style.LineColor = Color.Red;
style.MarkerSize = new Size2D(5, 5);
point.Style = style;
m_trackingLayer.Add(point, "playPoint");
//跟踪对象
m_count++;
if (m_count >= lineM.GetPart(0).Count)
{ m_count = 0; }
m_mapControl.Map.RefreshTrackingLayer(); }//刷新跟踪图层
catch (Exception ex)
{ Trace.WriteLine(ex.Message); } }
```

7. 窗体代码

(1) 添加 using 引用代码

添加 using 引用代码具体如下：

using System.Diagnostics; //添加 System.Diagnostics 命名空间，使其对应的类可以使用
using SuperMap.Data; //添加 SuperMap.Data 命名空间，使其对应的类可以使用
using SuperMap.UI; //添加 SuperMap.UI 命名空间，使其对应的类可以使用
using SuperMap.Mapping; //添加 SuperMap.Mapping 命名空间，使其对应的类可以使用
using System; //添加 System 命名空间，使其对应的类可以使用
using System.Threading; //添加 System.Threading 命名空间，使其对应的类可以使用

(2) 在 public FormMain() 函数前添加代码

在 public FormMain() 函数前添加如下代码：

```
//定义类对象
private SampleRun m_sampleRun;
private Workspace m_workspace;
private MapControl m_mapControl;
```

(3) 窗体的 Load 事件

窗体的 Load 事件，在主窗体加载时，初始化 SampleRun、DataGridView。在窗体的属性窗口中，选择 ≠ 按钮后，左键双击 Load，即可添加以下代码：

```
private void FormMain_Load(object sender, System.EventArgs e)
{ try { this.m_workspace = new SuperMap.Data.Workspace(); //定义新的对象
    this.m_mapControl = new SuperMap.UI.MapControl();
    //初始化控件
    dataGridView.AutoResizeColumns();
    dataGridView.AutoSizeColumnsMode = DataGridViewAutoSizeColumnsMode.Fill;
    dataGridView.Columns.Clear();
    dataGridView.Columns.Add("序号", "序号");
    dataGridView.Columns.Add("导引", "导引");
```

dataGridView.Columns.Add("耗费","耗费");
dataGridView.Columns.Add("距离","距离");
for(int i=0;i < dataGridView.Columns.Count;i++)
{dataGridView.Columns[i].ReadOnly=true;}
m_mapControl.Dock=DockStyle.Fill;
this.splitContainer2.Panel2.Controls.Add(m_mapControl);
this.splitContainer2.Panel2.Controls.SetChildIndex(m_mapControl,0);
m_sampleRun=new SampleRun(m_workspace,m_mapControl,dataGridView);
}//设置新对象
catch(Exception ex)
{Trace.WriteLine(ex.Message);}}

(4) 添加各个控件的代码

① 路径经过点 radioButtonPoint 控件的 CheckedChanged 事件。在 radioButtonPoint 控件的属性窗口双击 ≯ 按钮，即可添加以下代码，添加路径经过点：
private void radioButtonPoint_CheckedChanged(object sender, System.EventArgs e)
{ m_sampleRun.SetSelectMode(SampleRun.SelectMode.SelectPoint, false);}
//通过 sampleRun.SetSelectMode 添加经过点

② 障碍点 radioButtonBarrier 控件的 CheckedChanged 事件。在 radioButtonBarrier 控件的属性窗口双击 ≯ 按钮，即可添加以下代码，添加障碍点：
private void radioButtonBarrier_CheckedChanged(object sender, System.EventArgs e)
{m_sampleRun.SetSelectMode(SampleRun.SelectMode.SelectBarrier, true);}
//通过 sampleRun.SetSelectMode 添加障碍点

③ 分析 Button 控件的 Click 事件。左键双击分析 Button 控件，即可添加如下该按钮的 Click 事件代码：
private void toolStripButtonAnalyst_Click(object sender, System.EventArgs e)
{if(m_sampleRun.Analyst())//通过 sampleRun.Analyst 实现分析
{EnabledControl(false);}}

④ 开始导引 Button 控件的 Click 事件。左键双击分析 Button 控件，即可添加如下该按钮的 Click 事件代码：
private void toolStripButtonPlay_Click(object sender, System.EventArgs e)
{m_sampleRun.Play();}//通过 sampleRun.Play 实现导引

⑤ 停止导引 Button 控件的 Click 事件。左键双击分析 Button 控件，即可添加如下该按钮的 Click 事件代码：
private void toolStripButtonStop_Click(object sender, System.EventArgs e)
{m_sampleRun.Stop();}//通过 sampleRun.Stop 实现停止导引

⑥ 清除 Button 控件的 Click 事件。左键双击分析 Button 控件，即可添加如下该按钮的 Click 事件代码：
private void toolStripButtonClear_Click(object sender, EventArgs e)
{try{m_sampleRun.Clear();//通过 sampleRun.Clear 实现清除

```
SampleRun.SelectMode mode = SampleRun.SelectMode.SelectPoint;
m_sampleRun.SetSelectMode(mode,false);
if(radioButtonBarrier.Checked)
{mode = SampleRun.SelectMode.SelectBarrier;
m_sampleRun.SetSelectMode(mode,true);}
EnabledControl(true);}
catch(Exception ex)
{Trace.WriteLine(ex.Message);}}
```

(5) 窗体的 FormClose 事件

在退出程序时,要断开空间的关联,关闭地图窗口、工作空间。因此,在属性窗口点击 图标,左键双击 FormClose 事件,添加代码同前文的相应部分。

8. 运行结果

窗体设计完成后点击 ▶ 按钮,运行该程序,运行后弹出窗体,如图 7-6 所示。图中绿色圆点为路径必经点,蓝色线以外的红色圆点为障碍点,选好线路必经点及障碍点后即可分析出最佳路径,同时点击"导引"按钮,线路上面的红色圆点便会从点 1 移动至点 3,点击"停止",导引便会停止在蓝色线路上。

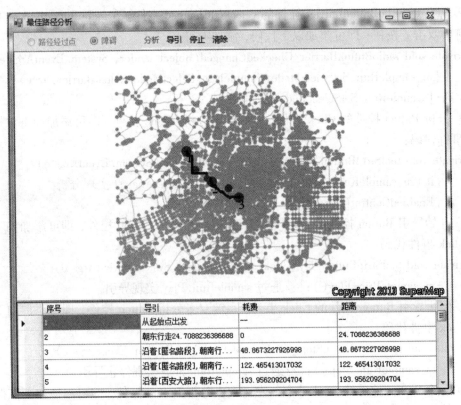

图 7-6　最佳路径分析图

7.1.4 批量建模与缓冲区分析

1. 数据

在安装目录\SampleData\ModelingAndAnalyst\ModelingAndAnalyst.smwu 路径下打开数据文件。

2. 新建文件夹和工程

由于二维缓冲区分析与三维缓冲区分析所用的接口是相同的，所以在此只介绍三维缓冲区分析。在【D:\MyProject\】文件中新建一个文件夹，命名为【第 7 章　空间分析】，并在其目录下建立一个文件夹命名为【三维分析】，在此文件夹里新建一个文件，夹命名为ModelingAndAnalyst，在此文件夹里新建一个工程将此工程命名为 ModelingAndAnalyst。

3. 关键类型/成员

关键类型/成员具体见表 7-7。

表 7-7　　　　　　　　　　　　　　关键类型/成员表

控件/类	方法	属　性	事件
SceneControl		Scene、Action	
Layer3DDataset	UpdateData、ToDatasetVector	AdditionalSetting、Selection、IsAlwaysRender、IsSelectable	
GeoStyle3D		ExtendedHeight、TopTextureFile、SideTextureFiles、TopTilingU、TopTilingV、TilingU、TilingV、FillMode	
Layer3DSettingVector		Style、ExtendedHeightField、TopTextureField、SideTextureField、TopTilingUField、TopTilingVField、TilingUField、TilingVField	
Selection	AddRange、UpdateData、Clear	Style	
BufferAnalyst	CreateBuffer		
BufferAnalystParameter		EndType、LeftDistance、RightDistance	

4. 窗体控件属性

窗体控件属性具体见表 7-8 和表 7-9。

表 7-8　　　　　　　　　　　窗体控件属性表 1

控　件	Name	Text
TabControl	m_tabMain	批量建模
	m_tabMain	缓冲区分析
folderBrowserDialog	folderBrowserDialog1	folderBrowserDialog1

表7-9　　　　　　　　　　　　窗体控件属性表2

Text	Label	ComboBox	GroupBox	Button
批量建模	label1（高度模式：）	m_cbboxAttitudMode	m_groupboxStyleSetting（风格设置）	m_buttonReload（重置数据）
	label2（拉伸高度：）	m_cbboxExtendHeight	groupBox2（侧面贴图）	
	label6（侧面纹理：）	m_cbboxSideImage	groupBox1（顶部贴图）	
	label13（重复模式：）	m_cbboxSideRepeatMode		
	label9（横向重复次数：）	m_cbboxTilingV（1）		
	label4（纵向重复次数：）	m_cbboxTilingU（1）		
	label5（顶部纹理：）	m_cbboxTopImage		
	label14（重复模式：）	m_cbboxTopRepeatMode		
	label7（横向重复次数：）	m_cbboxTopTilingV（1）		
	label8（纵向重复次数：）	m_cbboxTopTilingU（1）		
	label11（填充模式：）	m_cbboxFillMode		
缓冲区分析	label3（请在场景中选择一条道路线）	m_cbboxLeftRadius（50）	m_groupboxBufferSetting（缓冲设置）	
	label12（左半径：）			
	label10（右半径：）	m_cbboxRightRadius（50）	m_groupboxAnalyst（缓冲分析）	

5．窗体设计布局

窗体设计布局如图7-7所示。

6．创建SampleRun类

（1）添加using引用代码

添加using引用代码具体如下：

using SuperMap.Data；//添加SuperMap.Data命名空间，使其对应的类可以使用

using SuperMap.UI；//添加SuperMap.UI命名空间，使其对应的类可以使用

using SuperMap.Realspace；//添加SuperMap.Realspace命名空间，使其对应的类可以使用

using System.Drawing；//添加System.Drawing命名空间，使其对应的类可以使用

using System.Windows.Forms；//添加System.Windows.Forms命名空间，使其对应的类可以使用

using System.Diagnostics；//添加System.Diagnostics命名空间，使其对应的类可以使用

using SuperMap.Analyst.SpatialAnalyst；//添加SuperMap.Analyst.SpatialAnalyst命名空间，使其对应的类可以使用

using SuperMap.Data.Processing；//添加SuperMap.Data.Processing命名空间，使其对

7.1 网络分析

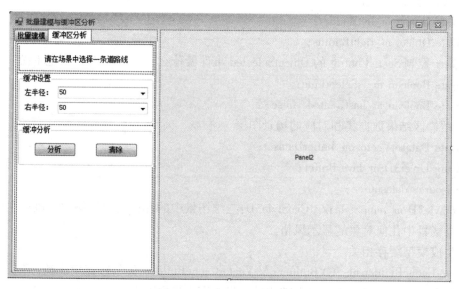

图 7-7 窗体设计布局图

应的类可以使用

using System. Drawing. Drawing2D；//添加 System. Drawing. Drawing2D 命名空间，使其对应的类可以使用

（2）在 public class SampleRun 中添加代码

在 public class SampleRun 中添加如下代码：

//定义类对象

private Workspace m_workspace;// Workspace:工作空间是用户的工作环境，主要完成数据的组织和管理，包括打开、关闭、创建、保存工作空间文件。

private SceneControl m_sceneControl;// SceneControl:三维地图控件。一个三维场景窗口（SceneControl）对应一个三维场景（Scene），即一个三维场景窗口中只能显示一个三维场景。

private Layer3DDataset m_layerRegion;// Layer3DDataset:三维数据集图层类。该类继承自 Layer3D 类。

private DatasetVector m_regionDataset;// DatasetVector:矢量数据集类。描述矢量数据集，并提供相应的管理和操作。对矢量数据集的操作主要包括数据查询、修改、删除、建立索引等。

//道路图层及道路数据集

private Layer3D m_lineLayer;

private DatasetVector m_dataset;

// 分析结果记录集

private Recordset m_lineRecordset;// Recordset:记录集类。通过此类，可以实现对矢量数据集中的数据进行操作。

// 缓冲区左右半径
private Object m_leftRadius;
private Object m_rightRadius;
//用于控制 SceneControl 的 ObjectSelected 事件做什么，为 false 时什么也不做
private Boolean m_isSelectLine;
private Boolean m_hasCreated=false;
// 缓冲区结果数据集和对应的场景图层
private DatasetVector m_bufferDataset;
private Layer3D m_layerBuffer;
Datasource datasource;
GeoStyle3D m_tempStyle;// GeoStyle3D:三维场景中的几何对象风格类。该类主要用于设置三维场景中几何对象的显示风格。
//生成模型缓存相关
VectorCacheBuilder m_SCVBuilder;// VectorCacheBuilder:矢量缓存生成类，该类主要用于为矢量数据生成供三维场景发布使用的矢量缓存，其配置文件为*.scv 格式。
private Dictionary<String, String> m_VectorRegionLayerFields;
private String _cacheName=String.Empty;
private String _outputFolder=String.Empty;
/// 矢量面字段集合
public Dictionary<String, String> VectorRegionLayerFields
    {get { return this.m_VectorRegionLayerFields; }}
///缓存名称
public String CacheName
    {get { return this._cacheName; }
    set { this._cacheName=value; }}
/// 缓存路径
public String OutputFolder
    {get { return this._outputFolder; }
    set { this._outputFolder=value; }}
///获取矢量面的高度模式
public Int32 AltitudeModeCode
    {get{
    try{Layer3DDataset layer3Dds=(Layer3DDataset)m_layerRegion;
        Layer3DSettingVector layer3DsettingVector = (Layer3DSettingVector) layer3Dds.AdditionalSetting;
AltitudeMode altitudeMode = layer3DsettingVector.Style.AltitudeMode;// AltitudeMode:该枚举定义了高度模式类型常量。高度模式用于指示 SuperMap 组件产品解析三维数据海拔高度值的方式。目前，SuperMap 组件产品提供了如下几种高度模式:贴地表模式（ClampToGround）、距地相对高度模式（RelativeToGround）、绝对海拔模式（Absolute）、相对

地下模式(RelativeToUnderground)、地下相对模式(RelativeUnderGround)、地下绝对模式(AbsoluteUnderGround),其中贴地表模式(ClampToGround)为 SuperMap 组件产品默认的高度模式。

```
        switch(altitudeMode)
        {case AltitudeMode.Absolute:
          return 2;
          case AltitudeMode.ClampToGround:
          return 0;
          case AltitudeMode.RelativeToGround:
          return 1;
          case AltitudeMode.RelativeToUnderground:
          return 3;
          default:
          return -1;}}
    catch(Exception e)
        {Trace.WriteLine(e.Message);
        return -1;}}}
        /// 获取矢量面的填充模式
    public Int32 FillModeCode
        {get
        {try{Layer3DDataset layer3Dds = (Layer3DDataset)m_layerRegion;
        Layer3DSettingVector layer3DsettingVector
        (Layer3DSettingVector)layer3Dds.AdditionalSetting;// Layer3DSettingVector:矢量数
```
据集三维图层扩展设置类。该类用于设置矢量数据集作为三维图层加入到三维窗口中所使用的一些显示风格或者获取相关的信息。

```
        FillMode3D fillMode = layer3DsettingVector.Style.FillMode;// FillMode3D:该枚举定
```
义了三维填充模式类型常量。该枚举提供了三种三维体对象的填充模式,分别为轮廓填充(Line)、区域填充(Fill)、轮廓和区域填充(LineAndFill)。

```
        switch(fillMode)
        {case FillMode3D.Fill:
          return 0;
          case FillMode3D.Line:
          return 1;
          case FillMode3D.LineAndFill:
          return 2;
          default:
          return -1;}}
    catch(Exception e)
        {Trace.WriteLine(e.Message);
```

```
        return -1;}}}
    public SampleRun(Workspace workspace, SceneControl sceneControl)
        {this.m_sceneControl = sceneControl;
        this.m_workspace = workspace;
WorkspaceConnectionInfo conInfo = new WorkspaceConnectionInfo(@"..\..\SampleData\
ModelingAndAnalyst\ModelingAndAnalyst.smwu");
        m_workspace.Open(conInfo);//打开工作空间
        m_sceneControl.Scene.Workspace = m_workspace;//打开三维地图场景
        datasource = m_workspace.Datasources[0];//返回数据源几何对象
        FlyToTarget();//初始化相机位置,飞到最初加载的位置
        m_VectorRegionLayerFields = new Dictionary<string,string>();
        m_SCVBuilder = new VectorCacheBuilder();}//构造新对象
    /// 初始化相机位置,飞到最初加载的位置
    public void FlyToTarget()
        {//创建相机
    Camera camera = new Camera();
    camera.Latitude = 39.883912845798775;//设置相机的纬度
    camera.Longitude = 116.35062907228864;//设置相机的经度
    camera.Altitude = 100.23800005298108;//设置相机的高度
    camera.AltitudeMode = AltitudeMode.RelativeToGround;//设置高度模式
    camera.Heading = 34.847028793470777;//设置相机的方位角
    camera.Tilt = 76.822802220392148;//设置相机的仰角
    m_sceneControl.Scene.Fly(camera,10);//将相机对象按照指定的时间飞行到指定
的目的地
    m_sceneControl.Scene.Refresh();}//刷新三维地图场景
    ///设置顶部纹理贴图横向重复次数
    public void SetTopTilingV(String value, Boolean isField)
        {try{if (String.IsNullOrEmpty(value))
    {return;}
    Layer3DDataset layer3Dds = (Layer3DDataset)m_layerRegion;
    Layer3DSettingVector layer3DsettingVector
    (Layer3DSettingVector)layer3Dds.AdditionalSetting;
    GeoStyle3D style3D = layer3DsettingVector.Style;
    if (isField)
        {style3D.TopTilingV = 0;//设置顶部纹理纵向重复次数
    layer3DsettingVector.Style = style3D;//设置矢量数据集作为三维图层加入到三维窗
口中所使用的显示风格
    layer3DsettingVector.TopTilingVField = value;}//设置顶部纹理纵向重复次数字段
    else
```

{layer3DsettingVector. TopTilingVField = " ";//设置顶部纹理纵向重复次数字段
layer3DsettingVector. Style. TopTilingV = Convert. ToInt32(value);}//设置顶部纹理纵向重复次数
RefreshScene();}
catch(Exception ex)
{MessageBox. Show(ex. Message);}}
///设置顶部纹理贴图纵向重复次数
public void SetTopTilingU(String value, Boolean isField)
{try{if(String. IsNullOrEmpty(value))
{return;}
Layer3DDataset layer3Dds = (Layer3DDataset)m_layerRegion;
Layer3DSettingVector layer3DsettingVector = (Layer3DSettingVector)layer3Dds. AdditionalSetting;
GeoStyle3D style3D = layer3DsettingVector. Style;
if(isField)
{style3D. TopTilingU = 0;//设置顶部纹理横向重复次数字段
layer3DsettingVector. Style = style3D;//设置矢量数据集作为三维图层加入到三维窗口中所使用的显示风格
layer3DsettingVector. TopTilingUField = value;}//设置顶部纹理横向重复次数字段
else
{layer3DsettingVector. TopTilingUField = " ";//设置顶部纹理横向重复次数字段
layer3DsettingVector. Style. TopTilingU = Convert. ToInt32(value);}//设置顶面纹理横向重复次数字段
RefreshScene();}
catch(Exception ex){
MessageBox. Show(ex. Message);}}
///设置侧面纹理贴图纵向重复次数
public void SetTilingU(String value, Boolean isField)
{try{if(String. IsNullOrEmpty(value))
{return;}
Layer3DDataset layer3Dds = (Layer3DDataset)m_layerRegion;
Layer3DSettingVector layer3DsettingVector = (Layer3DSettingVector)layer3Dds. AdditionalSetting;
GeoStyle3D style3D = layer3DsettingVector. Style;
if(isField)
{style3D. TilingU = 0;//设置顶部纹理横向重复次数
layer3DsettingVector. Style = style3D;//设置矢量数据集作为三维图层加入到三维窗口中所使用的显示风格
layer3DsettingVector. TilingUField = value;}//设置顶部纹理横向重复次数字段

```
            else{layer3DsettingVector.TilingUField="";//设置顶部纹理横向重复次数字段
                  layer3DsettingVector.Style.TilingU=Convert.ToInt32(value);}//设置顶部纹理
横向重复次数
            RefreshScene();}
            catch(Exception ex)
            {MessageBox.Show(ex.Message);}}
            /// 设置侧面纹理贴图横向重复次数
        public void SetTilingV(String value,Boolean isField)
              {try{if(String.IsNullOrEmpty(value))
            {return;}
            Layer3DDataset layer3Dds=(Layer3DDataset)m_layerRegion;
                  Layer3DSettingVector  layer3DsettingVector = ( Layer3DSettingVector ) layer3Dds.
     AdditionalSetting;
            GeoStyle3D style3D=layer3DsettingVector.Style;//设置矢量数据集作为三维图层加
入到三维窗口中所使用的显示风格
            if(isField)
            {style3D.TilingV=0;
            layer3DsettingVector.Style=style3D;//设置矢量数据集作为三维图层加入到三维窗
口中所使用的显示风格
            layer3DsettingVector.TilingVField=value;}//设置顶部纹理横向重复次数字段
            else{layer3DsettingVector.TilingVField="";//设置顶部纹理横向重复次数字段
                  layer3DsettingVector.Style.TilingV=Convert.ToInt32(value);}//设置顶面纹理横
向重复次数字段
            RefreshScene();}
            catch(Exception ex)
            {MessageBox.Show(ex.Message);}}
            /// 加载矢量面数据
        public void LoadRegionDataset()
              {try{m_regionDataset=datasource.Datasets["ModelRegion"] as DatasetVector;//获
取数据源所包含的数据集的几何对象
            GetRegionDatasetFields();//获取矢量面图层属性字段
            //设置矢量面初始样式
            GeoStyle3D style3D=new GeoStyle3D();
            style3D.AltitudeMode=AltitudeMode.ClampToGround;//设置三维几何对象的高度模
式
            style3D.FillForeColor=Color.White;//设置三维几何对象面区域的填充前景颜色
            style3D.FillMode=FillMode3D.LineAndFill;//设置三维几何体对象的填充模式
            Layer3DSettingVector layer3DSetting=new Layer3DSettingVector();
            layer3DSetting.Style=style3D;//设置矢量数据集作为三维图层加入到三维窗口中所
```

使用的显示风格

```
    m_layerRegion = m_sceneControl.Scene.Layers.Add(m_regionDataset, layer3DSetting,
true);//向三维图层集合中添加数据集类型的三维图层
    m_layerRegion.IsSelectable = false;//设置三维图层中的对象是否可以选择
    m_sceneControl.Scene.Refresh();}//刷新三维地图场景
    catch(Exception ex)
    {MessageBox.Show(ex.Message);}}
    /// 拉伸矢量面
    internal void ExtendRegion(String num_ExtendHeight)
    {try{if(String.IsNullOrEmpty(num_ExtendHeight))
    {return;}
    Layer3DDataset layer3Dds = (Layer3DDataset)m_layerRegion;
      Layer3DSettingVector layer3DsettingVector = (Layer3DSettingVector)layer3Dds.AdditionalSetting;
      if(VectorRegionLayerFields.ContainsKey(num_ExtendHeight.ToString()))
      {layer3DsettingVector.ExtendedHeightField = num_ExtendHeight.ToString();}//设置拉伸高度字段名称
      else{layer3DsettingVector.ExtendedHeightField = "";//设置拉伸高度字段名称
          layer3DsettingVector.Style.ExtendedHeight = Convert.ToDouble(num_ExtendHeight);}//设置拉伸高度值
    m_layerRegion.UpdateData();//更新渲染所需要的数据
    m_sceneControl.Scene.Refresh();//刷新三维地图场景}
    catch(Exception ex)
    {MessageBox.Show(ex.Message);}}
    /// 修改填充模式
    internal void SetFillMode(int p)
    {FillMode3D fillMode3D = new FillMode3D();
    switch(p)
    {case 0:
    fillMode3D = FillMode3D.Line;
    break;
    case 1:
    fillMode3D = FillMode3D.Fill;
    break;
    case 2:
    fillMode3D = FillMode3D.LineAndFill;
    break;
    default:
    break;}
```

```csharp
try{ Layer3DDataset layer3Dds=(Layer3DDataset)m_layerRegion;
    Layer3DSettingVector layer3DsettingVector = (Layer3DSettingVector)layer3Dds.AdditionalSetting;
    layer3DsettingVector.Style.FillMode=fillMode3D;//设置三维几何体对象的填充模式
    m_layerRegion.UpdateData();//更新渲染所需要的数据
    m_sceneControl.Scene.Refresh();}//刷新三维地图场景
catch(Exception ex)
{MessageBox.Show(ex.Message);}}
/// 改变高度模式
internal void SetAltitudeMode(int p)
{AltitudeMode altitudeMode=new AltitudeMode();
    switch(p)
    {case 0:
        altitudeMode=AltitudeMode.ClampToGround;
        break;
    case 1:
        altitudeMode=AltitudeMode.RelativeToGround;
        break;
    case 2:
        altitudeMode=AltitudeMode.Absolute;
        break;
    case 3:
        altitudeMode=AltitudeMode.RelativeUnderGround;
        break;
    case 4:
        altitudeMode=AltitudeMode.AbsoluteUnderGround;
        break;}
    try{ Layer3DDataset layer3Dds=(Layer3DDataset)m_layerRegion;
        Layer3DSettingVector layer3DsettingVector = (Layer3DSettingVector)layer3Dds.AdditionalSetting;
        layer3DsettingVector.Style.AltitudeMode=altitudeMode;//设置三维几何对象的高度模式
        RefreshScene();}//刷新
    catch(Exception ex)
    {MessageBox.Show(ex.Message);}}
/// 刷新场景更新图层数据
private void RefreshScene()
    {m_layerRegion.UpdateData();//更新渲染所需要的数据
    m_sceneControl.Scene.Refresh();}//刷新
```

```csharp
/// 加载分析使用的道路线数据集
public void LoadLineDataset()
    {try{Datasource datasource=m_workspace.Datasources[0];
m_dataset=datasource.Datasets["RoadLine"] as DatasetVector;//获取数据源所包含的数据集的集合对象
// 设置线路数据集的显示风格
Layer3DSettingVector layer3DSetting=new Layer3DSettingVector();
layer3DSetting.Style.LineColor=Color.FromArgb(255,255,153,51);//设置三维线几何对象的颜色
layer3DSetting.Style.LineWidth=2;//设置三维线几何对象的宽度
m_lineLayer=m_sceneControl.Scene.Layers.Add(m_dataset,layer3DSetting,true);//向三维图层集合中添加数据集类型的三维图层
m_lineLayer.Selection.Style.LineColor=Color.Red;//设置三维线几何对象的颜色
m_lineLayer.IsSelectable=false;//设置三维图层中的对象是否可以选择
m_sceneControl.Scene.Refresh();}//刷新
catch(Exception ex)
{Trace.WriteLine(ex.Message);}}
///设置缓冲区的左半径
public void SetLeftRadius(String text,Boolean isField)
    {try{if(String.Empty==text)
{m_leftRadius=null;
return;}
if(isField)
{m_leftRadius=text;}
else{m_leftRadius=Double.Parse(text);}}
catch(Exception ex)
{Trace.WriteLine(ex.Message);}}
/// 设置缓冲区的右半径
public void SetRightRadius(String text,Boolean isField)
    {try{if(String.Empty==text)
{m_leftRadius=null;
return;}
if(isField)
{m_rightRadius=text;}
else{m_rightRadius=Double.Parse(text);}}
catch(Exception ex)
{Trace.WriteLine(ex.Message);}}
/// 创建缓冲区
public void CreateBuffer()
```

```
{try{if(m_hasCreated)
{this.ClearBuffer();//清除缓冲区
this.ClearTempDataset();//清除执行过程中产生的数据集
m_hasCreated=false;}
 BufferAnalystParameter bufferAnalystParam = new BufferAnalystParameter();//
BufferAnalystParameter:缓冲区分析参数类。用于为缓冲区分析提供必要的参数信息。
 bufferAnalystParam.EndType = BufferEndType.Flat;  //设置缓冲端点类型
BufferEndType:该枚举定义了缓冲区端点类型常量。用以区分线对象缓冲区分析时的端点
是圆头缓冲还是平头缓冲。Flat:平头缓冲。平头缓冲区是在生成缓冲区时，在线段的端点
处作线段的垂线。
 bufferAnalystParam.LeftDistance=m_leftRadius;//设置缓冲区距离
 bufferAnalystParam.RightDistance=m_rightRadius;//设置缓冲区距离
 Datasource datasource=m_workspace.Datasources[0];
 String bufferName="bufferRegionDt";
 bufferName=datasource.Datasets.GetAvailableDatasetName(bufferName);//获取一个
数据源中未被使用的数据集的名称
 m_bufferDataset = datasource.Datasets.Create(new DatasetVectorInfo(bufferName,
DatasetType.Region));//根据指定的矢量数据集信息来创建矢量数据集
 Boolean isTrue = SuperMap.Analyst.SpatialAnalyst.BufferAnalyst.CreateBuffer(m_
lineRecordset, m_bufferDataset, bufferAnalystParam, false, true);//创建矢量记录集缓冲区
 //创建缓冲数据集
 Layer3DSettingVector layer3DSetting=new Layer3DSettingVector();
 GeoStyle3D style=new GeoStyle3D();
 style.FillForeColor=Color.FromArgb(100,255,128,64);//设置三维几何对象面区
域的填充前景颜色
 style.AltitudeMode=AltitudeMode.RelativeToGround;//设置三维几何对象的高度模
式
 style.FillMode=FillMode3D.Fill;//设置三维几何对象的填充模式
  Layer3DSettingVector layerRegion3DSetting = m_layerRegion.AdditionalSetting as
Layer3DSettingVector;
 double extendedHeight=layerRegion3DSetting.Style.ExtendedHeight;
 if(extendedHeight==0)
 {String extendedField=layerRegion3DSetting.ExtendedHeightField;
 extendedHeight=m_regionDataset.Statistic(extendedField,StatisticMode.Average);}
 style.ExtendedHeight=extendedHeight+1;//设置拉伸高度值
 layer3DSetting.Style=style;//设置矢量数据集作为三维图层加入到三维窗口中所使
用的显示风格
 m_layerBuffer = m_sceneControl.Scene.Layers.Add(m_bufferDataset, layer3DSetting,
true);//向三维图层集合中添加数据集类型的三维图层
```

```csharp
        m_layerBuffer.UpdateData();//更新渲染所需要的数据
        if(m_sceneControl.Action==Action3D.Select)
        {m_layerBuffer.IsSelectable=true;}//设置三维图层中的对象是否可以选择
        else
        {m_layerBuffer.IsSelectable=false;} //设置三维图层中的对象是否可以选择
        m_hasCreated=true;
        m_sceneControl.Scene.Refresh();}//刷新
        catch(Exception ex)
        {Trace.WriteLine(ex.Message);}}
    //使用创建出来的缓冲区进行叠加分析
    public void BufferQuery()
        {try{
        Datasource datasource=m_workspace.Datasources[0];
        QueryParameter para=new QueryParameter();//QueryParameter:查询参数类。用于描
```
述一个条件查询的限制条件,如所包含的 SQL 语句、游标方式、空间数据的位置关系条件
设定等。
```csharp
        para.SpatialQueryMode=SpatialQueryMode.Intersect;// SpatialQueryMode:该枚举定义
```
了空间查询操作模式常量。空间查询是通过几何对象之间的空间位置关系来构建过滤条件
的一种查询方式。例如:通过空间查询可以找到被包含在面中的空间对象,相离或者相邻
的空间对象等。Intersect:相交空间查询模式。
    //设置空间查询操作模式
```csharp
        para.SpatialQueryObject=m_bufferDataset;//设置空间查询中的搜索对象
        Recordset recordset=m_regionDataset.Query(para);
        List<Int32> ids=new List<int>(recordset.RecordCount);
        while(! recordset.IsEOF)
        {ids.Add(recordset.GetID());//添加对象
        recordset.MoveNext();}//移动当前记录位置到下一条记录
        m_layerRegion.Selection.AddRange(ids.ToArray());//根据指定的对象的系统 ID 值
```
数组,向三维选择集中批量加入对象,这些对象将从非选中状态变为选中状态
```csharp
        m_layerRegion.Selection.UpdateData();//更新渲染所需的数据
        m_layerRegion.Selection.Style.FillForeColor=Color.FromArgb(180,100,100,255);
```
//设置三维集合对象面区域的填充前景颜色
```csharp
        m_sceneControl.Scene.Refresh();//刷新
        recordset.Dispose();}//释放资源
        catch(Exception ex)
        {Trace.WriteLine(ex.Message);}}
    /// 清除缓冲区
    public void ClearBuffer()
        {try{m_lineLayer.Selection.Clear();//设置三维图层的选择集
```

m_layerRegion. Selection. Clear( );//设置三维图层的选择集
if (m_layerBuffer！=null)
{m_sceneControl. Scene. Layers. Remove(m_layerBuffer. Name）；//用于从三维图层集合中删除一个指定名称的三维图层。
m_layerBuffer=null;}
m_sceneControl. Scene. Refresh( );}//刷新
catch (Exception ex)
{Trace. WriteLine(ex. Message）；}} /// 清除程序执行过程中产生的数据集
public void ClearTempDataset( )
{try{ if (m_bufferDataset！=null)
{m_workspace. Datasources[0]. Datasets. Delete(m_bufferDataset. Name）；//根据指定的名称来删除数据集
m_bufferDataset=null;}
m_workspace. Save( );}//用于将现存的工作空间存盘，不改变原有名称，成功则返回true
catch (Exception ex)
{Trace. WriteLine(ex. Message）；}}
/// 通过鼠标右键切换sceneControl的Action状态，这个状态限于Pan和Select两个之间
public void ChangeAction( )
{try
{if (m_sceneControl. Action==Action3D. Pan)
{m_sceneControl. Action=Action3D. Select；//设置三维地图操作状态
m_lineLayer. IsSelectable=true；//设置三维图层中的对象是否可以选择
if (m_layerBuffer！=null)
{m_layerBuffer. IsSelectable=true；}} //设置三维图层中的对象是否可以选择
else{m_sceneControl. Action=Action3D. Pan；//设置三维地图操作状态
m_lineLayer. IsSelectable=false；//设置三维图层中的对象是否可以选择
if (m_layerBuffer！=null)
{m_layerBuffer. IsSelectable=false；}}//设置三维图层中的对象是否可以选择
m_sceneControl. Scene. Refresh( );}
catch (Exception ex)
{Trace. WriteLine(ex. Message）；}}
/// 获取选择的道路线，通过用户的鼠标选择，获取选中的道路线所对应的记录集
public void SelectLine( )
{try{ Int32 id=m_lineLayer. Selection[0]；
if (m_lineRecordset！=null)
{m_lineRecordset. Dispose( );//释放占有资源
m_lineRecordset=null；}

```
            m_lineRecordset = m_dataset.Query("SmID = " + id.ToString(), CursorType.Dynamic);
}//利用查询条件查询数据
    catch(Exception ex)
    {Trace.WriteLine(ex.Message);}}
/// 返回选择的道路线图层是否选中线
public Boolean HasSelectLine()
        {try{m_isSelectLine = (m_lineLayer.Selection.Count > 0);}
    catch(Exception ex)
    {Trace.WriteLine(ex.Message);}//将消息写入集合中的跟踪侦听器
    return m_isSelectLine;}
/// 渲染缓冲区分析结果
public void Twinkle()
        {try{m_layerRegion.Selection.Style.FillForeColor = Color.FromArgb(220, 102, 0,
51);//设置三维几何对象面区域的填充前景颜色
    m_sceneControl.Scene.Refresh();}//刷新
    catch(Exception ex)
    {Trace.WriteLine(ex.Message);}}
/// 重置数据
public void ReloadData()
            {m_sceneControl.Scene.Layers.Clear();//用于从三维图层集中删除所有三维图层
    LoadRegionDataset();//加载矢量面数据
    LoadLineDataset();//加载分析使用的道路线数据集
    FlyToTarget();}//初始化相机位置,飞到最初加载的位置
public void SetTopTexture(Int32 index, Object top)
            {try{Layer3DSettingVector layer3DSetting = (Layer3DSettingVector)m_layerRegion.
AdditionalSetting;
    GeoStyle3D style3D = layer3DSetting.Style;
    if(index <= 1)//用户输入或选择纹理文件
    {layer3DSetting.TopTextureField = "";//设置顶部纹理字段名称
    String topTexture = top.ToString();//
    style3D.TopTextureFile = topTexture;//设置顶部纹理文件全路径
    layer3DSetting.Style = style3D;}//设置矢量数据集作为三维图层加入到三维窗口中
所有使用的显示风格
    else//用户选择纹理字段
    {style3D.TopTextureFile = "";//设置顶部纹理文件全路径
    layer3DSetting.Style = style3D;//设置矢量数据集作为三维图层加入到三维窗口中所
有使用的显示风格
    String topTextureField = top.ToString();//
    layer3DSetting.TopTextureField = topTextureField;}//设置顶部纹理字段名称
```

```
m_layerRegion.AdditionalSetting=layer3DSetting;//设置三维图层扩展设置信息
m_layerRegion.UpdateData();//更新渲染所需要的数据
m_sceneControl.Scene.Refresh();}
catch(Exception ex)
{MessageBox.Show(ex.Message);}}
public void SetSideTexture(Int32 index,Object side)
{try{Layer3DSettingVector layer3DSetting=(Layer3DSettingVector)m_layerRegion.AdditionalSetting;
GeoStyle3D style3D=layer3DSetting.Style;
if(index<=1)//用户输入户选择纹理文件
{layer3DSetting.SideTextureField="";//设置侧面纹理字段名称
String sideTexture=side.ToString();
style3D.SideTextureFiles=new String[]{sideTexture};//设置用于侧面纹理渲染的一系列图片文件的全路径
layer3DSetting.Style=style3D;}//设置矢量数据集作为三维图层加入到三维窗口中所有使用的显示风格
else//用户选择纹理字段
{style3D.TopTextureFile="";
layer3DSetting.Style=style3D;
String sideTextureField=side.ToString();
layer3DSetting.SideTextureField=sideTextureField;}
m_layerRegion.AdditionalSetting=layer3DSetting;//设置三维图层扩展设置信息
m_layerRegion.UpdateData();
m_sceneControl.Scene.Refresh();}
catch(Exception ex)
{MessageBox.Show(ex.Message);}}
//设置纹理重复模式
public void SetTopRepeatMode(Int32 index)
{Layer3DSettingVector layer3DSetting=(Layer3DSettingVector)m_layerRegion.AdditionalSetting;
GeoStyle3D style3D=layer3DSetting.Style;
switch(index)
{case 0:
style3D.TopTextureRepeatMode=TextureRepeatMode.RepeatTimes;//TextureRepeatMode:纹理重复模式。1.按次数重复模式(repeattimes);2.按尺寸重复模式(realsize)。
break;
case 1:
style3D.TopTextureRepeatMode=TextureRepeatMode.RealSize;
break;
```

```
           default:
             style3D.TopTextureRepeatMode = TextureRepeatMode.RepeatTimes;
             break;}}
         public void SetSideRepeatMode(Int32 index)
             {Layer3DSettingVector layer3DSetting = (Layer3DSettingVector) m_layerRegion.AdditionalSetting;
             GeoStyle3D style3D = layer3DSetting.Style;
             switch (index)
             {case 0:
             style3D.TextureRepeatMode = TextureRepeatMode.RepeatTimes;
             break;
             case 1:
             style3D.TextureRepeatMode = TextureRepeatMode.RealSize;
             break;
             default:
             style3D.TextureRepeatMode = TextureRepeatMode.RepeatTimes;
             break;}}
         /// 设置底部高程
         public void SetButtomHeight(Object buttomHeight)
             {try{Layer3DDataset layer3Dds = (Layer3DDataset)m_layerRegion;
             Layer3DSettingVector layer3DsettingVector = (Layer3DSettingVector) layer3Dds.AdditionalSetting;
             layer3DsettingVector.Style.AltitudeMode = AltitudeMode.RelativeToGround;//设置三维几何对象的高度模式
             if (VectorRegionLayerFields.ContainsKey(buttomHeight.ToString()))
             {layer3DsettingVector.BottomAltitudeField = buttomHeight.ToString();}//设置底部高程字段的名称
             else
             {layer3DsettingVector.BottomAltitudeField = "";//设置底部高程字段的名称
             layer3DsettingVector.Style.BottomAltitude = Convert.ToDouble(buttomHeight);}//设置底部高程值
             m_layerRegion.UpdateData();
             m_sceneControl.Scene.Refresh();}
             catch (Exception ex){
             MessageBox.Show(ex.Message);}}
         /// 设置场景 Action 为 Pan
         public void SetActionToPan()
             {m_sceneControl.Action = Action3D.Pan;//设置三维地图操作状态
             m_lineLayer.IsSelectable = true;//设置三维图层中的对象是否可以选择
```

```csharp
            if(m_layerBuffer!=null)
            {m_layerBuffer.IsSelectable=true;} //设置三维图层中的对象是否可以选择
            m_sceneControl.Refresh();}
        /// 重置图层风格为矢量面拉伸样式
        public void LoadDefaultStyle()
                {try{Layer3DDataset layer3Dds=(Layer3DDataset)m_layerRegion;
            Layer3DSettingVector layerSettingVector=layer3Dds.AdditionalSetting as Layer3DSettingVector;
            if(m_tempStyle!=null)
                {layerSettingVector.Style.AltitudeMode=AltitudeMode.Absolute; //设置三维几何对象的高度模式
            layerSettingVector.ExtendedHeightField="建筑物高度"; //设置拉伸高度字段名称
            layerSettingVector.SideTextureField="SideTexture"; //设置侧面纹理字段名称
            layerSettingVector.Style.TextureRepeatMode=TextureRepeatMode.RepeatTimes; //
            layerSettingVector.Style.TilingV=10; //设置侧面纹理纵向重复次数
            layerSettingVector.TilingUField="层数"; //设置侧面纹理横向重复次数字段
            layerSettingVector.TopTextureField="TopTexture"; //设置顶部纹理字段名称
            layerSettingVector.Style.TopTextureRepeatMode=TextureRepeatMode.RepeatTimes; //
            layerSettingVector.Style.TopTilingV=1; //设置顶部纹理横向重复次数
            layerSettingVector.Style.TopTilingU=1; //设置顶部纹理横向重复次数字段
            layerSettingVector.Style.FillMode=FillMode3D.Fill;} //设置填充模式
            m_layerRegion.UpdateData();
            m_sceneControl.Scene.Refresh();}
            catch(Exception ex)
            {MessageBox.Show(ex.Message);}}
        public void SaveToDefaultStyle()
                {try{Layer3DDataset layer3Dds=(Layer3DDataset)m_layerRegion;
            m_tempStyle=new GeoStyle3D();
            m_tempStyle.ExtendedHeight=((Layer3DSettingVector)layer3Dds.AdditionalSetting).Style.ExtendedHeight; //设置拉伸高度值
            m_tempStyle.AltitudeMode=((Layer3DSettingVector)layer3Dds.AdditionalSetting).Style.AltitudeMode;} //设置对象高度模式
            catch(Exception ex)
            {MessageBox.Show(ex.Message);}}
        //私有方法
        private string[] array={"SMID","SMUSERID","SMAREA","SMPERIMETER","SMPERIMETER","SMGEOMETRYSIZE"};
        /// 判断给定的字段名称是否为系统字段
        private Boolean isSysField(String fieldName)
```

7.1 网络分析

```
{bool result=false;
try{for(int i=0; i < array.Length; i++)
{if(array[i]==fieldName)
{result=true;}}
return result;}
catch(Exception ex)
{Trace.WriteLine(ex.Message);
return result;}}
```
/// 获取矢量面图层属性字段
```
private void GetRegionDatasetFields()
{m_VectorRegionLayerFields.Clear();//删除所有值
for(int i=0; i < m_regionDataset.FieldCount; i++)
{if(! isSysField(m_regionDataset.FieldInfos[i].Name))
m_VectorRegionLayerFields.Add(m_regionDataset.FieldInfos[i].Name, m_regionDataset.FieldInfos[i].Type.ToString());//添加指定的值}}
```

7. 窗体代码

（1）添加 using 引用代码

添加 using 引用代码具体如下：

using SuperMap.UI; //添加 SuperMap.UI 命名空间，使其对应的类可以使用

using SuperMap.Data; //添加 SuperMap.Data 命名空间，使其对应的类可以使用

using SuperMap.Realspace; //添加 SuperMap.Realspace 命名空间，使其对应的类可以使用

using System.Diagnostics; //添加 System.Diagnostics 命名空间，使其对应的类可以使用

（2）实例化代码

实例化 SampleRun 以及 public FormMain() 函数部分的代码如下：

```
//定义类对象
private SceneControl m_SceneControl;
private Workspace m_Workspace;
private SampleRun m_SampleRun;
private OpenFileDialog m_OpenFileDialog;
private FolderBrowserDialog m_FolderBrowserDialog;
private Boolean isDefaultStyle=true;
public FormMain()
{// 开启全屏反走样，并设置反走样系数
SuperMap.Data.Environment.IsSceneAntialias=true;// Environment.:关于开发环境的一些配置信息管理类型，比如设置缓存目录、设置零值判断精度等功能，通过此类还可以设置像素与逻辑坐标的比例。
SuperMap.Data.Environment.SceneAntialiasValue=4;
InitializeComponent();
```

创建 SceneControl 和 Workspace 对象并放入窗体中

```
m_SceneControl = new SceneControl();
m_SceneControl.ObjectSelected += new SuperMap.UI.ObjectSelectedEventHandler(m_sceneControl_ObjectSelected);
m_SceneControl.Dock = DockStyle.Fill;
splitContainer1.Panel2.Controls.Add(m_SceneControl);
m_Workspace = new Workspace(this.components);
this.WindowState = FormWindowState.Maximized;//最大化窗口
//创建 SampleRun 实例,同时初始化 Workspace 和 SceneControl 信息
m_SampleRun = new SampleRun(m_Workspace, m_SceneControl);
//初始化地表高度枚举
this.m_cbboxAltitudeMode.Items.Add("贴地标");//添加项
this.m_cbboxAltitudeMode.Items.Add("相对地表");
this.m_cbboxAltitudeMode.Items.Add("绝对海拔");
this.m_cbboxAltitudeMode.Items.Add("地下相对");
this.m_cbboxAltitudeMode.Items.Add("地下绝对");
//初始化填充模式枚举
m_cbboxFillMode.Items.Add("轮廓填充");//添加项
m_cbboxFillMode.Items.Add("区域填充");
m_cbboxFillMode.Items.Add("轮廓和区域填充");
m_cbboxTopImage.Items.Add("");
m_cbboxTopImage.Items.Add("选择文件");
m_cbboxSideImage.Items.Add("");
m_cbboxSideImage.Items.Add("选择文件");
m_cbboxSideRepeatMode.Items.Add("重复次数");
m_cbboxSideRepeatMode.Items.Add("实际大小");
m_cbboxTopRepeatMode.Items.Add("重复次数");
m_cbboxTopRepeatMode.Items.Add("实际大小");
m_SampleRun.LoadRegionDataset();//加载矢量面数据
m_SampleRun.LoadLineDataset();//加载分析使用的道路线数据集
m_cbboxAltitudeMode.SelectedIndex = m_SampleRun.AltitudeModeCode;//设置指定当前选定项的索引
m_cbboxFillMode.SelectedIndex = m_SampleRun.FillModeCode;//设置指定当前选定项的索引
//填充可选择字段项到下拉列表
Dictionary<String, String> lstFields = m_SampleRun.VectorRegionLayerFields;
foreach (String fieldName in lstFields.Keys)
{
    String strType = lstFields[fieldName];
    if (strType == "Int32")
    {
        m_cbboxExtendHeight.Items.Add(fieldName);//添加项
```

```
           m_cbboxTilingV.Items.Add(fieldName);
           m_cbboxTilingU.Items.Add(fieldName);
           m_cbboxTopTilingV.Items.Add(fieldName);
           m_cbboxTopTilingU.Items.Add(fieldName);}
         if(strType=="Double")
           {m_cbboxExtendHeight.Items.Add(fieldName);}
          if(strType=="Text")
             {m_cbboxTopImage.Items.Add(fieldName);
              m_cbboxSideImage.Items.Add(fieldName);}}
            m_cbboxExtendHeight.Text="0";//设置文本
            m_OpenFileDialog=new OpenFileDialog();
            m_FolderBrowserDialog=new FolderBrowserDialog();
            InitSettingControl();}
```

(3)定义 InitSettingControl 初始化函数

定义 InitSettingControl 初始化函数的代码如下:

```
private void InitSettingControl()
     {//设置指定当前选定项的索引
   m_cbboxAttitudMode.SelectedIndex=2;
   m_cbboxExtendHeight.SelectedIndex=0;
   m_cbboxSideRepeatMode.SelectedIndex=0;
   m_cbboxTopRepeatMode.SelectedIndex=0;
   m_cbboxTopImage.SelectedIndex=3;
   m_cbboxSideImage.SelectedIndex=2;
   m_cbboxTilingV.SelectedIndex=9;
   m_cbboxTilingU.SelectedIndex=10;}
```

(4)其他控件的代码

① m_cbboxExtendHeight 的 TextChanged 事件。在控件的属性窗口单击 图标,左键双击 TextChanged,添加如下代码,实现拉伸高度:

```
private void m_cbboxExtendHeight_TextChanged(object sender,EventArgs e)
       {ComboBox comboBox=(ComboBox)sender;
   m_SampleRun.ExtendRegion(comboBox.Text);}//通过 SampleRun.ExtendRegion 实
```
现拉伸高度

② m_cbboxAltitudeMode 的 SelectedIndexChanged 事件。在控件的属性窗口单击 图标,左键双击 SelectedIndexChanged,添加如下代码,设置高度模式:

```
private void m_cbboxAltitudeMode_SelectedIndexChanged(object sender,EventArgs e)
       {ComboBox comboBox=(ComboBox)sender;
   m_SampleRun.SetAltitudeMode(comboBox.SelectedIndex);}//通过 SampleRun.
```
SetAltitudeMode 设置高度模式

③ m_cbboxFillMode 的 SelectedIndexChanged 事件。在控件的属性窗口单击 图标,左键双击 SelectedIndexChanged,添加如下代码,设置填充模式:

```
private void m_cbboxFillMode_SelectedIndexChanged(object sender, EventArgs e)
    {ComboBox comboBox = (ComboBox)sender;
    m_SampleRun.SetFillMode(comboBox.SelectedIndex);}//通过SampleRun.
```
SetFillMode设置填充模式

④ m_buttonReload 的 Click 事件。在控件的属性窗口单击 ≯ 图标，左键双击 Click，添加如下代码：

```
private void m_buttonReload_Click(object sender, EventArgs e)
    {m_SampleRun.ReloadData();//重置数据
    //设置控件是否可见
    m_groupboxStyleSetting.Enabled = true;
    m_cbboxAltitudeMode.Text = "";
    m_cbboxExtendHeight.Text = "";
    m_cbboxTopImage.Text = "";
    m_cbboxTopRepeatMode.Text = "";
    m_cbboxTopTilingV.Text = "";
    m_cbboxTopTilingU.Text = "";
    m_cbboxSideImage.Text = "";
    m_cbboxSideRepeatMode.Text = "";
    m_cbboxTilingV.Text = "";
    m_cbboxTilingU.Text = "";
    m_cbboxFillMode.Text = "";
    isDefaultStyle = true;}
```

⑤ m_cbboxTiling 的 TextChanged 事件。在控件的属性窗口单击 ≯ 图标，左键双击 TextChanged，添加如下代码，设置贴图重复次数：

```
private void m_cbboxTiling_TextChanged(object sender, EventArgs e)
    {ComboBox comboBox = (ComboBox)sender;
    Boolean isField = false;
    if(comboBox.SelectedIndex > 9)
    {isField = true;}
    switch(comboBox.Tag.ToString())
    {case "TopTilingV":
        m_SampleRun.SetTopTilingV(comboBox.Text, isField);//设置顶部纹理贴图横向
```
重复次数
```
        break;
    case "TopTilingU":
        m_SampleRun.SetTopTilingU(comboBox.Text, isField);//设置顶部纹理贴图纵向
```
重复次数
```
        break;
    case "TilingV":
        m_SampleRun.SetTilingV(comboBox.Text, isField);//设置侧面纹理贴图横向重复
```

次数

  break；
  case "TilingU"：
  m_SampleRun.SetTilingU(comboBox.Text,isField)；//设置侧面纹理贴图纵向重复次数
  break；
  default：
  break；}}

⑥ m_buttonClear 的 Click 事件。在控件的属性窗口单击图标，左键双击 Click，添加如下代码，实现清除结果：

  private void m_buttonClear_Click(object sender, EventArgs e)
   {m_SampleRun.ClearBuffer()；//清除缓冲区
   m_SampleRun.ClearTempDataset()；}//清除执行过程中产生的数据集

⑦ m_buttonAnalyst 的 Click 事件。在控件的属性窗口单击图标，左键双击 Click，添加如下代码，实现缓冲区分析：

  private void m_buttonAnalyst_Click(object sender, EventArgs e)
   {m_SampleRun.CreateBuffer()；//创建缓冲区
  m_SampleRun.BufferQuery()；}//使用创建出来的缓冲区进行叠加分析

⑧ m_cbboxLeftRadius 的 TextChanged 事件。在控件的属性窗口单击图标，左键双击 TextChanged，添加如下代码，设置左半径：

  private void m_cbboxLeftRadius_TextChanged(object sender, EventArgs e)
   {ComboBox comboBox=(ComboBox)sender；
  Boolean isField=false；
  if(comboBox.SelectedIndex！=-1)
   {isField=true；}
   m_SampleRun.SetLeftRadius(comboBox.Text,isField)；}//设置左半径

⑨ m_cbboxRightRadius 的 TextChanged 事件。在控件的属性窗口单击图标，左键双击 TextChanged，添加如下代码，设置右半径：

  private void m_cbboxRightRadius_TextChanged(object sender, EventArgs e)
   {ComboBox comboBox=(ComboBox)sender；
  Boolean isField=false；
  if(comboBox.SelectedIndex！=-1)
  {isField=true；}
   m_SampleRun.SetRightRadius(comboBox.Text,isField)；}//设置右半径

⑩ m_sceneControl 的 ObjectSelected 事件。在控件的属性窗口单击图标，左键双击 ObjectSelected，添加如下代码，通过该事件从场景中得到选择的道路线：

  private void m_sceneControl_ObjectSelected(object sender, SuperMap.UI.ObjectSelectedEventArgs e)
   {try
  {if(e.Count > 0 && m_SampleRun.HasSelectLine())

```
            {m_SampleRun.SelectLine();//获取选择的道路线
              if(m_cbboxLeftRadius.SelectedIndex==-1)
                {m_SampleRun.SetLeftRadius(m_cbboxLeftRadius.Text,false);}//设置缓冲区左
半径
              else
                {m_SampleRun.SetLeftRadius(m_cbboxLeftRadius.Text,true);}
              if(m_cbboxRightRadius.SelectedIndex==-1)
                {m_SampleRun.SetRightRadius(m_cbboxRightRadius.Text,false);}//设置缓冲
区右半径
              else
                {m_SampleRun.SetRightRadius(m_cbboxRightRadius.Text,true);}
              m_groupboxAnalyst.Enabled=true;}}
            catch(Exception ex)
              {Trace.WriteLine(ex.Message);}}
```

⑪ m_tabMain 的 SelectedIndexChanged 事件。在控件的属性窗口单击 图标,左键双击 SelectedIndexChanged,添加如下代码,初始化控件:

```
          private void m_tabMain_SelectedIndexChanged(object sender, EventArgs e)
              {if(m_tabMain.SelectedTab.Name=="m_tabpageBufferAnalyst" && !isDefaultStyle)
                {m_SampleRun.ReloadData();//重置数据
                 m_SampleRun.LoadDefaultStyle();//重置图层风格为矢量面拉伸样式
                 m_groupboxStyleSetting.Enabled=true;
                 m_cbboxAttitudMode.Text="绝对高度";
                 m_cbboxExtendHeight.Text="建筑物高度";
                 m_cbboxTopImage.Text="TopTexture";
                 m_cbboxTopRepeatMode.Text="重复模式";
                 m_cbboxTopTilingV.Text="1";
                 m_cbboxTopTilingU.Text="1";
                 m_cbboxSideImage.Text="SideTexture";
                 m_cbboxSideRepeatMode.Text="重复模式";
                 m_cbboxTilingV.Text="1";
                 m_cbboxTilingU.Text="层数";
                 m_cbboxFillMode.Text="区域填充";
                 isDefaultStyle=true;}
              else
                {m_SampleRun.ClearBuffer();
                 m_SampleRun.ClearTempDataset();}
                 m_SampleRun.SetActionToPan();}
```

⑫ m_cbboxTopImage 的 SelectedIndexChanged 事件,在控件的属性窗口单击 图标,左 键双击 SelectedIndexChanged,添加如下代码,设置顶部纹理贴图:

```csharp
private void m_cbboxTopImage_SelectedIndexChanged(object sender, EventArgs e)
    {if(m_cbboxTopImage.Text=="选择文件")
    {m_OpenFileDialog.Title="选择顶部纹理文件";
     if(m_OpenFileDialog.ShowDialog()==System.Windows.Forms.DialogResult.OK)
         { m_cbboxTopImage.Items[m_cbboxTopImage.Items.Count-3]=m_OpenFileDialog.FileName;}}//获取对象
     m_SampleRun.SetTopTexture(m_cbboxTopImage.SelectedIndex,m_cbboxTopImage.Text);}//设置顶部纹理贴图
```

⑬ m_cbboxSideImage 的 SelectedIndexChanged 事件。在控件的属性窗口单击 ⚡ 图标，左键双击 SelectedIndexChanged，添加如下代码，设置侧面贴图重复次数：

```csharp
private void m_cbboxSideImage_SelectedIndexChanged(object sender, EventArgs e)
    {if(m_cbboxSideImage.Text=="选择文件")
    {m_OpenFileDialog.Title="选择侧面纹理文件";
     if(m_OpenFileDialog.ShowDialog()==System.Windows.Forms.DialogResult.OK)
         {m_cbboxSideImage.Items[m_cbboxSideImage.Items.Count-3]=m_OpenFileDialog.FileName;}}
     m_SampleRun.SetSideTexture(m_cbboxSideImage.SelectedIndex,m_cbboxSideImage.Text);}//设置侧面贴图重复次数
```

⑭ m_cbboxSideRepeatMode 的 SelectedIndexChanged 事件。在控件的属性窗口单击 ⚡ 图标，左键双击 SelectedIndexChanged，添加如下代码，重置贴图 ComboBox 中项的内容，设置侧面纹理模式：

```csharp
private void m_cbboxSideRepeatMode_SelectedIndexChanged(object sender, EventArgs e)
    {m_SampleRun.SetSideRepeatMode(m_cbboxSideRepeatMode.SelectedIndex);}//设置侧面纹理模式
```

⑮ m_cbboxTopRepeatMode 的 SelectedIndexChanged 事件。在控件的属性窗口单击 ⚡ 图标，左键双击 SelectedIndexChanged，添加如下代码，设置顶部纹理模式：

```csharp
private void m_cbboxTopRepeatMode_SelectedIndexChanged(object sender, EventArgs e)
    {m_SampleRun.SetTopRepeatMode(m_cbboxTopRepeatMode.SelectedIndex);}//设置顶部纹理模式
```

⑯ m_cbboxSideImage 的 Click 事件。在控件的属性窗口单击 ⚡ 图标，左键双击 Click，添加如下代码：

```csharp
private void m_cbboxSideImage_Click(object sender, EventArgs e)
    {InitComboBox();}//重置 ComboBox 的内容
```

⑰ m_cbboxTopImage 的 Click 事件，在控件的属性窗口单击 ⚡ 图标，左键双击 Click，添加如下代码：

```csharp
private void m_cbboxTopImage_Click(object sender, EventArgs e)
    {
     InitComboBox();}//重置 ComboBox 的内容
```

(5) 窗体的 FormClose 事件

窗体的 FormClose 事件，代码同前文中的相应部分。

(6)定义 InitComboBox 初始化函数

定义 InitComboBox 初始化函数的代码如下：

```
private void InitComboBox()
{m_cbboxTopImage.Items.Clear();
m_cbboxSideImage.Items.Clear();
m_cbboxTopImage.Items.Add("");
m_cbboxTopImage.Items.Add("选择文件");
m_cbboxSideImage.Items.Add("");
m_cbboxSideImage.Items.Add("选择文件");
Dictionary<String,String> lstFields=m_SampleRun.VectorRegionLayerFields;
foreach (String fieldName in lstFields.Keys)
{String strType=lstFields[fieldName];
if (strType=="Text")
{m_cbboxTopImage.Items.Add(fieldName);
m_cbboxSideImage.Items.Add(fieldName);}}}
```

8．运行结果

窗体设计完成后点击 ▶ 按钮，运行该程序，运行后弹出窗体，如图7-8所示。如图可以根据不同的风格建模，还可以对图中的道路进行缓冲区分析。

图7-8　批量建模与缓冲区分析

7.2 拓扑分析

1. 数据

在安装目录\SampleData\Topo\TopoProcessing.smwu 路径下打开数据文件。

2. 新建文件夹和工程

在【D:\MyProject\】文件中新建一个文件夹，命名为【第 7 章 空间分析】，并在其目录下建立一个文件夹，命名为【拓扑分析】，在此文件夹里新建一个文件夹，命名为 Topology，在此文件夹中新建一个工程，将此工程命名为 Topology。

3. 关键类型/成员

关键类型/成员见表 7-10。

表 7-10　　　　　　　　　　关键类型/成员表

控件/类	方法	属性	事件
TopologyProcessing	BuildRegions、Clean		
TopologyValidator	Preprocess、Validate		

4. 窗体控件属性表

窗体控件属性具体见表 7-11。

表 7-11　　　　　　　　　　窗体控件属性表

控件	Name	Text
toolStrip	toolStrip1	toolStrip1
Button	topoLineToRegion	拓扑构面
Button	topoProcess	拓扑处理
Button	topoCheck	拓扑检查
StatusBar	StatusBar1	等待中…

5. 窗体设计布局

窗体设计布局如图 7-9 所示。

6. 创建 SampleRun 类

（1）添加 using 引用代码

添加 using 引用代码具体如下：

using System.Diagnostics; //添加 System.Diagnostics 命名空间，使其对应的类可以使用

using System.Drawing; //添加 System.Drawing 命名空间，使其对应的类可以使用

using System.Windows.Forms; //添加 System.Windows.Forms 命名空间，使其对应的类可以使用

图 7-9　窗体设计布局图

using SuperMap.Data；//添加 SuperMap.Data 命名空间，使其对应的类可以使用

using SuperMap.UI；//添加 SuperMap.UI 命名空间，使其对应的类可以使用

using SuperMap.Data.Topology；//添加 SuperMap.Data.Topology 命名空间，使其对应的类可以使用

using SuperMap.Mapping；//添加 SuperMap.Mapping 命名空间，使其对应的类可以使用

（2）在 public class SampleRun 中添加代码

在 public class SampleRun 中添加如下代码：

//定义类对象

private Workspace m_workspace；// Workspace：工作空间是用户的工作环境，主要完成数据的组织和管理，包括打开、关闭、创建、保存工作空间文件。

private MapControl m_mapControl；// MapControl：地图控件类。该类是用于为地图的显示提供界面的，同时为地图与数据的互操作提供了途径。

private Dataset m_dataset；// Dataset：所有数据集类型（如矢量数据集，栅格数据集等）的基类。提供各数据集共有的属性、方法和事件。

private String m_datasetName；

private String m_createRegionName；

private String m_processDatasetName；

private String m_checkDataName；

private String m_bufDatasetName；

private DatasetVector m_bufDataset；// DatasetVector：矢量数据集类。描述矢量数据集，并提供相应的管理和操作。对矢量数据集的操作主要包括数据查询、修改、删除、建立索引等。

```csharp
private DatasetVector m_resultDataset;
private TopologyProcessingOptions m_topoOptions;// TopologyProcessingOptions:拓扑预处
理选项类。该类提供了关于拓扑预处理的相关设置信息。
/// 根据 workspace 和 map 构造 SampleRun 对象
public SampleRun(Workspace workspace, MapControl mapControl)
    {try{ m_workspace = workspace;
        m_mapControl = mapControl;
        Initialize();}
    catch (Exception ex)
        {Trace.WriteLine(ex.Message);}}
private void Initialize()
    {try{//打开工作空间及地图
        WorkspaceConnectionInfo conInfo = new WorkspaceConnectionInfo(@"../../SampleData/Topo/TopoProcessing.smwu");// WorkspaceConnectionInfo:工作空间连接信息
类。包括工作空间连接的所有信息,如所要连接的服务器名称、数据库名称、用户名、密
码等。
            m_datasetName = "RoadLine";
            m_createRegionName = "LineToRegion";
            m_processDatasetName = "copyDataset";
            m_checkDataName = "checkTopo";
            m_bufDatasetName = "bufDataset";
            m_workspace.Open(conInfo);
            Datasets datasets = m_workspace.Datasources[0].Datasets;
            if (datasets.Contains(m_datasetName))
            {this.m_mapControl.Map.Layers.Add(datasets[m_datasetName], true);
            m_mapControl.Map.Refresh();
            m_dataset = datasets[m_datasetName];}
            // 调整 mapControl 的状态
            m_mapControl.Action = SuperMap.UI.Action.Pan;
            SetOptions();}
    catch (Exception ex)
        {Trace.WriteLine(ex.Message);}}
    /// 构造一个全为 true 的 TopologyProcessingOption
private void SetOptions()
    {m_topoOptions = new TopologyProcessingOptions();
    m_topoOptions.AreAdjacentEndpointsMerged = true;//设置是否进行邻近端点合并
    m_topoOptions.AreDuplicatedLinesCleaned = true;//设置是否去除重复线
    m_topoOptions.AreLinesIntersected = true;//设置是否进行弧段求交
    m_topoOptions.ArePseudoNodesCleaned = true;//设置是否去除假节点
```

```
m_topoOptions.AreRedundantVerticesCleaned=true;//设置是否去除多余点
m_topoOptions.AreUndershootsExtended=true;//设置是否进行长悬线延伸
m_topoOptions.AreOvershootsCleaned=true;}//设置是否去除短悬线
/// 设置图层的一些属性
private void SetLayerStyle(Layer layer,Color color,Double width)
{try{LayerSettingVector layerSetting=new LayerSettingVector();//LayerSettingVector:矢
量图层设置类。该类主要用来设置矢量图层的显示风格。矢量图层用单一的符号或风格绘
制所有的要素。当用户只想可视化地显示空间数据,只关心空间数据中各要素在什么位
置,而不关心各要素在数量或性质上的不同时,可以用普通图层来显示要素数据。
    GeoStyle style=new GeoStyle();
    style.FillForeColor=Color.SkyBlue;//填充符号的前景色
    style.LineColor=color;//设置线的颜色
    style.LineWidth=width;//设置线宽
    layerSetting.Style=style;//设置风格
    layer.AdditionalSetting=layerSetting;}
    catch(Exception ex)
    {Trace.WriteLine(ex.Message);}}
/// 清除所有图层并添加新的图层
private void ResetDatasetAddMap()
    {try{m_mapControl.Map.Layers.Clear();
    m_mapControl.Map.Layers.Add(m_dataset,true);
    // 拓扑构面、处理等会改变数据,并要求数据为关闭状态,所以每次都使用
新的数据
    m_dataset.Datasource.Datasets.Delete(m_bufDatasetName);
    m_bufDataset=(DatasetVector)m_dataset.Datasource.CopyDataset(m_dataset,m
_bufDatasetName,m_dataset.EncodeType);}
    catch(Exception ex)
    {Trace.WriteLine(ex.Message);}}
/// 拓扑构面
public Boolean LineToRegion()
    {Boolean result=false;
    // 数据不为空才执行下面的操作
    if(m_dataset!=null)
    {try{this.ResetDatasetAddMap();
    m_dataset.Datasource.Datasets.Delete(m_createRegionName);
    m_resultDataset=TopologyProcessing.BuildRegions(m_bufDataset,m_workspace.
Datasources[0],m_createRegionName,m_topoOptions);// TopologyProcessing:拓扑处理类。
用于进行拓扑处理。
    m_mapControl.Map.Layers.Add(m_resultDataset,true);
```

```
            SetLayerStyle(m_mapControl.Map.Layers[0], Color.Red, 0.2);
            m_mapControl.Map.Refresh();
             result = true;}
            catch(Exception ex)
            {Trace.Write(ex.Message);
             result = false;}}
            return result;}
        /// 拓扑处理
    public Boolean TopoProcess()
        {Boolean result = false;
        if(m_dataset ! = null)
        {try{this.ResetDatasetAddMap();
            m_dataset.Datasource.Datasets.Delete(m_processDatasetName);
            if (TopologyProcessing.Clean(m_bufDataset, m_topoOptions)) //
TopologyProcessing:拓扑处理类。用于进行拓扑处理。
            {m_mapControl.Map.Layers.Add(m_bufDataset, true);
            SetLayerStyle(m_mapControl.Map.Layers[0], Color.Red, 0.2);
            m_mapControl.Map.Refresh();}
            result = true;}
            catch(Exception ex)
            {Trace.Write(ex.Message);
            result = false;}}}
        /// 拓扑检查
    public Boolean TopoCheck()
        {Boolean result = false;
        if(m_dataset ! = null)
        {try{this.ResetDatasetAddMap();
                TopologyDatasetRelationItem topoItem = new TopologyDatasetRelationItem(m_
bufDataset);// TopologyDatasetRelationItem:拓扑预处理项类,用于为拓扑预处理提供必要
的参数信息,包括用于预处理的数据集、用于预处理数据集的精度序号。
            TopologyDatasetRelationItem[] items = {topoItem};
            // 拓扑预处理
            TopologyValidator.Preprocess(items, 2);// TopologyValidator:拓扑检查类。用
于进行拓扑预处理、拓扑检查和拓扑错误自动修复。
            // 检查线相叠
            m_dataset.Datasource.Datasets.Delete(m_checkDataName);
                m_resultDataset = TopologyValidator.Validate(m_bufDataset, m_bufDataset,
TopologyRule.LineNoOverlap, 2, null, m_dataset.Datasource, m_checkDataName);
            m_mapControl.Map.Layers.Add(m_resultDataset, true);
```

```
            this.SetLayerStyle(m_mapControl.Map.Layers[0],Color.Red,0.5);
            m_mapControl.Map.Refresh();
            result=true;}
        catch(Exception ex)
          {Trace.Write(ex.Message);
             result=false;}}
            return result;}
```

7. 窗体代码

(1) 添加 using 引用代码

添加 using 引用代码具体如下:

using System.Diagnostics;   //添加 System.Diagnostics 命名空间,使其对应的类可以使用
using SuperMap.Data;   //添加 SuperMap.Data 命名空间,使其对应的类可以使用
using SuperMap.UI;   //添加 SuperMap.UI 命名空间,使其对应的类可以使用
using SuperMap.Data.Topology;   //添加 SuperMap.Data.Topology 命名空间,使其对应的类可以使用

(2) 在 public FormMain() 函数前添加代码

在 public FormMain() 函数前添加如下代码:

```
//定义类对象
private SampleRun m_sampleRun;
private Workspace m_workspace;
private MapControl m_mapControl;
```

(3) public FormMain() 函数部分的代码

public FormMain() 函数部分的代码如下:

```
public FormMain()
    {try{InitializeComponent();
        m_workspace=new Workspace();
        m_mapControl=new MapControl(m_workspace);
        m_mapControl.Dock=DockStyle.Fill;
        this.Controls.Add(m_mapControl);}
        catch(Exception ex)
        {Trace.WriteLine(ex.Message);}}
```

(4) 窗体的 Load 事件

在窗体的属性窗口中,选择 ≠ 按钮后,左键双击 Load,即可添加以下代码,实现加载窗口时加载数据:

```
private void FormMain_Load(object sender,EventArgs e)
     {try{stateText.Text="打开数据中,请等待…";
        statusBar.Update();
        m_sampleRun=new SampleRun(m_workspace,m_mapControl);//定义新对象
        topoLineToRegion.Click+=new EventHandler(topoLineToRegion_Click);
```

```
    topoProcess.Click+=new EventHandler(topoProcess_Click);
    topoCheck.Click+=new EventHandler(topoCheck_Click);
    stateText.Text="请？选？择?";}
    catch(Exception ex)
    {Trace.WriteLine(ex.Message);}}
```

(5) 添加各个控件的代码

① 拓扑构面 Button 控件的 Click 事件。左键双击漫游 Button 控件，即可添加该按钮的 Click 事件代码，实现拓扑构面：

```
    private void topoLineToRegion_Click(object sender,EventArgs e)
        {try{stateText.Text="拓扑构面中,请等待…";
        statusBar.Update();
        this.Cursor=Cursors.WaitCursor;
    if(m_sampleRun.LineToRegion())//通过 sampleRun.LineToRegion 实现拓扑构面
        {stateText.Text="拓扑构面成功完成";}
    else{stateText.Text="拓扑构面失败";}
        this.Cursor=Cursors.Arrow;}
        catch(Exception ex)
        {Trace.WriteLine(ex.Message);}}
```

② 拓扑处理 Button 控件的 Click 事件。左键双击全屏 Button 控件，即可添加该按钮的 Click 事件代码，实现拓扑处理：

```
    private void topoProcess_Click(object sender,EventArgs e)
        {try{stateText.Text="拓扑处理中,请等待…";
            statusBar.Update();
            this.Cursor=Cursors.WaitCursor;
        if(m_sampleRun.TopoProcess())//通过 sampleRun.TopoProcess 实现拓扑处理
            {stateText.Text="拓扑处理成功完成";}
        else{stateText.Text="拓扑处理失败";}
            this.Cursor=Cursors.Arrow;}
        catch(Exception ex)
        {Trace.WriteLine(ex.Message);}}
```

③ 拓扑检查 Button 控件的 Click 事件。左键双击分析 Button 控件，即可添加该按钮的 Click 事件代码，实现拓扑检查：

```
    private void topoCheck_Click(object sender,EventArgs e)
        {try{stateText.Text="拓扑检查中,请等待…";
            statusBar.Update();
            this.Cursor=Cursors.WaitCursor;
            if(m_sampleRun.TopoCheck())//通过 sampleRun.TopoCheck 实现拓扑检查
            {stateText.Text="拓扑检查成功完成";}
            else{stateText.Text="拓扑检查失败";}
```

```
            this.Cursor=Cursors.Arrow;}
        catch(Exception ex)
            {Trace.WriteLine(ex.Message);}}
```

(6)窗体的 FormClose 事件

窗体的 FormClose 事件，添加代码同前文中的相应部分。

8. 运行结果

窗体设计完成后点击▶按钮，运行该程序，运行后弹出窗体，如图 7-10 所示。图 7-10 是经过拓扑构面、拓扑处理、拓扑检查的结果显示。

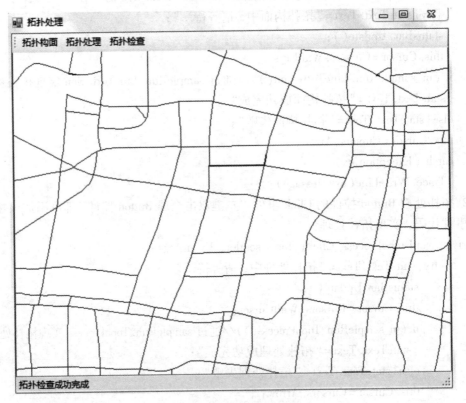

图 7-10　最近设施查找分析图

第 8 章 三维浏览

8.1 气　　泡

1. 数据

在安装目录\SampleData\Beijing\Beijing.smwu 路径下打开数据文件。

2. 新建文件夹和工程

在【D:\MyProject\】文件中新建一个文件夹，命名为【第 8 章　三维浏览】，并在其目录下建立一个文件夹，命名为【Bubble】，在此文件夹里新建一个文件夹，命名为 Bubble，在此文件夹里新建一个工程，将此工程命名为 Bubble。

3. 关键类型/成员

关键类型/成员见表 8-1。

表 8-1　　　　　　　　　　　　　关键类型/成员表

控件/类	方法	属性	事件
SceneControl		Scene、Bubbles	MouseDown、ObjectSelected、BubbleInitialize、BubbleResize、BubbleClose
Scene			BubbleChanged
Bubble		ClientWidth、ClientHeight、Title、TitleTextStyle、RoundQuality、IsAutoHide、FrameWidth、BackColor	
Bubbles	Add、Count、Clear		

4. 窗体控件属性

窗体控件属性具体见表 8-2。

表 8-2　　　　　　　　　　　　　窗体控件属性表

控件	Name	Text
GroupBox	GroupBox2	气泡设置
CheckBox	m_checkBoxAutoHide	自动隐藏

续表

控件	Name	Text
DomainUpDown	m_numericUpDownRoundQuality	
Label	m_labelTitleName	气泡名称
	m_labelFontName	标题字体
	m_labelFontColour	字体颜色
	m_labelAlignment	对齐方式
	m_labelFrameWidth	边框宽度
	m_labelBackColor	边框背景
	m_labelRoundQuality	圆角程度
	label2	设置气泡任一属性后，需要关闭当前气泡，选择模型对象，弹出新气泡，修改属性生效
Button	m_buttonFontColor	
	m_buttonBackColor	
	m_buttonWeather	天气预报气泡
	m_buttonModel	模型气泡
TextBox	m_textBoxTitleName	
	m_textBoxFrameWidth	
ComboBox	m_comboBoxAlignment	
	m_comboBoxFontName	

5. 窗体设计布局

窗体设计布局如图 8-1 所示。

图 8-1 窗体设计布局图

6. 创建 SampleRun 类

（1）添加 using 引用代码

添加 using 引用代码具体如下：

using System.Windows.Forms；//添加 System.Windows.Forms 命名空间，使其对应的类可以使用

using System.Diagnostics；//添加 System.Diagnostics 命名空间，使其对应的类可以使用

using SuperMap.Realspace；//添加 SuperMap.Realspace 命名空间，使其对应的类可以使用

using SuperMap.Data；//添加 SuperMap.Data 命名空间，使其对应的类可以使用

using SuperMap.UI；//添加 SuperMap.UI 命名空间，使其对应的类可以使用

using System.Drawing；//添加 System.Drawing 命名空间，使其对应的类可以使用

（2）在 public class SampleRun 中添加代码

在 public class SampleRun 中添加如下代码：

//定义类对象

private Workspace m_workspace；// Workspace：工作空间是用户的工作环境，主要完成数据的组织和管理，包括打开、关闭、创建、保存工作空间文件。

private SceneControl m_sceneControl；// SceneControl：三维地图控件。一个三维场景窗口（SceneControl）对应一个三维场景（Scene），即一个三维场景窗口中只能显示一个三维场景。

private Selection3D m_selection；// Selection3D：三维选择集类。

private DatasetVector m_datasetRegion；// DatasetVector：矢量数据集类。描述矢量数据集，并提供相应的管理和操作。对矢量数据集的操作主要包括数据查询、修改、删除、建立索引等。

private DatasetVector m_datasetCAD；

private BubbleControlModel m_bubbleControlModel；//：

private BubbleControlModel1 m_bubbleControlModel1；

private BubbleControlWeather m_bubbleControlWeather；

private Boolean m_flag=false；

List<Int32> m_smids；

String m_path=String.Empty；

private Bubble m_bubble；

public Bubble CurrentBubble

　　{get { return m_bubble；} }

private String m_bubbleTitle；

public String BubbleTitle

　　{get { return m_bubbleTitle；}

　set { m_bubbleTitle=value；} }

private Int32 m_frameWidth；

public Int32 BubbleFrameWidth

```csharp
        {get { return m_frameWidth;} set { m_frameWidth=value;}}
    private Color m_backColor;
    public Color BackColor
        {get { return m_backColor;}
        set { m_backColor=value;}}
    private Int32 m_roundQuality;
    public Int32 RoundQuality
        {get { return m_roundQuality;}
        set { m_roundQuality=value;}}
    private Boolean m_isHide;
    public Boolean IsHide
        {get { return m_isHide;}
        set { m_isHide=value;}}
    private TextStyle m_titleStyle;
    public TextStyle TitleStyle
        {get { return m_titleStyle;}
        set { m_titleStyle=value;}}
    /// 根据 workspace 和 sceneControl 构造 SampleRun 对象
    public SampleRun(Workspace workspace, SceneControl sceneControl)
    {try{m_workspace=workspace;
        m_sceneControl=sceneControl;
        m_sceneControl.Scene.Workspace=m_workspace;
        Initialize();}
        catch(Exception ex)
        {Trace.WriteLine(ex.Message);}}
    /// 打开需要的工作空间文件及场景
    private void Initialize()
        {try{
            WorkspaceConnectionInfo conInfo=new WorkspaceConnectionInfo(@"..\..\SampleData\Beijing\Beijing.smwu");//
WorkspaceConnectionInfo:工作空间连接信息类。包括了工作空间连接的所有信息,如所要连接的服务器名称、数据库名称、用户名、密码等。
            m_workspace.Open(conInfo);//打开工作空间
            m_sceneControl.Scene.Open("Beijing");//打开三维地图场景
            Layer3DDataset layer3DDatasetCAD = m_sceneControl.Scene.Layers[1] as Layer3DDataset;// Layer3DDataset:三维数据集图层类。
            Layer3DDataset layer3DDatasetRegion = m_sceneControl.Scene.Layers[3] as Layer3DDataset;
            layer3DDatasetRegion.IsSelectable=false;//设置三维图层中的对象是否可以选择
```

```
m_datasetCAD = layer3DDatasetCAD.Dataset as DatasetVector;
m_datasetRegion = layer3DDatasetRegion.Dataset as DatasetVector;
//创建自定义气泡并添加到 m_sceneControl 中
m_bubbleControlModel = new BubbleControlModel();
m_bubbleControlWeather = new BubbleControlWeather();//定义新对象
m_bubbleControlModel1 = new BubbleControlModel1();
m_sceneControl.Controls.Add(m_bubbleControlModel);//将指定的控件添加到控件集合中
m_sceneControl.Controls.Add(m_bubbleControlWeather);
m_sceneControl.Controls.Add(m_bubbleControlModel1);
m_bubbleControlWeather.Visible = false;//设置一个值，该值指示是否该控件及其所有子控件
m_bubbleControlModel.Visible = false;
m_bubbleControlModel1.Visible = false;
m_path = @"..\..\SampleData\Beijing\";
//将特殊模型的 ID 存储到 m_smids 中
m_smids = new List<Int32>();
m_smids.Add(155);//添加对象
m_smids.Add(156);
m_smids.Add(159);
m_smids.Add(160);
m_smids.Add(162);
m_smids.Add(206);
//调整 sceneControl 的状态
m_sceneControl.Action = Action3D.Pan;
FlyToWeather();
m_sceneControl.ObjectSelected += new ObjectSelectedEventHandler(m_sceneControl_ObjectSelected);
m_sceneControl.MouseDown += new MouseEventHandler(m_sceneControl_MouseDown);//定义新的对象
m_sceneControl.BubbleResize += new BubbleResizeEventHandler(m_sceneControl_BubbleResize);
m_sceneControl.BubbleClose += new BubbleCloseEventHandler(m_sceneControl_BubbleClose);
m_sceneControl.BubbleInitialize += new BubbleInitializeEventHandler(m_sceneControl_BubbleInitialize);
m_titleStyle = new TextStyle();}
catch (Exception ex)
{Trace.WriteLine(ex.Message);}}
```

```csharp
/// 关闭气泡时，触发的事件
private void m_sceneControl_BubbleClose(object sender, BubbleEventArgs e)
{//关闭气泡
    m_bubbleControlWeather.Visible=false; //设置一个值，该值指示是否该控件及其所有子控件
    m_bubbleControlModel.Visible=false;
    m_bubbleControlModel1.Visible=false;}
/// 气泡改变时，触发的事件
private void m_sceneControl_BubbleResize(object sender, BubbleEventArgs e)
{//选择不同的地物弹出不同的气泡
    System.Drawing.Point point=new Point(e.Bubble.ClientLeft, e.Bubble.ClientTop);
    if (m_selection!=null)
    {if ((m_selection.Layer as Layer3DDataset).Dataset.Type==DatasetType.Point)
        {m_bubbleControlWeather.Location=point;
         m_bubbleControlWeather.Visible=true;}
        else if ((m_selection.Layer as Layer3DDataset).Dataset.Type==DatasetType.CAD)
        {if (m_flag)
            {m_bubbleControlModel.Location=point;
             m_bubbleControlModel.Visible=true;}
            else
            {m_bubbleControlModel1.Location=point;
             m_bubbleControlModel1.Visible=true;}}}
/// 气泡初始化时触发的事件
private void m_sceneControl_BubbleInitialize(object sender, BubbleEventArgs e)
{//选择不同的地物弹出不同的气泡
    System.Drawing.Point point = new Point(e.Bubble.ClientLeft, e.Bubble.ClientTop);
    if (m_selection!=null)
    {if ((m_selection.Layer as Layer3DDataset).Dataset.Type==DatasetType.Point)
        {m_bubbleControlWeather.Location=point;
         m_bubbleControlWeather.Visible=true;}
        else if ((m_selection.Layer as Layer3DDataset).Dataset.Type==DatasetType.CAD)
        {if (m_flag)
            {m_bubbleControlModel.Location=point;
```

```
                    m_bubbleControlModel.Visible=true;}
                else{
                    m_bubbleControlModel1.Location=point;
                    m_bubbleControlModel1.Visible=true;}}}}
                /// 鼠标右键，销毁气泡和清除选择集
private void m_sceneControl_MouseDown(object sender,MouseEventArgs e)
    {try{if(e.Button==MouseButtons.Right)
        {if(m_selection.Count>0)
            {m_selection.Clear();}
         if(m_sceneControl.Bubbles.Count>0)
            {m_sceneControl.Bubbles.Clear();}}}
         catch
            {}}
```

/// 选中模型或点对象时弹出气泡
```
private void m_sceneControl_ObjectSelected(object sender,ObjectSelectedEventArgs e)
    {Recordset recordset=null;
        try
        {if(m_sceneControl.Bubbles.Count>0)
            {m_sceneControl.Bubbles.Clear();}
         if(m_sceneControl.Scene.FindSelection(true)[0]!=null)
            {m_selection=m_sceneControl.Scene.FindSelection(true)[0];
             recordset=m_selection.ToRecordset();
             Bubble bubble=null;
             Layer3DDataset layer3DDataset=m_selection.Layer as Layer3DDataset;
             if(layer3DDataset.Dataset.Type==DatasetType.Point)// Dataset.
```
Type：该枚举定义了数据集类型常量。数据集一般为存储在一起的相关数据的集合；根据数据类型的不同，分为矢量数据集、栅格数据集和影像数据集，以及为了处理特定问题而设计的如拓扑数据集、网络数据集等。根据要素的空间特征的不同，矢量数据集又分为点数据集、线数据集、面数据集、CAD 数据集、文本数据集、纯属性数据集等。Point：点数据集。
```
{FillBubbleControlWeather(recordset.GetFieldValue("Name").ToString());
                bubble=new Bubble();
                m_sceneControl.Bubbles.Add(bubble);
                    Point3D point3D=new Point3D(recordset.GetGeometry().Inner
Point.X,
                recordset.GetGeometry().InnerPoint.Y,0);
                // 设置气泡指向点
                bubble.Pointer=point3D;
                // 设置气泡标题
```

```
                    bubble.Title=m_bubbleTitle；
                    //设置气泡圆角的程度
                    bubble.RoundQuality=m_roundQuality；
                    //设置气泡是否自动隐藏
                    bubble.IsAutoHide=m_isHide；
                    //设置气泡边框宽度
                    bubble.FrameWidth=m_frameWidth；
                    //设置气泡边框背景
                    bubble.BackColor=m_backColor；
                    //设置气泡绘图区宽度
                    bubble.ClientWidth=m_bubbleControlWeather.Width；
                    //设置气泡绘图区高度
                    bubble.ClientHeight=m_bubbleControlWeather.Height；
                    //设置气泡标题文本风格
                    bubble.TitleTextStyle=m_titleStyle；}
              else if (layer3DDataset.Dataset.Type==DatasetType.CAD)
              {bubble=new Bubble()；
              m_sceneControl.Bubbles.Add(bubble)；
              if (m_smids.Contains((Int32)(recordset.GetFieldValue("SMID"))))
                    {FillBubbleControlModel(recordset)；//填充特殊模型气泡上的
信息
                    bubble.ClientWidth=m_bubbleControlModel.Width；
                    bubble.ClientHeight=m_bubbleControlModel.Height；
                    m_flag=true；}
              else
                    {FillBubbleControlModel1(recordset)；//填充特殊模型气泡上的
信息
                    bubble.ClientWidth=m_bubbleControlModel1.Width；
                    bubble.ClientHeight=m_bubbleControlModel1.Height；
                    m_flag=false；}
                    Geometry3D geometry3D=recordset.GetGeometry() as Geometry3D；
// Geometry3D:所有三维几何类的基类，提供了基本的三维几何类的属性和方法。通过本
类可以对三维几何对象的姿态进行控制，包括对象的位置、旋转角度、缩放比例和内点；还
可以对三维几何对象进行偏移；还可以获取三维模型几何对象。
                       Point3D point3D=geometry3D.BoundingBox.Center；// Point3D:
用于表示精度为Double 的三维点结构。
                    bubble.Pointer=point3D；
                    bubble.Title=m_bubbleTitle；
                    bubble.RoundQuality=m_roundQuality；
```

```
                    bubble.IsAutoHide=m_isHide;
                    bubble.FrameWidth=m_frameWidth;
                    bubble.BackColor=m_backColor;
                    bubble.TitleTextStyle=m_titleStyle;}}}
            catch(System.Exception ex)
            {Trace.WriteLine(ex.Message);}
            finally
            {if(recordset!=null)
                {recordset.Close();
                recordset.Dispose();}}}
///填充天气预报气泡上的信息
private void FillBubbleControlWeather(String name)
    {Recordset recordset=null;
        try
        {//查询指定名称的记录集
        recordset=m_datasetRegion.Query("Name='"+name+"'",CursorType.Static);
//填写气泡
                m_bubbleControlWeather.LabelName.Text=name;//获取名称
                    m_bubbleControlWeather.TodayWeather.Text=recordset.GetFieldValue
("TodayWeather") as String;//填写今天天气
                m_bubbleControlWeather.TodayWindLevel.Text=recordset.GetFieldValue
("TodayWindLevel") as String;//填写今天风级
                m_bubbleControlWeather.TomorrowWindLevel.Text=recordset.GetFieldValue
("TomorrowWindLevel") as String;//填写明天风级
                    m_bubbleControlWeather.AfterTomorrowWindLevel.Text=recordset.
GetFieldValue("AfterTomorrowWindLevel") as String;//填写后天风级
                    m_bubbleControlWeather.TodayTemperature.Text=recordset.GetFieldValue
("TodayTemperature") as String;//填写今天气温
                    m_bubbleControlWeather.TomorrowTemperature.Text=recordset.GetFieldValue
("TomorrowTemperature") as String;//填写明天气温
                    m_bubbleControlWeather.AfterTomorrowTemperature.Text=recordset.
GetFieldValue("AfterTomorrowTemperature") as String;//填写后天气温
                        SetBitmap(m_bubbleControlWeather.TodayPath1,recordset.GetFieldValue
("TodayPath1") as String);
                        SetBitmap(m_bubbleControlWeather.TodayPath2,recordset.GetFieldValue
("TodayPath2") as String);
                        SetBitmap(m_bubbleControlWeather.TomorrowPath1,recordset.GetFieldValue
("TomorrowPath1") as String);
                        SetBitmap(m_bubbleControlWeather.TomorrowPath2,recordset.GetFieldValue
```

```
("TomorrowPath2") as String);
            SetBitmap(m_bubbleControlWeather.AfterTomorrowPath1,recordset.GetFieldValue
("AfterTomorrowPath1") as String);
            SetBitmap(m_bubbleControlWeather.AfterTomorrowPath2,recordset.GetFieldValue
("AfterTomorrowPath2") as String);
                Int32 centerX = m_bubbleControlWeather.LabelName.Location.X + m_bubble
ControlWeather.LabelName.Width / 2;
                Int32 locationX = centerX - m_bubbleControlWeather.TodayWeather.Width / 2;
                Point location = new Point(locationX, m_bubbleControlWeather.TodayWeather.
Location.Y);
                m_bubbleControlWeather.TodayWeather.Location = location;}
            catch(System.Exception ex)
            {Trace.WriteLine(ex.Message);}
            finally
            {if(recordset ! = null)
            {recordset.Close();
             recordset.Dispose();}}}
/// 填充特殊模型气泡上的信息
private void FillBubbleControlModel(Recordset recordset)
    {// 存放特殊模型临近点的记录集
            Recordset recordsetNeighbor1 = null;
            Recordset recordsetNeighbor2 = null;
        try{// 获取选中模型对象所在的记录集中 Name 字段值,并赋值给特殊气泡
控件的 SelectModelName 属性
            m_bubbleControlModel.ModelName.Text = recordset.GetFieldValue("Name").
ToString();
            //获取选中模型对象所在的记录集中 Address 字段值,并赋值给特殊气泡控件
的 SelectedModelAddress 属性
                m_bubbleControlModel.ModelAddress.Text = recordset.GetFieldValue
("Address").ToString();
            //获取选中模型对象所在的记录集中 Introduction 字段值,并赋值给特殊气泡
控件的 ModelIntroduction 属性
                m_bubbleControlModel.ModelIntroduction.Text = recordset.GetFieldValue
("Introduction").ToString();
                //获取选中模型对象所在的记录集中 Path 字段值,并赋值给特殊气泡控件的
PicturePath 属性
            SetBitmap(m_bubbleControlModel.PicturePath, recordset.GetFieldValue
("Path").ToString());
            // 遍历 m_smids 数组
```

```
for(int i=0; i < m_smids.Count; i++)
    {// 获取 m_smids 数组中与当前记录中的 SMID 字字段值相等的值所在的索引位置
    if(recordset.GetFieldValue("SMID").ToString().Equals(m_smids[i].ToString()))
        {// 如果选中的模型在 m_smids 数组的起始位置,则临近点为 am_smids 数组中的第二个和最后一个点
        if(i==0)
            {recordsetNeighbor1 = m_datasetCAD.Query("SMID = "+m_smids[m_smids.Count - 1].ToString(), CursorType.Static);
            recordsetNeighbor2 = m_datasetCAD.Query("SMID = "+m_smids[i+1].ToString(), CursorType.Static);}
        // 如果选中的模型在 m_smids 数组的结束位置,则临近点为 am_smids 数组中的倒数第二个点和起始点
        else if(i==m_smids.Count - 1)
            {recordsetNeighbor1 = m_datasetCAD.Query("SMID = "+m_smids[i - 1].ToString(), CursorType.Static);
            recordsetNeighbor2 = m_datasetCAD.Query("SMID = "+m_smids[0].ToString(), CursorType.Static);}
        // 其他位置,则临近点为该位置的前一个点和后一个点
        else
            {recordsetNeighbor1 = m_datasetCAD.Query("SMID = "+m_smids[i - 1].ToString(), CursorType.Static);
            recordsetNeighbor2 = m_datasetCAD.Query("SMID = "+m_smids[i+1].ToString(), CursorType.Static);}
        // 获取第一个临近点所载及陆基的 Name 字段值,并赋值给特殊气泡空间的 NeighborModelName1 属性
        m_bubbleControlModel.ModelName1.Text = recordsetNeighbor1.GetFieldValue("Name").ToString();
        //获取第一个临近点所载及陆基的 Address 字段值,并赋值给特殊气泡空间的 NeighborModelName1 属性
        m_bubbleControlModel.ModelAddress1.Text = recordsetNeighbor1.GetFieldValue("Address").ToString();
        //获取第二个临近点所载及陆基的 Name 字段值,并赋值给特殊气泡空间的 NeighborModelName2 属性
        m_bubbleControlModel.ModelName2.Text = recordsetNeighbor2.GetFieldValue("Name").ToString();
        //获取第二个临近点所载及陆基的 Address 字段值,并赋值给特殊气泡空间的
```

NeighborModelName2 属性

```
            m_bubbleControlModel.ModelAddress2.Text = recordsetNeighbor2.GetFieldValue("Address").ToString();}}}
        catch(System.Exception ex)
        {Trace.WriteLine(ex.Message);}
        finally
        {if(recordsetNeighbor1! = null)
        {recordsetNeighbor1.Close();
        recordsetNeighbor1.Dispose();}
        if(recordsetNeighbor2! = null)
        {recordsetNeighbor2.Close();
        recordsetNeighbor2.Dispose();}}}
/// 填充普通模型气泡上的信息
private void FillBubbleControlModel1(Recordset recordset)
    {try
        {String name = String.Empty;
        Object obj = recordset.GetFieldValue("Name");
        if(obj! = null)
        {name = obj.ToString();}
        else
        {name = "暂无信息";}
        m_bubbleControlModel1.ModelName.Text = name;}
        catch(System.Exception e)
        {Trace.WriteLine(e.Message);}}
/// 将图片的大小旋转到与气泡控件的大小一致,保证能正常显示图片
private void SetBitmap(Label label, String path)
    {Bitmap bitmap = new Bitmap(m_path+path);//定义新对象
    Bitmap bitmap1 = new Bitmap(bitmap, label.Size);
    label.Image = bitmap1;}
/// 飞到模型图层
public void FlyToModel()
    {Layer3DDataset layer3DDataset = m_sceneControl.Scene.Layers[1] as Layer3DDataset;
        Camera camera = new Camera();//Camera:相机类。该对象用照相机的道理来模拟观察者通过视锥来观察场景内的三维物体。
        camera.Longitude = 116.383863205411;//设置相机的经度
        camera.Latitude = 40.007987058504071;//设置相机的纬度
        camera.Altitude = 502.11111768707633;//设置相机的高度
```

```csharp
            camera.Heading = 347.15071992962771;//设置相机的方位角
            camera.Tilt = 80.536373338903346;//设置相机的仰角
            m_sceneControl.Scene.Fly(camera, 5);
            DisplayBubble(false);//切换图层时,弹出一个默认气泡}
///飞到面图层
public void FlyToWeather()
    {Layer3DDataset layer3DDataset = m_sceneControl.Scene.Layers[0] as Layer3DDataset;
        m_sceneControl.Scene.EnsureVisible(layer3DDataset.Bounds, 5);
        DisplayBubble(true);}
///切换图层时,弹出一个默认的气泡
private void DisplayBubble(Boolean enable)
    {Recordset recordset = null;
        try{
            if (m_sceneControl.Bubbles.Count > 0)
            {m_sceneControl.Bubbles.Clear();}
            if (m_selection != null)
            {m_selection.Clear();}
            //默认的天气气泡
            Bubble bubble = null;
            if (enable)
            {(m_sceneControl.Scene.Layers[0] as Layer3DDataset).Selection.Add(2);
                m_selection = m_sceneControl.Scene.FindSelection(true)[0];
                recordset = m_selection.ToRecordset();
                Layer3DDataset layer3DDataset = m_selection.Layer as Layer3DDataset;
                if (layer3DDataset.Dataset.Type == DatasetType.Point)
                {String name = recordset.GetFieldValue("Name") as String;
                    FillBubbleControlWeather(name);
                    bubble = new Bubble();
                    m_sceneControl.Bubbles.Add(bubble);
                    Point3D point3D = new Point3D(recordset.GetGeometry().InnerPoint.X,
                    recordset.GetGeometry().InnerPoint.Y, 0);
                    bubble.Pointer = point3D;
                    bubble.Title = "天气预报";//设置标题
                    bubble.RoundQuality = 1;
                    bubble.IsAutoHide = true;
                    bubble.FrameWidth = 10;
                    bubble.BackColor = Color.FromArgb(100, 185, 230, 253);//设置背景色
```

```
            bubble.ClientWidth = m_bubbleControlWeather.Width;//设置宽度
            bubble.ClientHeight = m_bubbleControlWeather.Height;//设置高度
            TextStyle textStyle = new TextStyle();//设置字体风格
            textStyle.Alignment = TextAlignment.MiddleCenter;
            textStyle.FontName = "宋体";
            textStyle.ForeColor = Color.Blue;//设置前景色
            bubble.TitleTextStyle = textStyle;//设置主题风格}}
        //默认的模型气泡
        else
        {   (m_sceneControl.Scene.Layers[1] as Layer3DDataset).Selection.Add(206);
            m_selection = m_sceneControl.Scene.FindSelection(true)[0];
            recordset = m_selection.ToRecordset();
            bubble = new Bubble();
            m_sceneControl.Bubbles.Add(bubble);
            FillBubbleControlModel(recordset);//填充气泡信息
            bubble.ClientWidth = m_bubbleControlModel.Width;//设置宽度
            bubble.ClientHeight = m_bubbleControlModel.Height;//设置高度
            m_flag = true;
            Geometry3D geometry3D = recordset.GetGeometry() as Geometry3D;
            Point3D point3D = geometry3D.BoundingBox.Center;
            bubble.Pointer = point3D;}
            m_bubble = bubble;}
        catch(System.Exception ex)
        {Trace.WriteLine(ex.Message);}
        finally
        {if(recordset! = null)
        {recordset.Close();
        recordset.Dispose();}}}
```

7. 创建用户控件 BubbleControlModel

创建用户控件,并命名为 BubbleControlModel。

(1) BubbleControlModel 的控件

BubbleControlModel 的控件见表 8-3 和表 8-4。

表 8-3 **BubbleControlModel 控件表 1**

控件	Name	Text
TapControl		位置
		简介

表8-4　　　　　　　　　　　　**BubbleControlModel 控件表2**

Text	Label	Toolstrip	TextBox
位置	m_labelModelName(北京国家体育馆)	Toolstrip1	
	label3(地址:)		
	toolStripLabel1(邻近地点)		
	m_labelModelName1(北京国家体育场)		
	label1(地址:)		
	m_labelModelAddress1(北京市朝阳区安翔路1号)		
	label2(label2)		
	label2(地址:)		
	Label5(m_labelModelAddress2)		
	m_labelAddress(北京市朝阳区安翔路1号)		
简介	m_toolStripLabelIntroduction(简介)	Toolstrip1	m_textBoxIntroduction

(2) BubbleControlModel 的窗体分布

BubbleControlModel 的窗体分布如图8-2所示。

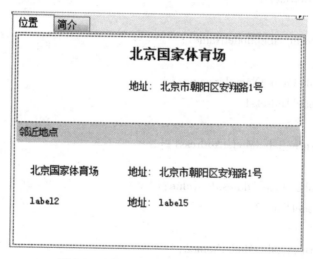

图8-2　BubbleControlModel 窗体分布图

(3) 在 public partial class BubbleControlModel 中添加代码

在 public partial class BubbleControlModel 中添加如下代码:
private void m_tabControl_SelectedIndexChanged(object sender, EventArgs e)
　　{m_tabControl.TabPages[m_tabControl.SelectedIndex].Controls.Add(m_panel);}
public Label PicturePath
　　{get

```csharp
            { return m_labelPicture; }
            set
            { m_labelPicture = value; } }
    public Label ModelName
        { get
            { return m_labelModelName; }
            Set
            { m_labelModelName = value; } }
    public Label ModelAddress
        { get
            { return m_labelAddress; }
            set
            { m_labelAddress = value; } }
    public TextBox ModelIntroduction
        { get
            { return m_textBoxIntroduction; }
            set
            { m_textBoxIntroduction = value; } }
    public Label ModelName1
        { get
            { return m_labelModelName1; }
            set
            { m_labelModelName1 = value; } }
    public Label ModelAddress1
        { get
            { return m_labelModelAddress1; }
            set
            { m_labelModelAddress1 = value; } }
    public Label ModelName2
        { get
            { return m_labelModelName2; }
            set
            { m_labelModelName2 = value; } }
    public Label ModelAddress2
        { get
            { return m_labelModelAddress2; }
            set
            { m_labelModelAddress2 = value; } }
```

8. 创建用户控件 BubbleControlModel1

创建用户控件,并命名为 BubbleControlModel1。

(1) BubbleControlModel1 的控件

BubbleControlModel1 的控件见表 8-5。

表 8-5　　　　　　　　　　**BubbleControlModel 1 控件表**

控件	Name	Text
Label	m_labelModelName	朝阳区大屯街道办为民服务大厅

(2) BubbleControlModel1 的窗体分布

BubbleControlModel1 的窗体分布如图 8-3 所示。

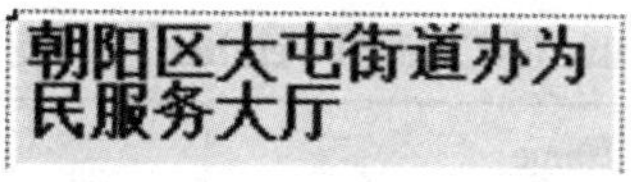

图 8-3　BubbleControlModel1 窗体分布图

(3) 添加代码

① 在 private void InitializeComponent() 之前添加如下代码:

/// 必需的设计器变量

private System. ComponentModel. IContainer components = null;

/// 清理所有正在使用的资源

/// 如果应释放托管资源,为 true;否则为 false

protected override void Dispose(bool disposing)

　　　　{if (disposing && (components ! = null))

　　　　{components. Dispose();}

　　　　base. Dispose(disposing);}

② 在 private void InitializeComponent() 中添加如下代码:

private void InitializeComponent()

　　　{this. m_labelModelName = new System. Windows. Forms. Label();

　　　　　this. SuspendLayout();

　　　　　// m_labelModelName

　　　　　this. m_labelModelName. Dock = System. Windows. Forms. DockStyle. Fill;

　　　　　this. m _ labelModelName. Font = new System. Drawing. Font ("宋体", 12F, System. Drawing. FontStyle. Bold);

　　　　　this. m_labelModelName. Location = new System. Drawing. Point(0, 0);

　　　　　this. m_labelModelName. Name = "m_labelModelName";

　　　　　this. m_labelModelName. Size = new System. Drawing. Size(161, 39);

　　　　　this. m_labelModelName. TabIndex = 0;

```
this.m_labelModelName.Text = "朝阳区大屯街道办为民服务大厅";
// BubbleControlModel1
this.AutoScaleDimensions = new System.Drawing.SizeF(6F, 12F);
this.AutoScaleMode = System.Windows.Forms.AutoScaleMode.Font;
this.Controls.Add(this.m_labelModelName);
this.Name = "BubbleControlModel1";
this.Size = new System.Drawing.Size(161, 39);
this.ResumeLayout(false);}
```

9. 创建用户控件 BubbleControlWeather

创建用户控件,并命名为 BubbleControlWeather。

(1) BubbleControlWeather 的控件。

BubbleControlWeather 的控件具体见表 8-6。

表 8-6 **BubbleControlWeather 控件表**

控件	Name	Text
Label	m_labelName	北京
	m_labelDatatime	2014 年 4 月 21 日星期一
	m_labelWind	风力风向:
	m_labelWindLevelToday	南风 3—4 级转西北风 4—4 级
	m_labelUltraviolet	紫外线指数:
	m_labelUltravioletLevel	弱
	m_labelDegreeTomorrow	7℃ ~ 13℃
	m_labelDegreeAfterTomorrow	7℃ ~ 13℃
	m_labelWindLevelTomorrow	微风
	m_labelWindLevelAfterTomorrow	微风
	m_labelWeatherToday	晴
	m_labelDegreeToday	7℃ ~ 13℃
	m_labelTomorrow	明天
	m_labelAfterTomorrow	后天

(2) BubbleControlModel1 的窗体分布

BubbleControlModel1 的窗体分布如图 8-4 所示。

(3) 在 public BubbleControlWeather() 后面添加代码

在 public BubbleControlWeather() 后面添加如下代码:

```
private void m_linkLabel_LinkClicked(object sender, LinkLabelLinkClickedEventArgs e)
    {System.Diagnostics.Process.Start("http://www.weather.com.cn/weather/101010100.shtml");}
```

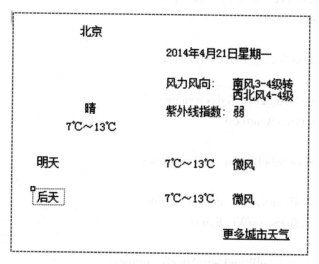

图 8-4 BubbleControlWeather 窗体分布图

```
public Label LabelName
    {get
        {return m_labelName;}
    set{
        m_labelName = value;}}
public Label TodayTemperature
    {get
        {return m_labelDegreeToday;}
    set
        {m_labelDegreeToday = value;}}
public Label TomorrowTemperature
    {get
        {return m_labelDegreeTomorrow;}
    set{
        m_labelDegreeTomorrow = value;}}
public Label AfterTomorrowTemperature
    {get
        {return m_labelDegreeAfterTomorrow;}
    set
        {m_labelDegreeAfterTomorrow = value;}}
public Label TodayWeather
    {get
        {return m_labelWeatherToday;}
    set
```

```csharp
            { m_labelWeatherToday = value; } }
    public Label TodayWindLevel
        { get
            { return m_labelWindLevelToday; }
        set
            { m_labelWindLevelToday = value; } }
    public Label TomorrowWindLevel
        { get
            { return m_labelWindLevelTomorrow; }
        set
            { m_labelWindLevelTomorrow = value; } }
    public Label AfterTomorrowWindLevel
        { get
            { return m_labelWindLevelAfterTomorrow; }
        set
            { m_labelWindLevelAfterTomorrow = value; } }
    public Label TodayPath1
        { get
            { return m_labelIconTodayDay; }
        set
            { m_labelIconTodayDay = value; } }
    public Label TodayPath2
        { get
            { return m_labelIconTodayNight; }
        set
            { m_labelIconTodayNight = value; } }
    public Label TomorrowPath1
        { get
            { return m_labelIconTomorrowDay; }
        set
            { m_labelIconTomorrowDay = value; } }
    public Label TomorrowPath2
        { get
            { return m_labelIconTomorrowNight; }
        set
            { m_labelIconTomorrowNight = value; } }
    public Label AfterTomorrowPath1
        { get
            { return m_labelIconAfterTomorrowDay; }
        set
```

```
        {m_labelIconAfterTomorrowDay = value;}}
public Label AfterTomorrowPath2
        {get
            {return m_labelIconAfterTomorrowNight;}
        set
            {m_labelIconAfterTomorrowNight = value;}}
```

10. 主窗体代码

(1) 添加 using 引用代码

添加 using 引用代码具体如下：

using SuperMap.Data; //添加 SuperMap.Data 命名空间，使其对应的类可以使用

using SuperMap.Realspace; //添加 SuperMap.Realspace 命名空间，使其对应的类可以使用

using SuperMap.UI; //添加 SuperMap.UI 命名空间，使其对应的类可以使用

(2) 在 public FormMain() 函数前添加代码

在 public FormMain() 函数前添加如下代码：

```
//定义类对象
private SceneControl m_sceneControl;
private Workspace m_workspace;
private SampleRun m_sampleRun;
private TextStyle m_titleStyle;
```

(3) 窗体的 Load 事件

窗体的 Load 事件，在窗体的属性窗口单击 图标，左键双击 Load，添加如下代码：

```
private void MainForm_Load(object sender, EventArgs e)
    {this.m_workspace = new SuperMap.Data.Workspace(this.components);//定义新的对象
        m_sceneControl = new SuperMap.UI.SceneControl();
        m_sceneControl.Dock = DockStyle.Fill;//指定控件的停靠位置及方式
        splitContainer1.Panel2.Controls.Add(m_sceneControl);//添加指定控件
        m_titleStyle = new TextStyle();
        m_sampleRun = new SampleRun(m_workspace, m_sceneControl);
        InitControls();}
```

(4) 初始化气泡设置区域的控件

初始化气泡设置区域的控件，需添加如下代码：

```
private void InitControls()
    {Bubble bubble = m_sampleRun.CurrentBubble;
        m_textBoxFrameWidth.Text = bubble.FrameWidth.ToString();//设置文本
        m_textBoxTitleName.Text = bubble.Title;//设置主题名称
        m_numericUpDownRoundQuality.Value = bubble.RoundQuality;
        m_checkBoxAutoHide.Checked = bubble.IsAutoHide;
        m_buttonBackColor.BackColor = bubble.BackColor;//设置背景颜色
```

```
m_sampleRun.TitleStyle = m_titleStyle;//设置主题风格
m_sampleRun.BubbleTitle = m_textBoxTitleName.Text;//设置气泡主题
m_sampleRun.RoundQuality = (Int32)m_numericUpDownRoundQuality.Value;
m_sampleRun.BubbleFrameWidth = Convert.ToInt32(m_textBoxFrameWidth.Text);//设置宽度
m_sampleRun.BackColor = m_buttonBackColor.BackColor;//设置背景色
if(! m_textBoxTitleName.Text.Equals(String.Empty))
    {m_comboBoxAlignment.SelectedIndex = GetTitleAlignment(bubble.TitleTextStyle.Alignment);
     m_comboBoxFontName.SelectedItem = bubble.TitleTextStyle.FontName;//设置名称
     m_buttonFontColor.BackColor = bubble.TitleTextStyle.ForeColor;}//设置习题颜色
else
{m_comboBoxAlignment.Text = "";
m_comboBoxFontName.Text = "";
m_buttonFontColor.BackColor = SystemColors.Control;}
m_titleStyle.ForeColor = m_buttonFontColor.BackColor;}
```

(5) 获取气泡标题的对齐方式

添加如下代码，可获取气泡标题的对齐方式：

```
private Int32 GetTitleAlignment(TextAlignment textAlignment)
   {String alignment = textAlignment.ToString();
    Int32 index = 0;
    if(alignment.Contains("Left"))
    {index = 0;}
    else if(alignment.Contains("Center"))
    {index = 1;}
    else if(alignment.Contains("Right"))
    {index = 2;}
    return index;}
```

(6) 其他控件的代码

① m_checkBoxAutoHide 的 CheckedChanged 事件。在控件的属性窗口单击 图标，左键双击 CheckedChanged，添加如下代码，实现自动隐藏：

```
private void m_checkBoxAutoHide_CheckedChanged(object sender, EventArgs e)
    {m_sampleRun.IsHide = m_checkBoxAutoHide.Checked;}//通过 sampleRun.IsHide 实现自动隐藏
```

② m_textBoxTitleName 的 TextChanged 事件。在控件的属性窗口单击 图标，左键双击 TextChanged，添加如下代码，设置气泡标题名称：

```
private void m_textBoxTitleName_TextChanged(object sender, EventArgs e)
    {m_sampleRun.BubbleTitle = m_textBoxTitleName.Text;//通过 sampleRun.
```

BubbleTitle 设置气泡标题名称

```
            if(m_textBoxTitleName.Text.Equals(String.Empty))
            {m_comboBoxAlignment.Text="";
            m_comboBoxFontName.Text="";
            m_buttonFontColor.BackColor=SystemColors.Control;}}
```

③ m_textBoxFrameWidth 的 TextChanged 事件。在控件的属性窗口单击 ✶ 图标，左键双击 TextChanged，添加如下代码，设置气泡边框宽度：

```
private void m_textBoxFrameWidth_TextChanged(object sender, EventArgs e)
    {m_sampleRun.BubbleFrameWidth=Convert.ToInt32(m_textBoxFrameWidth.Text);
}//通过 sampleRun.BubbleFrameWidth 设置气泡边框宽度
```

④ m_numericUpDownRoundQuality 的 ValueChanged 事件。在控件的属性窗口单击 ✶ 图标，左键双击 ValueChanged，添加如下代码，设置气泡圆角程度：

```
private void m_numericUpDownRoundQuality_ValueChanged(object sender, EventArgs e)
    {m_sampleRun.RoundQuality=(Int32)m_numericUpDownRoundQuality.Value;}//
通过 sampleRun.RoundQuality 设置气泡圆角程度
```

⑤ m_buttonWeather 的 Click 事件。在控件的属性窗口单击 ✶ 图标，左键双击 Click，添加如下代码，设置天气预报：

```
private void m_buttonWeather_Click(object sender, EventArgs e)
    {m_sampleRun.FlyToWeather();
        InitControls();}
```

⑥ m_buttonModel 的 Click 事件。在控件的属性窗口单击 ✶ 图标，左键双击 Click，添加如下代码，显示模型气泡：

```
private void m_buttonModel_Click(object sender, EventArgs e)
    {m_sampleRun.FlyToModel();//通过 sampleRun.FlyToModel 显示模型气泡
        InitControls();}
```

⑦ m_comboBoxFontName 的 SelectedIndexChanged 事件。在控件的属性窗口单击 ✶ 图标，左键双击 SelectedIndexChanged，添加如下代码，设置标题字体：

```
private void m_comboBoxFontName_SelectedIndexChanged(object sender, EventArgs e)
    {m_titleStyle.FontName=m_comboBoxFontName.SelectedItem.ToString();}//通过
titleStyle.FontName 设置标题字体
```

⑧ m_comboBoxAlignment 的 SelectedIndexChanged 事件。在控件的属性窗口单击 ✶ 图标，左键双击 SelectedIndexChanged，添加如下代码，设置气泡对齐方式：

```
private void m_comboBoxAlignment_SelectedIndexChanged(object sender, EventArgs e)
    {if(m_comboBoxAlignment.SelectedIndex==0)
        {m_titleStyle.Alignment=TextAlignment.MiddleLeft;}
    else if(m_comboBoxAlignment.SelectedIndex==1)
        {m_titleStyle.Alignment=TextAlignment.MiddleCenter;}
    else if(m_comboBoxAlignment.SelectedIndex==2)
        {m_titleStyle.Alignment=TextAlignment.MiddleRight;}}
```

⑨ m_buttonFontColor 的 Click 事件。在控件的属性窗口单击 ✶ 图标，左键双击 Click，

添加如下代码，设置气体颜色：
```
private void m_buttonFontColor_Click(object sender, EventArgs e)
{ColorDialog colorDialog=new ColorDialog();
    if(colorDialog.ShowDialog()==DialogResult.OK)
    {m_buttonFontColor.BackColor=colorDialog.Color;}
    m_titleStyle.ForeColor=m_buttonFontColor.BackColor;}
```

⑩ m_buttonFrameColor 的 Click 事件。在控件的属性窗口单击 图标，左键双击 Click，添加如下代码，设置边框颜色：
```
private void m_buttonFrameColor_Click(object sender, EventArgs e)
{ColorDialog colorDialog=new ColorDialog();
    if(colorDialog.ShowDialog()==DialogResult.OK)
    {m_buttonBackColor.BackColor=colorDialog.Color;}
    m_sampleRun.BackColor=m_buttonBackColor.BackColor;}
```

(7) 窗体的 FormClose 事件

在退出程序时，要断开空间的关联，关闭地图窗口、工作空间。具体代码同前文中的相应部分。

11. 运行结果

窗体设计完成后点击 按钮，运行该程序，运行后弹出窗体，如图 8-5 所示。

图 8-5 气泡

8.2 飞行管理

1. 数据

分别在安装目录\OlympicGreenSCV\OlympicGreen.sxwu 和安装目录\SampleData\OlympicGreen\flyroute.fpf 路径下打开数据文件。

2. 新建文件夹和工程

在【D:\MyProject\】文件中新建一个文件夹,命名为【第8章 三维浏览】,并在其目录下建立一个文件夹,命名为【FlyManager】,在此文件夹里新建一个文件夹,命名为 FlyManager,在此文件夹里新建一个工程,将此工程命名为 FlyManager。

3. 关键类型/成员

关键类型/成员见表 8-7。

表 8-7 关键类型/成员表

控件/类	方法	属性	事件
SceneControl		Scene、Action	MouseEvent
GeoCardinal	ConvertToLine		
FlyManager	Play、Pause、Stop	Scene、Routes、PlayRate	
Routes	FromFile、ToFile、Add、Remove	CurrentRoute	
Route	FromGeoLine3D	IsFlyAlongTheRoute、IsHeadingFixed、IsAltitudeFixed、IsLinesVisible、IsStopsVisible	

4. 窗体控件属性

窗体控件属性具体见表 8-8。

表 8-8 窗体控件属性表

控件	Name	Text
GroupBox	m_groupBoxSetting	飞行路线设置
	m_groupBoxFlyControl	飞行控制
Label	m_labelLongitude	经度:
	m_labelLatitude	纬度:
	m_labelCurrentStopHeight	高程:
	label1	速度:
	label2	m/s

续表

控件	Name	Text
Button	m_buttonLoadRoute	加载默认飞行路线
	m_buttonCustomRoute	自定义飞行路线
	m_buttonSaveRoute	保存飞行路线
	m_buttonClearRoute	清除飞行路线
	m_buttonFly	飞行
	m_buttonPause	暂停
	m_buttonStop	停止
	m_buttonFaster	+
	m_buttonSlower	-
Panel	m_panel	
	m_splitContainer.Panel2	
splitContainer	splitContainer	

5. 窗体设计布局

窗体设计布局如图 8-6 所示。

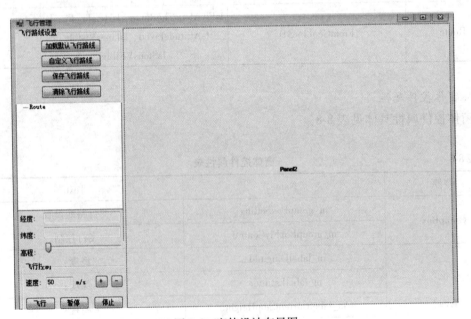

图 8-6　窗体设计布局图

6. 创建 SampleRun 类

(1) 添加 using 引用代码

添加 using 引用代码具体如下：
using System.Diagnostics;　//添加 System.Diagnostics 命名空间，使其对应的类可以使用
using System.Windows.Forms;　//添加 System.Windows.Forms 命名空间，使其对应的类可以使用
using SuperMap.Realspace;　//添加 SuperMap.Realspace 命名空间，使其对应的类可以使用
using SuperMap.UI;　//添加 SuperMap.UI 命名空间，使其对应的类可以使用
using SuperMap.Data;　//添加 SuperMap.Data 命名空间，使其对应的类可以使用

（2）在 public class SampleRun 中添加代码

在 public class SampleRun 中添加如下代码：

```
//定义类函数
private SceneControl m_sceneControl;// SceneControl:三维地图控件。一个三维场景窗口(SceneControl)对应一个三维场景(Scene)，即一个三维场景窗口中只能显示一个三维场景。
private Point3Ds m_point3Ds;// Point3Ds:用于表示精度为 Double 的三维点对象集合。
private Point3Ds m_point3DsAll;
private TreeView m_treeView;
private RouteStops m_routeStops;// RouteStops:站点集合类。该类提供了对站点对象的添加、移除、导入、导出等管理功能。
//用于存取当前显示的坐标点在 m-point 中的位置
private Int32 m_index=-1;
private FlyManager m_flyManager;// FlyManager:三维场景的飞行管理类。该类提供飞行过程中的停止、暂停、继续中断飞行等方法。
private String m_pointName;
private GeoLine3D m_geoLine3D;// GeoLine3D:三维线几何对象类。该类继承自 Geometry3D 类。三维线几何对象是由三维点串构成的线对象，这些点可以不在同一平面上。
private Workspace m_workspace;// Workspace:工作空间是用户的工作环境，主要完成数据的组织和管理，包括打开、关闭、创建、保存工作空间文件。
private Boolean m_customButtonEnable=true;
///获取绘制站点
public Point3Ds Point3Ds
    {get
        {return m_point3Ds;}}
///获取飞行管理器
public FlyManager FlyManager
    {get
        {return m_flyManager;}}
/// 获取飞行路线上所有点，包括插值点和绘制站点
```

```csharp
public Point3Ds Point3DsAll
{get
    {return m_point3DsAll;}}
/// 获取或设置站点的索引值
public Int32 Index
    {get
        {return m_index;}
    set
        {m_index = value;}}
/// 根据 workspace 和 map 构造 SampleRun 对象
public SampleRun(SceneControl sceneControl, TreeView treeView)
    {try
        {m_sceneControl = sceneControl;
        m_treeView = treeView;
        m_workspace = new Workspace();
        m_flyManager = m_sceneControl.Scene.FlyManager;
        m_point3Ds = new Point3Ds();
        m_point3DsAll = new Point3Ds();
        m_geoLine3D = new GeoLine3D();
        m_geoLine3D.Style3D = GetLineGeoStyle3D();
        m_pointName = "Point";
        Initialize();}
    catch (Exception ex)
    {Trace.Write(ex.Message);}}
private void Initialize()
    {try
        {AddData();}
    catch (Exception ex)
    {Trace.Write(ex.Message);}}
/// 加载数据
private void AddData()
    {try
        {WorkspaceConnectionInfo conInfo = new WorkspaceConnectionInfo(@"..\..\SampleData\OlympicGreenSCV\OlympicGreen.sxwu");
        onInfo.Type = WorkspaceType.SXWU;
        m_workspace.Open(conInfo);//打开工作空间
        m_sceneControl.Scene.Workspace = m_workspace;
        m_sceneControl.Scene.Open("OlympicGreen");//打开三维地图场景
        for (int i = 0; i < m_sceneControl.Scene.Layers.Count; i++)
```

```
                {Layer3D layer = m_sceneControl.Scene.Layers[i];
                layer.IsSelectable = false;}//设置三维图层中的对象是否可以选择
                Layer3D olympicGreenLayer = m_sceneControl.Scene.Layers[0];//获取 3D 图
层集合对象
                m_sceneControl.Scene.EnsureVisible(olympicGreenLayer.Bounds, 10);}//根
据指定的经纬度范围显示场景
            catch
            {}}
    ///   鼠标移动时绘制飞行路线
    private void m_sceneControl_MouseMove(object sender, System.Windows.Forms.
MouseEventArgs e)
        {if (m_customButtonEnable && m_point3Ds.Count > 0
            && !Double.IsNaN(m_sceneControl.Scene.PixelToGlobe(e.Location).X))
            {Point2Ds point2Ds = m_point3Ds.ToPoint2Ds();//Point2Ds:点集合对象。此
类管理线对象或线对象的子对象上的所有节点。由于线对象或线对象的子对象都是有向
的,所以其点集合对象为有序的点的集合。
            Point3D point3D = m_sceneControl.Scene.PixelToGlobe(e.Location);//
Point3D:用于表示精度为 Double 的三维点结构。将屏幕点对象转换成三维点对象,该三维
点对象包含了经纬度和海拔高度信息。
            point3D.Z = m_sceneControl.Scene.GetAltitude(point3D.X, point3D.Y);//获
取地面上某点的海拔高度
            point2Ds.Add(new Point2D(point3D.X, point3D.Y));//将指定的点对象添
加到点集合对象的末尾
            SetGeoLine3DToTrackingLayer(point2Ds, 20);}}
    /// 鼠标点击时绘制飞行路线经过的站点
    private void m_sceneControl_MouseDown(object sender, System.Windows.Forms.
MouseEventArgs e)
        {// 鼠标左键
        if (e.Button == System.Windows.Forms.MouseButtons.Left
            && !Double.IsNaN(m_sceneControl.Scene.PixelToGlobe(e.Location).X))
            {TreeNode treeNode = new TreeNode();
            m_treeView.Nodes[0].Nodes.Add(treeNode);//将先前创建的树节点添加到
树节点的末尾
                if (treeNode.Index > 0)
                {treeNode.Text = GetStopName(treeNode.PrevNode.Text);}//获取站点名称
                else
                {treeNode.Text = "Stop1";}
            m_treeView.Nodes[0].ExpandAll();//展开所有子树节点
            Point3D point3D = m_sceneControl.Scene.PixelToGlobe(e.Location);//将屏幕
```

点对象转换成三维点对象，该三维点对象包含了经纬度和海拔高度信息

  point3D. Z = m_sceneControl. Scene. GetAltitude(point3D. X, point3D. Y) + 30;
//获取地面上的某点的海拔高度

  m_point3Ds. Add(point3D); //向集合中添加三维点对象,添加成功,返回被添加对象的序号

  GeoPoint3D geoPoint3D = new GeoPoint3D(point3D);

  geoPoint3D. Style3D = GetPointGeoStyle3D(geoPoint3D. Z); //获取绘制到场景跟踪图层上的站点的样式

  geoPoint3D. Z = 0; //设置三维点集合对象的 Z 坐标,单位为米

  m _ sceneControl. Scene. TrackingLayer. Add (geoPoint3D, m _ pointName + treeNode. Index. ToString( ));}//该方法用于向三维跟踪图层中添加一个几何对象,并给出标签信息

  // 鼠标右键,结束绘制,将实时生成的路线加到 FlyManager

  else if (e. Button == System. Windows. Forms. MouseButtons. Right) //按下鼠标右键

  {if (m_point3Ds. Count > 1)

  {SetGeoLine3DToTrackingLayer(m_point3Ds. ToPoint2Ds( ), 20);

  m_flyManager. Routes. Add(GetRoute( ));}//向路线集合中添加一个指定的路线对象

  else

  {ResumeDefault( );}//重置数据

  RegisterEvents(false); }}//注销事件

  ///打开默认飞行数据并将各个站点显示到路线树

  internal void OpenData( )

  {m_flyManager. Routes. Clear( );//清除路线集合中所有路线

  m_flyManager. Routes. FromFile(
"../../SampleData/OlympicGreen/flyroute. fpf");//从指定的文件中导入路线几何对象

  m_flyManager. Scene = m_sceneControl. Scene; //获取三维地图场景

  for (Int32 i = 0; i < m_flyManager. Routes[0]. Stops. Count; i++)

  {TreeNode treeNode = new TreeNode( );

  m_treeView. Nodes[0]. Nodes. Add(treeNode); //将先前创建的树节点添加到树节点集合的末尾

  treeNode. Text = m_flyManager. Routes[0]. Stops[i]. Name; } //获取路线站点的名称

  m_treeView. Nodes[0]. ExpandAll( );}//展开所有树节点

  /// 飞行

  internal void Fly( )

  {if (m_flyManager ! = null && m_flyManager. Routes. CurrentRoute ! = null)

```
        {m_flyManager.Play();}}//按照返回的路线集合开始飞行
/// 暂停
        internal void Pause()
        {if(m_flyManager! =null)
        {m_flyManager.Pause();}}//暂停当前飞行
/// 停止
        internal void Stop()
        {if(m_flyManager! =null)
        {m_flyManager.Stop();}}//停止当前飞行
            /// 注册或注销 m_sceneControl 的事件
    public void RegisterEvents(Boolean register)
        {if(register)
            {m _ sceneControl. MouseDown + = new System. Windows. Forms. Mouse
EventHandler(m_sceneControl_MouseDown);
             m _ sceneControl. MouseMove + = new System. Windows. Forms. Mouse
EventHandler(m_sceneControl_MouseMove);
            Route route=new Route();//Route:路线对象类。该类用于对飞行路线进行
设置,同时还提供了路线对象与XML字符串对象、路线对象与 GeoLine3D 对象之间的转
换。
            m_sceneControl. Action=Action3D. Null;}// Action3D:该枚举定义了三维地
图的操作状态类型常量。设置三维地图操作状态
            else
                {m _ sceneControl. MouseDown - = new System. Windows. Forms. MouseEvent
Handler(m_sceneControl_MouseDown);
             m _ sceneControl. MouseMove - = new System. Windows. Forms. MouseEvent
Handler(m_sceneControl_MouseMove);
            m_sceneControl. Action=Action3D. Pan;}}//设置三维地图操作状态 Pan:平
移,包括绕球旋转。注意:如果操作中拖动比较快,平移的过程会有惯性效果。
/// 获取设置 m_customButtonEnable
    public Boolean CustomButtonEnable
        {get
            {return m_customButtonEnable;}
        set
            {m_customButtonEnable=value;}}
///将绘制的站点组成的 GeoCardinal 曲线的所有节点转化成三维点,设置同一高度
30m
    private Point3Ds ConvertPoint2DsToPoint3Ds(Point2Ds point2Ds)
        {m_point3DsAll. Clear();//清空集合中的三维点集合对象
            for(int i=0; i < point2Ds. Count; i++)
```

```csharp
            {m_point3DsAll.Add(new Point3D(point2Ds[i].X, point2Ds[i].Y, //向集合
中添加三维点对象,添加成功,返回被添加对象的序号
                m_sceneControl.Scene.GetAltitude(point2Ds[i].X, point2Ds[i].Y)+30));
}//获取地面某点的海拔高度
            return m_point3DsAll;}
        /// 获取绘制到场景跟踪图层上的飞行路线的样式
        private GeoStyle3D GetLineGeoStyle3D()
            {GeoStyle3D geoStyle3D = new GeoStyle3D();
                geoStyle3D.AltitudeMode = AltitudeMode.Absolute;//设置三维几何对象的高度
模式为绝对高度
                geoStyle3D.LineColor = System.Drawing.Color.FromArgb(255, 128, 255, 128);
//设置三维几何对象的颜色
                return geoStyle3D;}
        /// 获取绘制到场景跟踪图层上的站点的样式
        private GeoStyle3D GetPointGeoStyle3D(Double z)
            {GeoStyle3D geoStyle3D = new GeoStyle3D();
                geoStyle3D.AltitudeMode = AltitudeMode.Absolute;
                geoStyle3D.MarkerColor = System.Drawing.Color.FromArgb(255, 255, 255, 0);
                geoStyle3D.MarkerSize = 5;//设置三维点几何对象的大小,单位为0.1毫米
                geoStyle3D.ExtendedHeight = z;//设置拉伸高度
                return geoStyle3D;}
        ///   设置选择站点后改变站点的样式
        private GeoStyle3D SelectPointGeoStyle3D(Double z)
            {GeoStyle3D geoStyle3D = new GeoStyle3D();
                geoStyle3D.AltitudeMode = AltitudeMode.Absolute;
                geoStyle3D.MarkerColor = System.Drawing.Color.FromArgb(255, 255, 0, 0);
                geoStyle3D.MarkerSize = 5;
                geoStyle3D.ExtendedHeight = z;
                return geoStyle3D;}
        public void SetPointStyle3D(String name, Boolean enable)
            {Int32 index = m_sceneControl.Scene.TrackingLayer.IndexOf(m_pointName+name);
                Int32 indexPoints = 0;
                if (index >= 0)
                {GeoPoint3D geoPoint3D = m_sceneControl.Scene.TrackingLayer.Get(index) as
GeoPoint3D;//获取三维跟踪图层中指定索引的几何对象
                    //在跟踪图层中索引为的是线对象
                    if (index > 0)
                    {indexPoints = index - 1;}
                    else
```

```
            {indexPoints = index;}
            if(enable)
              {geoPoint3D.Style3D = SelectPointGeoStyle3D(m_point3Ds[indexPoints].Z);
}//设置三维几何对象的风格
            else
              {geoPoint3D.Style3D = GetPointGeoStyle3D(m_point3Ds[indexPoints].Z);}
            m_sceneControl.Scene.TrackingLayer.Set(index, geoPoint3D);}}//将三维跟
踪图层中的指定色索引出的几何对象替换为指定的几何对象
    private void SetGeoLine3DToTrackingLayer(Point2Ds point2Ds, Int32 count)
        {Int32 index = m_sceneControl.Scene.TrackingLayer.IndexOf("line");
            GeoCardinal geoCardinal = new GeoCardinal(point2Ds); // GeoCardinal:二维
Cardinal 样条曲线几何对象类。该类主要用于 CAD 图层,是 Geometry 对象的子对象。
            GeoLine geoLine = geoCardinal.ConvertToLine(count);
            if(m_geoLine3D.Length > 0)
              {m_geoLine3D.RemovePart(0);}//删除三维线几何对象中指定序号的子对
象
              m_geoLine3D.AddPart(ConvertPoint2DsToPoint3Ds(geoLine[0]));//向三维
线几何对象追加一个子对象
            if(index > 0)
              {m_sceneControl.Scene.TrackingLayer.Set(index, m_geoLine3D);} //将三维
跟踪图层中的指定色索引出的几何对象替换为指定的几何对象
            else
              {m_sceneControl.Scene.TrackingLayer.Add(m_geoLine3D, "line");}}//该方
法用于向三维跟踪图层中添加一个集合对象,并给出标签信息
    /// 根据所绘制的飞行站点插值
    public void InterpolateHeight(Int32 count)
        {Point2Ds curPoint2Ds = new Point2Ds();
            Point2Ds point2Ds = new Point2Ds();
            for(int i = 0; i < m_point3Ds.Count; i++)
              {curPoint2Ds.Add(new Point2D(m_point3Ds[i].X, m_point3Ds[i].Y));//
将指定的点对象添加到集合对象的末尾
            GeoCardinal geoCardinal = new GeoCardinal(curPoint2Ds);
            GeoLine geoLine = geoCardinal.ConvertToLine(count);//
            double dLength = geoLine.Length;
            Point2D point2D = new Point2D(dLength, m_point3Ds[i].Z);
            point2Ds.Add(point2D);}//将指定的点对象添加到点集合对象的末尾
            GeoCardinal geoCardinalHeight = new GeoCardinal(point2Ds);
            GeoLine geoLineHeight = geoCardinalHeight.ConvertToLine(count);
            for(Int32 i = 0; i < m_point3DsAll.Count; i++)
```

```
            {Point3D point3D = new Point3D(m_point3DsAll[i].X, m_point3DsAll[i].Y,
geoLineHeight[0][i].Y);
                m_point3DsAll[i] = point3D;}
            Int32 indexLine = m_sceneControl.Scene.TrackingLayer.IndexOf("line");
            m_geoLine3D.SetEmpty();
            m_geoLine3D.AddPart(m_point3DsAll);//向三维线几何对象追加一个子对
象
            if(indexLine > 0)
            {m_sceneControl.Scene.TrackingLayer.Set(indexLine, m_geoLine3D);}//将
三维跟踪图层中的指定色索引出的几何对象替换为指定的几何对象
            else
            {m_sceneControl.Scene.TrackingLayer.Add(m_geoLine3D, "line");}//该方
法用于向三维跟踪图层中添加一个几何对象，并给出标签信息
            SetPointStyle3D(m_index.ToString(), true);
            m_flyManager.Routes.Remove(0);//从飞行管理中移除给定索引出的路线
对象
            m_flyManager.Routes.Add(GetRoute());}//向路线集合中添加一个置顶的
录像对象
    /// 重置数据
    public void ResumeDefault()
        {m_point3Ds.Clear();//清空集合中的三维点集合对象
        m_point3DsAll.Clear();//清空集合中所有的三维点集合对象
        m_treeView.Nodes[0].Nodes.Clear();//从集合中删除所有树节点
        m_index = -1;
        if(m_geoLine3D.Length > 0)
        {m_geoLine3D.SetEmpty();}
        if(m_flyManager.Routes.Count > 0)
        {m_flyManager.Routes.Clear();}//清空路线集合中所有路线
        m_sceneControl.Scene.TrackingLayer.Clear();}//清空三维跟踪图层中的所
有几何对象
    ///获取飞行路线
    private Route GetRoute()
        {Route route = new Route();
        Point3Ds point3Ds = m_geoLine3D[0];
        for(int i = 0; i < point3Ds.Count; i++)
        {Point3D point3D = new Point3D(point3Ds[i].X, point3Ds[i].Y, point3Ds
[i].Z);
        point3Ds[i] = point3D;}
        GeoLine3D geoLine3D = new GeoLine3D(point3Ds);
```

```
        route.FromGeoLine3D(geoLine3D);//初始化路线
        route.IsFlyAlongTheRoute=true;//设置是否沿路线飞行
        route.IsHeadingFixed=true;//设置是否锁定方位角
        route.IsAltitudeFixed=true;//设置是否锁定高程
        route.IsTiltFixed=true;//设置是否锁定仰俯角
        route.IsLinesVisible=false;//设置路线是否可见
        route.IsStopsVisible=false;//设置站点是否可见
        return route;}
/// 获取站点名字
private String GetStopName(String name)
    {String lastName=name.Substring(4);
        Int32 index=Convert.ToInt32(lastName);
        lastName="Stop"+(++index).ToString();
        return lastName;}
/// 定位到站点
public void FlyTo(Int32 index, Boolean enable)
    {if(enable)
        {GeoPoint3D geoPoint3D=new GeoPoint3D(m_point3Ds[index]);
        m_sceneControl.Scene.Fly(geoPoint3D,500);}//根据指定的目标对象和飞行时间进行飞行
        else
        {m_sceneControl.Scene.Fly(m_flyManager.Routes[0].Stops[index].Camera,500);}}//将相机对象按照指定的时间飞行到指定的目的地
public void TransferCamera()
    {Camera camera=new Camera();
        camera.Altitude=285.05341279041;//设置相机高度
        camera.Longitude=116.391305696988;//设置相机经度
        camera.Latitude=39.9933447121584;//设置相机纬度
        camera.Heading=2.76012171129487;//设置相机方位角
        camera.Tilt=75.2282529563474;//设置相机仰角
        m_sceneControl.Scene.Fly(camera,10);}//将相机对象按照指定的时间飞行到指定的目的地
```

7. 窗体代码

(1)添加 using 引用代码

添加 using 引用代码具体如下:

using SuperMap.Data;//添加 SuperMap.Data 命名空间,使其对应的类可以使用

using SuperMap.Realspace;//添加 SuperMap.Realspace 命名空间,使其对应的类可以使用

using SuperMap.UI;//添加 SuperMap.UI 命名空间,使其对应的类可以使用

(2) 实例化代码

实例化 SampleRun 在 public FormMain() 函数前添加的代码如下:

```
//定义类对象
private SceneControl m_sceneControl;
private SampleRun m_sampleRun;
//记录上一次选中的节点的位置
private Int32 m_preSelectNodeIndex = -1;
private Boolean m_preTrackBarEnable = false;
Double m_speed;
// 用于记录当前的速度变化率
Double curPlayRate = 0.0;
// 用于记录速度改变后的速度变化率
Double updatedPlayRate = 0.0;
// 用于记录速度改变后的速度
Double updatedSpeed = 0.0;
```

(3) 窗体 Load 事件

窗体的 Load 事件,在窗体的属性窗口单击 图标,左键双击 Load,添加如下代码:

```
private void FormMain_Load(object sender, EventArgs e)
{ m_sceneControl = new SuperMap.UI.SceneControl();
    //设置 m_sceneControl 的属性
    m_sceneControl.Dock = DockStyle.Fill;
    m_splitContainer.Panel2.Controls.Add(m_sceneControl);
    m_buttonClearRoute.Enabled = false;
    m_buttonSaveRoute.Enabled = false;
    m_groupBoxFlyControl.Enabled = false;
    m_sampleRun = new SampleRun(m_sceneControl, m_treeView);
    m_sampleRun.FlyManager.StatusChanged += new StatusChangedEventHandler(FlyManager_StatusChanged);
    m_speed = Convert.ToInt32(m_textBoxSpeed.Text);}
```

(4) 其他控件的代码

① m_buttonLoadRoute 的 Click 事件。在控件的属性窗口单击 图标,左键双击 Click,添加如下代码,打开飞行路线:

```
private void m_buttonLoadRoute_Click(object sender, EventArgs e)
    {//设置控件是否可用
        m_buttonCustomRoute.Enabled = false;
        m_buttonLoadRoute.Enabled = false;
        m_buttonClearRoute.Enabled = true;
        m_buttonSaveRoute.Enabled = true;
        m_trackBar.Enabled = false;
```

　　　　m_groupBoxFlyControl.Enabled=true;
　　　　m_buttonFly.Enabled=true;
　　　　m_buttonPause.Enabled=false;
　　　　m_buttonStop.Enabled=false;
　　　　m_sampleRun.OpenData();}//打开默认飞行数据,并将各个站点显示到路线树中
　　② m_buttonCustomRoute 的 Click 事件。在控件的属性窗口单击🗲图标,左键双击 Click,添加如下代码,自定义飞行路线:
　　private void m_buttonCustomRoute_Click(object sender, EventArgs e)
　　　　{//设置控件是否可用
　　　　m_buttonCustomRoute.Enabled=false;
　　　　m_buttonLoadRoute.Enabled=false;
　　　　m_buttonClearRoute.Enabled=true;
　　　　m_buttonSaveRoute.Enabled=true;
　　　　m_trackBar.Enabled=false;
　　　　m_groupBoxFlyControl.Enabled=true;
　　　　m_buttonPause.Enabled=false;
　　　　m_buttonStop.Enabled=false;
　　　　m_buttonFly.Enabled=true;
　　　　m_preSelectNodeIndex=-1;
　　　　m_sampleRun.RegisterEvents(true);
　　　　m_sampleRun.TransferCamera();}//自定义飞行路线
　　③ m_buttonClearRoute 的 Click 事件。在控件的属性窗口单击🗲图标,左键双击 Click,添加如下代码,清除飞行路线:
　　private void m_buttonClearRoute_Click(object sender, EventArgs e)
　　　　{//设置控件是否可用及其属性
　　　　m_buttonCustomRoute.Enabled=true;
　　　　m_buttonLoadRoute.Enabled=true;
　　　　m_buttonClearRoute.Enabled=false;
　　　　m_buttonSaveRoute.Enabled=false;
　　　　m_groupBoxFlyControl.Enabled=false;
　　　　m_trackBar.Enabled=false;
　　　　m_trackBar.Value=0;
　　　　m_labelAltitude.Text="";
　　　　m_textBoxLatitude.Text="";
　　　　m_textBoxLongitude.Text="";
　　　　m_sampleRun.ResumeDefault();//重置数据
　　　　m_sampleRun.RegisterEvents(false);}//注销事件
　　④ m_treeView 的 NodeMouseDoubleClick 事件。在控件的属性窗口单击🗲图标,左键双

击 NodeMouseDoubleClick，添加如下代码，定位到站点：

```
private void m_treeView_NodeMouseDoubleClick（object sender, TreeNodeMouseClickEventArgs e）
    {if（e.Button==MouseButtons.Left && e.Node.Parent！=null）
        {if（m_sampleRun.Point3Ds.Count>0）
            {m_sampleRun.FlyTo（e.Node.Index,true）;}
        else
            {m_sampleRun.FlyTo（e.Node.Index,false）;}}}//定位到站点
```

⑤ m_treeView 的 NodeMouseClick 事件。在控件的属性窗口单击 ⚡ 图标，左键双击 NodeMouseClick，添加如下代码，设置背景色：

```
private void m_treeView_NodeMouseClick（object sender, TreeNodeMouseClickEventArgs e）
    {if（e.Button==MouseButtons.Left && e.Node.Parent！=null && m_sampleRun.Point3Ds.Count>0）
        {if（m_sampleRun.Index！=-1）
            {m_sampleRun.SetPointStyle3D（m_sampleRun.Index.ToString（）,false）;}
        if（m_preSelectNodeIndex！=-1）
            {m_treeView.Nodes[0].Nodes[m_preSelectNodeIndex].BackColor=Color.White;}//设置背景颜色
        m_preSelectNodeIndex=e.Node.Index;
        e.Node.BackColor=Color.FromArgb（255,201,201,201）;
        m_trackBar.Scroll-=new EventHandler（this.m_trackBar_Scroll）;//在通过或键盘操作移动滚动框时发生
        m_textBoxLongitude.Text=m_sampleRun.Point3Ds[e.Node.Index].X.ToString（）;//设置文本
        m_textBoxLatitude.Text=m_sampleRun.Point3Ds[e.Node.Index].Y.ToString（）;
        m_trackBar.Value=（Int32）m_sampleRun.Point3Ds[e.Node.Index].Z;//设置表示跟踪条上滚动框的当前位置的数值
        if（！m_buttonStop.Enabled）
            {m_trackBar.Enabled=true;}//设置看空间是否可用
        m_labelAltitude.Text=m_sampleRun.Point3Ds[e.Node.Index].Z.ToString（）;
        m_sampleRun.Index=e.Node.Index;设置站点索引值
        m_sampleRun.SetPointStyle3D（e.Node.Index.ToString（）,true）;
        m_trackBar.Scroll+=new EventHandler（this.m_trackBar_Scroll）;}}
```

⑥ m_trackBar 的 Scroll 事件，在控件的属性窗口单击 ⚡ 图标，左键双击 Scroll，添加如下代码，设置高程：

```
private void m_trackBar_Scroll（object sender, EventArgs e）
    {Point3D point3D=new Point3D（）;
    point3D.X=m_sampleRun.Point3Ds[m_sampleRun.Index].X;//获取绘制
```

站点

　　　　　point3D.Y = m_sampleRun.Point3Ds[m_sampleRun.Index].Y;//

　　　　　point3D.Z = (Double)m_trackBar.Value;//设置表示跟踪条上滚动框的当前位置的数值

　　　　　m_labelAltitude.Text = m_trackBar.Value.ToString();

　　　　　m_sampleRun.Point3Ds[m_sampleRun.Index] = point3D;

　　　　　m_sampleRun.InterpolateHeight(20);}//根据所绘制的飞行站点插值

⑦ m_buttonFly 的 Click 事件。在控件的属性窗口单击 🗲 图标，左键双击 Click，添加如下代码，实现飞行：

　　private void m_buttonFly_Click(object sender, EventArgs e)

　　　　{m_sampleRun.Fly();//通过 sampleRun.Fly 实现飞行

　　　　　m_buttonFly.Enabled = false;

　　　　　m_buttonPause.Enabled = true;

　　　　　m_buttonStop.Enabled = true;

　　　　　m_preTrackBarEnable = m_trackBar.Enabled;

　　　　　m_trackBar.Enabled = false;

　　　　　m_buttonClearRoute.Enabled = false;}

⑧ m_buttonPause 的 Click 事件。在控件的属性窗口单击 🗲 图标，左键双击 Click，添加如下代码，实现暂停：

　　private void m_buttonPause_Click(object sender, EventArgs e)

　　　　{m_sampleRun.Pause();//通过 sampleRun.Pause 实现赞同

　　　　　m_buttonFly.Enabled = true;

　　　　　m_buttonPause.Enabled = false;

　　　　　m_buttonStop.Enabled = true;}

⑨ m_buttonStop 的 Click 事件。在控件的属性窗口单击 🗲 图标，左键双击 Click，添加如下代码，实现停止：

　　private void m_buttonStop_Click(object sender, EventArgs e)

　　　　{m_sampleRun.Stop();//通过 sampleRun.Stop 实现停止

　　　　　m_buttonFly.Enabled = true;

　　　　　m_buttonPause.Enabled = false;

　　　　　m_buttonStop.Enabled = false;

　　　　　m_trackBar.Enabled = m_preTrackBarEnable;

　　　　　m_buttonClearRoute.Enabled = true;

　　　　　m_sceneControl.Scene.IsFirstPersonView = false;

　　　　　m_sampleRun.TransferCamera();}

⑩ m_buttonFaster 的 Click 事件，在控件的属性窗口单击 🗲 图标，左键双击 Click，添加如下代码，实现加速：

　　private void m_buttonFaster_Click(object sender, EventArgs e)

　　　　{curPlayRate = m_sampleRun.FlyManager.PlayRate;//获取飞行管理器

m_sampleRun.FlyManager.PlayRate = curPlayRate * 1.2;
updatedPlayRate = m_sampleRun.FlyManager.PlayRate;
updatedSpeed = Math.Round(50 * updatedPlayRate, 2);
m_textBoxSpeed.Text = updatedSpeed.ToString();}

⑪ m_buttonSlower 的 Click 事件。在控件的属性窗口单击 ❡ 图标，左键双击 Click，添加如下代码，实现减速：

```
private void m_buttonSlower_Click(object sender, EventArgs e)
    {curPlayRate = m_sampleRun.FlyManager.PlayRate;//获取飞行管理器
    m_sampleRun.FlyManager.PlayRate = curPlayRate * 0.8;
    updatedPlayRate = m_sampleRun.FlyManager.PlayRate;
    updatedSpeed = Math.Round(50 * updatedPlayRate, 2);
    m_textBoxSpeed.Text = updatedSpeed.ToString();}
```

⑫ m_buttonSaveRoute 的 Click 事件。在控件的属性窗口单击 ❡ 图标，左键双击 Click，添加如下代码，保存飞行路线：

```
private void m_buttonSaveRoute_Click(object sender, EventArgs e)
    {SaveFileDialog saveFileDialog = new SaveFileDialog();
    saveFileDialog.Title = "保存";
    saveFileDialog.Filter = "SuperMap 三维飞行路径文件(*.fpf)|*.fpf";
    saveFileDialog.FileName = "NewSceneRoutes.fpf";
    if(saveFileDialog.ShowDialog() == DialogResult.OK)
    {m_sampleRun.FlyManager.Routes.ToFile(saveFileDialog.FileName);}}//将
```
路线集合对象输出成 xml 文件

(5) 窗体的 FormClose 事件

在退出程序时，要断开空间的关联，关闭地图窗口、工作空间。因此在属性窗口点击 ❡ 图标，左键双击 FormClose 事件，添加以下代码：

```
private void FormMain_FormClosing(object sender, FormClosingEventArgs e)
    {try
        {m_sceneControl.Dispose();}
    catch
    {}}
```

(6) 设置控件的属性

设置控件的属性需添加以下代码：

```
private void FlyManager_StatusChanged(object sender, StatusChangedEventArgs e)
    {if(e.PreStatus == FlyStatus.Play && e.CurrentStatus == FlyStatus.Stop)
        {m_buttonFly.Enabled = true;
        m_buttonPause.Enabled = false;
        m_buttonStop.Enabled = false;
        m_trackBar.Enabled = m_preTrackBarEnable;
        m_buttonClearRoute.Enabled = true;
```

m_sceneControl. Scene. IsFirstPersonView = false;}}

8. 运行结果

窗体设计完成后点击 ▶ 按钮，运行该程序，运行后弹出窗体，如图 8-7 所示。如图可以根据默认路线或者自定义路线进行三维场景的飞行浏览。

图 8-7　飞行管理

第9章 水文分析

1. 数据

在安装目录\SampleData\SurfaceAnalyst\SurfaceAnalyst.swmu 路径下打开数据文件。

2. 新建文件夹和工程

在【D:\MyProject\】文件中新建一个文件夹,命名为【第9章 水文分析】,并在其目录下建立一个文件夹,命名为【水文分析】,在此文件夹里新建一个工程,将此工程命名为 HydrologyAnalyst。

3. 关键类型/成员

关键类型/成员见表9-1。

表9-1 关键类型/成员表

控件/类	方法	属性	事件
Workspace	Open		
MapControl		Map	MouseMove
HydrologyAnalyst	FillSink、FlowDirection、Basin、FlowAccumulation、StreamToLine		
MathAnalyst	Execute		

4. 窗体控件属性

窗体控件属性具体见表9-2。

表9-2 窗体控件属性表

控件	Name	Text
toolStrip	toolStrip	toolStrip1
StatusStrip	StatusStrip1	StatusStrip1
Button	buttonFillSink	填充伪洼地
	buttonFlowDirection	计算流向
	buttonBasin	计算流域盆地
	buttonFlowAccumulation	计算累积汇水量
	buttonStreamGrid	提取栅格水系
	buttonStreamToLine	提取矢量水系
	buttonReset	重置

续表

控件	Name	Text
labelStatus	labelStatus	labelStatus
labelValue	labelValue	labelValue

5. 窗体设计布局

窗体设计布局如图 9-1 所示。

图 9-1　窗体设计布局图

6. 创建 SampleRun 类

（1）添加 using 引用代码

添加 using 引用代码具体如下：

using SuperMap.Data;　//添加 SuperMap.Data 命名空间，使其对应的类可以使用
using SuperMap.UI;　//添加 SuperMap.UI 命名空间，使其对应的类可以使用
using SuperMap.Analyst.TerrainAnalyst;　//添加 SuperMap.Analyst.TerrainAnalyst 命名空间，使其对应的类可以使用
using System.Drawing;　//添加 System.Drawing 命名空间，使其对应的类可以使用
using System.Windows.Forms;　//添加 System.Drawing 命名空间，使其对应的类可以使用
using System.Diagnostics;　//添加 System.Diagnostics 命名空间，使其对应的类可以使用
using SuperMap.Analyst.SpatialAnalyst;　//添加 System.Diagnostics 命名空间，使其对应的类可以使用
using SuperMap.Mapping;　//添加 SuperMap.Mapping 命名空间，使其对应的类可以使用

（2）在 public class SampleRun 中添加代码

在 public class SampleRun 中添加如下代码：

//定义各个结果数据集名称变量　定义类对象
private const String FILL_SINK_NAME = "FillSink";

```csharp
private const String FLOW_DIRECTION_NAME = "Direction";
private const String FLOW_ACCUMULATION_NAME = "Accumulation";
private const String BASIN_NAME = "Basin";
private const String STREAM_GRID_NAME = "StreamGrid";
private const String STREAM_LINE_NAME = "StreamLine";
private Workspace m_workspace;// Workspace:工作空间是用户的工作环境,主要完成数据的组织和管理,包括打开、关闭、创建、保存工作空间文件。
private MapControl m_mapControl;// MapControl:地图控件类。该类是用于为地图的显示提供界面的,同时为地图与数据的互操作提供途径。
private Datasource m_datasource;// Datasource:数据源类。该类管理投影信息、数据源与数据库的连接信息和对其中的数据集的相关操作,如通过已有数据集复制生成新的数据集等。
private ToolStripStatusLabel m_labelValue;
public SampleRun(Workspace workspace, MapControl mapControl, ToolStripStatusLabel labelValue)
{
    m_workspace = workspace;
    m_mapControl = mapControl;
    m_labelValue = labelValue;
    Initialize();
}
/// 打开工作空间获取数据源
private void Initialize()
{
    try
    {
        WorkspaceConnectionInfo info = new WorkspaceConnectionInfo(@"..\..\SampleData\SurfaceAnalyst\SurfaceAnalyst.smwu");
        info.Type = WorkspaceType.SMWU;
        m_workspace.Open(info);//打开工作空间
        m_datasource = m_workspace.Datasources["SurfaceAnalyst"];
        //添加 DEM 栅格数据到当前地图上
        AddDEM();
        m_mapControl.MouseMove += new MouseEventHandler(m_mapControl_MouseMove);
    }
    catch (Exception ex)
    {
        Trace.WriteLine(ex.Message);
    }
}
/// 地图控件的鼠标移动事件 当鼠标在地图上移动时,获取并显示栅格值
private void m_mapControl_MouseMove(object sender, System.Windows.Forms.MouseEventArgs e)
{
    try
    {
        Dataset currentDataset = m_mapControl.Map.Layers[0].Dataset;//获取当前地图所包含的图层集合对象
        if (currentDataset.Type == DatasetType.Grid)
        {
            Point2D mapPoint = m_mapControl.Map.PixelToMap(new Point(e.X, e.Y));//将地图中指定点的像素坐标转换为地图坐标
```

```
            DatasetGrid currentGrid = currentDataset as DatasetGrid;
            //点在当前数据集的范围内才能获取栅格
            Rectangle2D bounds = m_mapControl.Map.Layers[0].Bounds;
            if(bounds.Contains(mapPoint))
            {Point position = currentGrid.XYToGrid(mapPoint);//将地理坐标系下的点(X,Y)
转换为栅格数据集中对应的栅格
            Double value = currentGrid.GetValue((Int32)position.X,(Int32)position.Y);//根
据给定的行数和列数获取栅格数据集的栅格所对应的栅格值
            m_labelValue.Text = "栅格值" + value.ToString();}
            else
            {m_labelValue.Text = "栅格值:";}}
            else{m_labelValue.Text = " ";}}//设置要显示在顶上的文本
          catch(Exception ex)
          {Trace.WriteLine(ex.Message);}}
     public Boolean CalculateSink()
          {return false;}
          ///填充洼地 true,否则返回 false
     public Boolean FillSink()
            {try{DatasetGrid surfaceGrid = m_datasource.Datasets["T1"] as DatasetGrid;//获取
数据源所包含的数据集的集合对象
            if(m_datasource.Datasets.Contains(FILL_SINK_NAME))//检查当前数据源中是
否包含指定名称的数据集
            {m_datasource.Datasets.Delete(FILL_SINK_NAME);}//根据指定的名称来删除
数据集
            DatasetGrid resultGrid = HydrologyAnalyst.FillSink(surfaceGrid, m_datasource, FILL_
SINK_NAME);
            if(resultGrid ! = null)
            {AddToMap(resultGrid,true);//将数据添加到当前地图上
            return true;}
            else{return false;}}
          catch(Exception ex)
          {Trace.WriteLine(ex.Message);
            return false;}}
          /// 对无伪洼地的地形栅格数据集计算流向
          /// <returns>成功生成流向栅格返回 true,否则返回 false</returns>
     public Boolean FlowDirection()
            {try{DatasetGrid surfaceGrid = m_datasource.Datasets[FILL_SINK_NAME] as
DatasetGrid;//获取数据源所包含的数据集的集合对象
            if(m_datasource.Datasets.Contains(FLOW_DIRECTION_NAME))//检查当前数据
```

源中是否包含指定名称的数据集

    {m_datasource.Datasets.Delete(FLOW_DIRECTION_NAME);} //根据指定的名称来删除数据集

    DatasetGrid resultGrid = HydrologyAnalyst.FlowDirection(surfaceGrid, true, m_datasource, FLOW_DIRECTION_NAME);

    if(resultGrid！=null)

    {AddToMap(resultGrid, true); //将数据集添加到当前地图上

    return true;}

    else{return false;}}

    Trace.WriteLine(ex.Message);

    return false;}}

/// 根据流向栅格计算累积汇水量

///成功生成累积汇水量返回 true, 否则返回 false

public Boolean FlowAccumulation()

    {try{DatasetGrid directionGrid=m_datasource.Datasets[FLOW_DIRECTION_NAME] as DatasetGrid; //获取数据源所包含的数据集的集合对象

    if(m_datasource.Datasets.Contains(FLOW_ACCUMULATION_NAME)) //检查当前数据源中是否包含指定名称的数据集

    {m_datasource.Datasets.Delete(FLOW_ACCUMULATION_NAME);} //根据指定的名称来删除数据集

    DatasetGrid resultFlowAccumulationGrid = HydrologyAnalyst.FlowAccumulation(directionGrid, null, m_datasource, FLOW_ACCUMULATION_NAME);

    if(resultFlowAccumulationGrid！=null)

    {AddToMap(resultFlowAccumulationGrid, true); //将数据集添加到当前地图上

    return true;}

    else{

    return false;}}

    catch(Exception ex)

    {Trace.WriteLine(ex.Message);

    return false;}}

/// 提取栅格水系，是一个栅格值由 0 和 1 构成的二值栅格

///成功获得栅格水系返回 true, 否则返回 false

public Boolean StreamGrid()

    {try{DatasetGrid accumulationGrid=m_datasource.Datasets[FLOW_ACCUMULATION_NAME] as DatasetGrid; //获取数据源所包含的数据集的集合对象

    if(m_datasource.Datasets.Contains(STREAM_GRID_NAME)) //检查当前数据源中是否包含指定名称的数据集

    {m_datasource.Datasets.Delete(STREAM_GRID_NAME);} //根据指定的名称来删除数据集

```
            String expression = "Con([" +m_datasource.Alias+"." +accumulationGrid.Name+"] >
1000, 1, 0)";
            DatasetGrid resultStreamGrid = MathAnalyst.Execute(expression, null, PixelFormat.
Bit16, true, true, m_datasource, STREAM_GRID_NAME);// DatasetGrid:栅格数据集类,用
于描述栅格数据,例如高程数据集和土地利用图。
            if(resultStreamGrid ! =null)
            {AddToMap(resultStreamGrid, true); //将数据集添加到当前地图上
            return true;}
            else
            {return false;}}
        catch (Exception ex)
            {Trace.WriteLine(ex.Message);
            return false;}}
        /// 根据流向计算流域盆地
        ///成功生成流域盆地栅格返回 true,否则返回 false
    public Boolean Basin()
            {try{DatasetGrid directionGrid=m_datasource.Datasets[FLOW_DIRECTION_NAME]
as DatasetGrid; //获取数据源所包含的数据集的集合对象
            if (m_datasource.Datasets.Contains(BASIN_NAME)) //检查当前数据源中是
否包含指定名称的数据集
                {m_datasource.Datasets.Delete(BASIN_NAME);} //根据指定的名称来删除
数据集
                DatasetGrid resultBasinGrid = HydrologyAnalyst.Basin(directionGrid, m_
datasource, BASIN_NAME);
            if(resultBasinGrid ! =null)
            {AddToMap(resultBasinGrid, true); //将数据集添加到当前地图上
            return true;}
            else{return false;}}
        catch (Exception ex)
            {Trace.WriteLine(ex.Message);
            return false;}}
            /// 根据栅格水系和流向栅格提取矢量水系
            //成功获取矢量水系返回 true,否则返回 false
    public Boolean StreamToLine()
            {try{DatasetGrid streamGrid = m_datasource.Datasets[STREAM_GRID_NAME] as
DatasetGrid; //获取数据源所包含的数据集的集合对象
                DatasetGrid directionGrid=m_datasource.Datasets[FLOW_DIRECTION_NAME]
as DatasetGrid; //获取数据源所包含的数据集的集合对象
            if (m_datasource.Datasets.Contains(STREAM_LINE_NAME)) //检查当前数
```

第9章 水文分析

据源中是否包含指定名称的数据集

　　　　{m_datasource. Datasets. Delete(STREAM_LINE_NAME);} //根据指定的名称来删除数据集

　　　　DatasetVector resultStreamLine = HydrologyAnalyst. StreamToLine(streamGrid, directionGrid, m_datasource, STREAM_LINE_NAME, StreamOrderType. Shreve); // HydrologyAnalyst：水文分析类。用于填充洼地、流向分析、洼地计算、累积汇水量计算、流域计算、流长分析及矢量河网提取等水文分析功能，还提供网格剖分功能。

　　　　if (resultStreamLine ! = null)
　　　　{//将原始 DEM 和提取的矢量水系叠加显示
　　　　AddDEM( );
　　　　Layer streamLineLayer = m_mapControl. Map. Layers. Add(resultStreamLine, true); //添加图层
　　　　//设置图层风格
　　　　GeoStyle style = new GeoStyle( );
　　　　style. LineColor = Color. FromArgb(0, 0, 255); //线状符号或点状符号的颜色
　　　　style. LineWidth = 0.4; //设置线状符号的宽度
　　　　LayerSettingVector layerSetting = new LayerSettingVector( );
　　　　layerSetting. Style = style; //设置矢量图层的风格
　　　　streamLineLayer. AdditionalSetting = layerSetting; //设置普通图层的风格设置
　　　　m_mapControl. Map. Refresh( );//刷新
　　　　return true;}
　　　　else{return false;}}
　　　　catch(Exception ex)
　　　　{Trace. WriteLine(ex. Message);
　　　　return false;}}
　　　　/// 添加地形数据到地图上
　　public void AddDEM( )
　　　　{try{Dataset dataset = m_datasource. Datasets["T1"]; //获取数据源所包含的数据集的集合对象
　　　　AddToMap(dataset, true);}//将数据集添加到当前地图上
　　catch(Exception ex)
　　　　{Trace. WriteLine(ex. Message);}}
　　　　/// 将数据集添加到当前地图上
　　private void AddToMap(Dataset dataset, Boolean isClearLayers)
　　　　{if (isClearLayers)
　　　　{m_mapControl. Map. Layers. Clear( );}//删除此图层集合对象中所有的图层
　　　　m_mapControl. Map. Layers. Add(dataset, true); //添加图层
　　　　m_mapControl. Map. ViewEntire( );//全幅显示地图
　　　　m_mapControl. Map. Refresh( );}//刷新

7. 窗体代码

(1) 添加 using 引用代码

添加 using 引用代码具体如下：

using SuperMap.UI;　//添加 SuperMap.UI 命名空间，使其对应的类可以使用

using SuperMap.Data;　//添加 SuperMap.Data 命名空间，使其对应的类可以使用

(2) 实例化代码

实例化 SampleRun、MapControl、Workspace，设置控件的初始状态代码如下：

```
//定义类对象
private MapControl m_mapControl;
private Workspace m_workspace;
private SampleRun m_sampleRun;
public FormMain()
    {InitializeComponent();
    Initialize();}
private void Initialize()
    {m_mapControl = new MapControl();
    this.Controls.Add(m_mapControl);　//将指定控件添加到控件集合中
    if(m_mapControl! = null)
    {m_mapControl.Dock = DockStyle.Fill;}　//设置控件的停靠方式
    m_workspace = new Workspace();
    m_mapControl.Map.Workspace = m_workspace;
    m_mapControl.Map.IsAntialias = true;　//设置一个布尔值是否反走样地图
    //设置控件的初始状态
    ControlsInitialStatus();
    //实例化 SampleRun
    m_sampleRun = new SampleRun(m_workspace, m_mapControl, labelValue);
    labelStatus.Text = "当前图层为:地形栅格数据集";
    labelValue.Text = "";}
```

(3) 设置控件的初始状态

设置工具条、状态条上控件的初始状态需添加以下代码：

```
private void ControlsInitialStatus()
    {//初始化控件属性
    buttonFillSink.Enabled = true;
    buttonFlowDirection.Enabled = false;
    buttonFlowAccumulation.Enabled = false;
    buttonBasin.Enabled = false;
    buttonStreamGrid.Enabled = false;
    buttonStreamToLine.Enabled = false;
    buttonReset.Enabled = false;
```

labelStatus.Text="当前图层为:地形栅格数据集";

labelStatus.Text="";}

(4)添加各个控件的代码

① 填充洼地 Button 控件的 Click 事件。左键双击漫游 Button 控件,即可添加如下该按钮的 Click 事件代码,实现填充伪洼地:

private void buttonFillSink_Click(object sender, EventArgs e)

{

if (m_sampleRun.FillSink())//实现填伪洼地

{labelStatus.Text="当前图层为:无伪洼地栅格数据集";

buttonFlowDirection.Enabled=true;

buttonReset.Enabled=true;

buttonFillSink.Enabled=false;}}

② 计算流向 Button 控件的 Click 事件。左键双击全屏 Button 控件,即可添加如下该按钮的 Click 事件代码,实现计算流向:

private void buttonFlowDirection_Click(object sender, EventArgs e)

{if (m_sampleRun.FlowDirection())//通过 sampleRun.FlowDirection 实现计算流向

{labelStatus.Text="当前图层为:流向栅格数据集";

buttonFlowAccumulation.Enabled=true;

buttonBasin.Enabled=true;

buttonFlowDirection.Enabled=false;}}

③ 计算累积汇水量 Button 控件的 Click 事件。左键双击清除 Button 控件,即可添加如下该按钮的 Click 事件代码,实现计算累积汇水量:

private void buttonFlowAccumulation_Click(object sender, EventArgs e)

{if (m_sampleRun.FlowAccumulation())//通过 sampleRun.FlowAccumulation 实现计算累积汇水量

{labelStatus.Text="当前图层为:累积汇水量栅格数据集";

buttonStreamGrid.Enabled=true;

buttonFlowAccumulation.Enabled=false;}}

④ 提取栅格水系 Button 控件的 Click 事件。左键双击清除 Button 控件,即可添加如下该按钮的 Click 事件代码,实现提取栅格水系:

private void buttonStreamGrid_Click(object sender, EventArgs e)

{if (m_sampleRun.StreamGrid())//通过 sampleRun.StreamGrid 实现提取栅格水系

{labelStatus.Text="当前图层为:栅格水系数据集";

buttonStreamToLine.Enabled=true;

buttonStreamGrid.Enabled=false;}}

⑤ 计算流域盆地 Button 控件的 Click 事件。左键双击清除 Button 控件,即可添加如下该按钮的 Click 事件代码,实现计算流域盆地:

private void buttonBasin_Click(object sender, EventArgs e)

{ if (m_sampleRun.Basin())//通过 sampleRun.Basin 实现计算流域盆地

{labelStatus.Text=" 当前图层为:流域盆地栅格数据集";
　　buttonBasin.Enabled=false;}}

⑥ 提取矢量水系 Button 控件的 Click 事件,左键双击清除 Button 控件,即可添加如下该按钮的 Click 事件代码,实现提取矢量水系:
private void buttonStreamToLine_Click(object sender, EventArgs e)
　　{
　　if(m_sampleRun.StreamToLine())//通过 sampleRun.StreamToLine 提取矢量水系
　　{labelStatus.Text=" 当前图层为:地形栅格数据集、矢量水系数据集";
　　buttonStreamToLine.Enabled=false;}}

⑦ 重置 Button 控件的 Click 事件,左键双击清除 Button 控件,即可添加如下该按钮的 Click 事件代码,实现重置:
private void buttonReset_Click(object sender, EventArgs e)
　　{ControlsInitialStatus();
　　m_sampleRun.AddDEM();}//通过 sampleRun.AddDEM 实现重置

(5)窗体的 FormClose 事件
窗体的 FormClose 事件,添加代码同前文中的相应部分。

8. 运行结果
窗体设计完成后点击▶按钮,运行该程序,运行后弹出窗体,如图 9-2 所示,可以实现填充伪洼地、计算流向、计算流域盆地、计算累积汇水量、提取栅格水系、重置的功能。

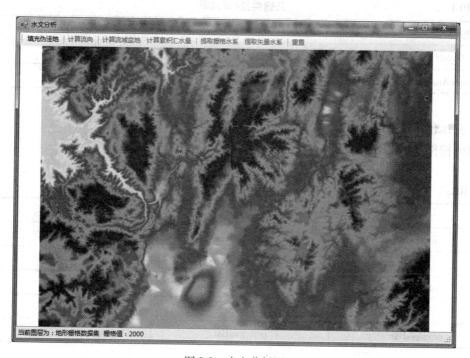

图 9-2　水文分析图

第 10 章　表面分析与动态分段

10.1　表面分析

1. 数据

在安装目录\SampleData\SurfaceAnalyst\SurfaceAnalyst.smwu 路径下打开数据文件。

2. 新建文件夹和工程

在【D:\MyProject\】文件中新建一个文件夹，命名为【第 10 章　表面分析与动态分段】，并在其目录下建立一个文件夹，命名为【SurfaceAnalyst】，在此文件夹里新建一个工程，将此工程命名为 SurfaceAnalyst。

3. 关键类型/成员

关键类型/成员见表 10-1。

表 10-1　关键类型/成员表

控件/类	方法	属性	事件
Layers	Add、Clear		
SurfaceAnalyst	ExtractIsoline、ExtractIsoregion		

4. 窗体控件属性

窗体控件属性见表 10-2。

表 10-2　窗体控件属性表

控件	Name	Text
toolStrip	toolStrip1	toolStrip1
Button	ExtractIsoline	等值线
Button	ExtractIsoregion	等值面
Button	Review	复原
progressBar	progressBar	

5. 窗体设计布局

窗体设计布局如图 10-1 所示。

10.1 表面分析

图 10-1　窗体设计布局图

6. 创建 SampleRun 类

（1）添加 using 引用代码

添加 using 引用代码具体如下：

using System. Diagnostics；//添加 System. Diagnostics 命名空间，使其对应的类可以使用
using System. Drawing；//添加 System. Drawing 命名空间，使其对应的类可以使用
using System. Windows. Forms；//添加 System. Windows. Forms 命名空间，使其对应的类可以使用
using SuperMap. Data；//添加 SuperMap. Data 命名空间，使其对应的类可以使用
using SuperMap. UI；//添加 SuperMap. UI 命名空间，使其对应的类可以使用
using SuperMap. Analyst. SpatialAnalyst；//添加 SuperMap. Analyst. SpatialAnalyst 命名空间，使其对应的类可以使用
using SuperMap. Mapping；//添加 SuperMap. Mapping 命名空间，使其对应的类可以使用
using System. ComponentModel；//添加 System. ComponentModel 命名空间，使其对应的类可以使用

（2）在 public class SampleRun 中添加代码

在 public class SampleRun 中添加如下代码：

//定义类对象

private Workspace m_workspace；// Workspace：工作空间是用户的工作环境，主要完成数据的组织和管理，包括打开、关闭、创建、保存工作空间文件。

private MapControl m_mapControl；// MapControl：地图控件类。该类是用于为地图的显示提供界面的，同时为地图与数据的互操作提供途径。

private Datasource m_datasource；// Datasource：数据源类。该类管理投影信息、数据源与数据库的连接信息和对其中的数据集的相关操作，如通过已有数据集复制生成新的数据集等。

private DatasetGrid m_datasetGrid；// DatasetGrid：栅格数据集类，用于描述栅格数据，

例如高程数据集和土地利用图。

```csharp
    private DatasetVector m_datasetVectorLine;// DatasetVector:矢量数据集类。描述矢量数
据集,并提供相应的管理和操作。对矢量数据集的操作主要包括数据查询、修改、删除、
建立索引等。
    private DatasetVector m_datasetVectorRegion;
    private Double m_minValue;
    private Double m_maxValue;
    /// 根据 workspace 和 map 构造 SampleRun 对象
    public SampleRun(Workspace workspace, MapControl mapControl)
        {try{m_workspace = workspace;
    m_mapControl = mapControl;
    m_mapControl.Map.Workspace = workspace;
    Initialize();}
    catch(Exception ex){
    Trace.WriteLine(ex.Message);}}
    private void Initialize()
        {try{//打开工作空间
WorkspaceConnectionInfo conInfo = new WorkspaceConnectionInfo(@"..\..\SampleData
\SurfaceAnalyst\SurfaceAnalyst.smwu");
    conInfo.Type = WorkspaceType.SMWU;
    m_workspace.Open(conInfo);//打开工作空间
    m_datasource = m_workspace.Datasources[0];
    //得到栅格数据集
    m_datasetGrid = (DatasetGrid)m_datasource.Datasets["T1"];//获取数据源所包含的
数据集的集合对象
    //得到数据集的最大值和最小值
    m_minValue = m_datasetGrid.MinValue;//获取栅格数据集中栅格值的最小值
    m_maxValue = m_datasetGrid.MaxValue;//获取栅格数据集中栅格值的最大值
    m_mapControl.Map.Layers.Add(m_datasetGrid, true);//添加图层
    m_mapControl.Map.Refresh();}//刷新
    catch(Exception ex)
    {Trace.WriteLine(ex.Message);}}
    //打开栅格数据集
    public void OpenDataset()
        {try{m_mapControl.Map.Layers.Clear();//删除此图层集合对象中所有的图层
    m_mapControl.Map.Layers.Add(m_datasetGrid, true);//添加图层
    m_mapControl.Map.Refresh();}//刷新
    catch(Exception ex)
        {Trace.WriteLine(ex.Message);}}
```

```
//提取等值线
public void ExtrsctIsoLine()
    {try{m_mapControl.Map.Layers.Clear();//删除此图层集合对象中所有的图层
//设置表面分析,提取操作参数
SurfaceExtractParameter surfaceExtractParameter = new SurfaceExtractParameter();
surfaceExtractParameter.DatumValue = m_minValue;//设置等值线的基准值
surfaceExtractParameter.Interval = (m_maxValue - m_minValue) / 10;//设置等值距
surfaceExtractParameter.SmoothMethod = SmoothMethod.Polish;//设置光滑处理所使用的方法
surfaceExtractParameter.Smoothness = 3;//设置等值线或面的光滑度
//提取等值线
String name = m_datasource.Datasets.GetAvailableDatasetName("Del_Line");//获取数据集
m_datasetVectorLine = SuperMap.Analyst.SpatialAnalyst.SurfaceAnalyst.ExtractIsoline
(surfaceExtractParameter, m_datasetGrid, m_datasource, name);// SurfaceAnalyst:表面分析类。该类用于从表面栅格数据、具有高程信息的点数据中提取等值线或等值面,用于从栅格数据集中提取等值线。
m_mapControl.Map.Layers.Clear();//删除图层集合对象中所有的图层
m_mapControl.Map.Layers.Add(m_datasetGrid, false);//添加图层
//设置线图层颜色
LayerSettingVector layerSettingVector = new LayerSettingVector();
layerSettingVector.Style.LineColor = Color.Black;//设置颜色
layerSettingVector.Style.LineWidth = 0.2;//设置宽度
m_mapControl.Map.Layers.Add(m_datasetVectorLine, layerSettingVector, true);//添加图层
//制作标签专题图
ThemeLabel themeLabel = new ThemeLabel();//ThemeLabel:标签专题图类。用文本的形式在图层上直接显示属性表中的数据,其实质就是对图层的标注。
themeLabel.LabelExpression = "dZValue";//设置标注字段表达式
themeLabel.NumericPrecision = 2;//设置标签中数字的精度
themeLabel.UniformStyle.ForeColor = Color.Black;//设置文本的前景色
themeLabel.UniformStyle.BackColor = Color.White;//设置文本的背景色
themeLabel.UniformStyle.Outline = true;//设置是否以轮廓的方式来显示文本的背景
Layer themeLabelLayer = m_mapControl.Map.Layers.Add(m_datasetVectorLine, themeLabel, true);//添加图层
m_mapControl.Map.Refresh();}//刷新
catch (Exception ex)
{Trace.WriteLine(ex.Message);}}
//提取等值面
```

```csharp
public void ExtractIsoRegion()
    {try{m_mapControl.Map.Layers.Clear();//删除图层集合对象中所有的图层
//设置表面分析提取操作参数
    SurfaceExtractParameter surfaceExtractParameter = new SurfaceExtractParameter();
    surfaceExtractParameter.Interval = 100;//设置等值距
    surfaceExtractParameter.SmoothMethod = SmoothMethod.None;//设置光滑处理所使用的方法
    //提取等值面
    String name = m_datasource.Datasets.GetAvailableDatasetName("Del_Region1");//获取数据集
    m_datasetVectorRegion =
     SuperMap.Analyst.SpatialAnalyst.SurfaceAnalyst.ExtractIsoregion(surfaceExtractParameter,
    m_datasetGrid, m_datasource, name, null);//提取等值面
    m_mapControl.Map.Layers.Clear();//删除此图层集合对象中所有的图层
    m_mapControl.Map.Layers.Add(m_datasetVectorRegion, true);//添加图层
    //制作标签专题图
    ThemeLabel themeLabel = new ThemeLabel();
    themeLabel.LabelExpression = "dZValue";//设置标注字段表达式
    themeLabel.NumericPrecision = 2;//设置标签中数字的精度
    themeLabel.UniformStyle.ForeColor = Color.Black;//设置文本的前景色
    themeLabel.UniformStyle.BackColor = Color.White;//设置文本的背景色
    themeLabel.UniformStyle.Outline = true;//设置是否以轮廓的方式来显示文本的背景
    Layer themeLabelLayer = m_mapControl.Map.Layers.Add(m_datasetVectorRegion, themeLabel, true);//添加图层
    m_mapControl.Map.Refresh();}//刷新
    catch(Exception ex)
    {Trace.WriteLine(ex.Message);}}
```

7. 窗体代码

(1) 添加 using 引用代码

添加 using 引用代码具体如下:

using System.Diagnostics;//添加 System.Diagnostics 命名空间,使其对应的类可以使用

(2) 实例化代码

实例化 SampleRun 以及 public Form1() 函数部分的代码如下:

```csharp
//定义类对象
private SampleRun m_sampleRun;
private SuperMap.Data.Workspace m_workspace;
private SuperMap.UI.MapControl m_mapControl;
public Form1()
```

```
{try{InitializeComponent();
    this.m_workspace = new SuperMap.Data.Workspace(this.components);//定义一个新
对象
    this.m_mapControl = new SuperMap.UI.MapControl();}
catch(Exception ex)
{Trace.WriteLine(ex.Message);}}
//初始化窗体
private void FormMain_Load(object sender,EventArgs e)
    {try{//指定控件的停靠方式 m_mapControl.Dock=DockStyle.Fill;
//实例化 SampleRun
    m_sampleRun = new SampleRun(m_workspace, m_mapControl);
    base.Controls.Add(m_mapControl);//将指定的控件添加到控件集合中
    progressBar.Maximum = 100;//设置范围上限
    progressBar.Minimum = 5;}//设置范围下限
catch(Exception ex)
{Trace.WriteLine(ex.Message);}
```

(3) 添加各个控件的代码

① 等值线 Button 控件的 Click 事件。左键双击漫游 Button 控件，即可添加如下该按钮的 Click 事件代码，实现生成等值线：

```
private void ExtractIsoline_Click(object sender,EventArgs e)
    {try{progressBar.Value = 10;
    m_sampleRun.ExtrsctIsoLine();//通过 sampleRun.ExtrsctIsoLine 生成等值线
    progressBar.Value = 100;}
catch(Exception ex)
{Trace.WriteLine(ex.Message);}}
```

② 等值面 Button 控件的 Click 事件。左键双击全屏 Button 控件，即可添加如下该按钮的 Click 事件代码，实现生成等值面：

```
private void ExtractIsoregion_Click(object sender,EventArgs e)
    {try{progressBar.Value = 10;
    m_sampleRun.ExtractIsoRegion();//通过 sampleRun.ExtractIsoRegion 生成等值面
    progressBar.Value = 100;}
catch(Exception ex)
{Trace.WriteLine(ex.Message);}}
```

③ 复原 Button 控件的 Click 事件。左键双击分析 Button 控件，即可添加如下该按钮的 Click 事件代码,实现复原：

```
private void ReView_Click(object sender,EventArgs e)
    {try{progressBar.Value = 10;
    m_sampleRun.OpenDataset();//通过 sampleRun.OpenDataset 实现复原
    progressBar.Value = 100;}
```

catch（Exception ex）

{Trace.WriteLine(ex.Message);}}

（4）窗体的 FormClose 事件

窗体的 FormClose 事件，代码同前文中的相应部分。

8. 运行结果

窗体设计完成后点击 ▶ 按钮，运行该程序，运行后弹出窗体，如图 10-2 所示，可以表示等值线、等值面。

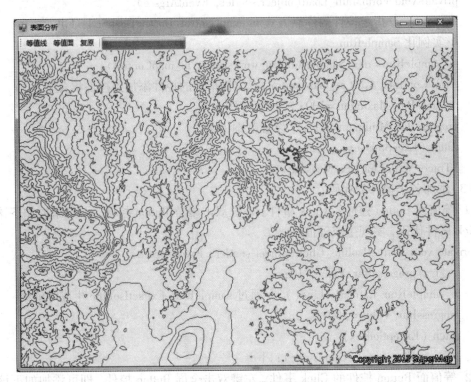

图 10-2　表面分析图

10.2　动态分段

1. 数据

在安装目录\SampleData\LinearReferencing\LinearReferencing.udb 路径下打开数据文件。

2. 新建文件夹和工程

在【D:\MyProject\】文件中新建一个文件夹，命名为【第 10 章　表面分析与动态分段】，并在其目录下建立一个文件夹，命名为【LinearReferencing】，在此文件夹里新建一个工程，将此工程命名为 LinearReferencing。

3. 关键类型/成员

关键类型/成员见表 10-3。

表 10-3　　　　　　　　　　　　　　　关键类型/成员表

控件/类	方法	属性	事件
Layers	Add、Clear		
MapControl		Map、Action	GeometrySelected
Workspace	Open		
DatasetVector	GetRecordset		
Recordset	GetGeometry、GetFieldValue、SetFieldValue		
LinearReferencing	GenerateRoutes、GenerateSpatialData		
GenerateRoutes-Parameter		Type、DatasetPoint、PointRouteIDField、LineRouteIDField、MeasureField、MeasureStartField、MeasureEndField、Tolerance、IgnoringGaps、MeasureOffset、FieldError、OutputDatasource、OutputDatasetName	
GenerateSpatialData-Parameter		EventTable、ReferenceLineM、EventRouteIDField、RouteIDField、MeasureStartField、MeasureEndField、ErrorInfoField、OutputDatasource、OutputDatasetName	
DynamicSegment-Manager	RemoveDynamicSegmentInfos		

4. 窗体控件属性

窗体控件属性具体见表 10-4。

表 10-4　　　　　　　　　　　　　　　窗体控件属性表

控件	Name	Text
toolStrip	toolStrip1	toolStrip1
DataGridview	dataGridView	
Button	buttonLoadLine	加载线数据
	buttonGenerateRoutes	生成路由数据集
	buttonLoadEvent	加载事件表
	buttonGenerateSpatialData	生成空间数据
	buttonModify	修改事件表
	buttonOK	完成修改
	buttonReset	重置
	buttonEntireView	全幅显示

5. 窗体设计布局

窗体设计布局如图 10-3 所示。

图 10-3　窗体设计布局图

6. 创建 SampleRun 类

(1) 添加 using 引用代码

添加 using 引用代码具体如下：

using SuperMap. Data；//添加 SuperMap. Data 命名空间，使其对应的类可以使用

using System. Windows. Forms；//添加 System. Windows. Forms 命名空间，使其对应的类可以使用

using SuperMap. UI；//添加 SuperMap. UI 命名空间，使其对应的类可以使用

using SuperMap. Mapping；//添加 SuperMap. Mapping 命名空间，使其对应的类可以使用

using System. Drawing；//添加 System. Drawing 命名空间，使其对应的类可以使用

using SuperMap. Analyst. SpatialAnalyst；//添加 SuperMap. Analyst. SpatialAnalyst 命名空间，使其对应的类可以使用

using System. Diagnostics；//添加 System. Diagnostics 命名空间，使其对应的类可以使用

using System. IO；//添加 System. IO 命名空间，使其对应的类可以使用

(2) 在 public class SampleRun 中添加代码

在 public class SampleRun 中添加如下代码：

//定义各个数据集的名称常量　　定义类对象

private const String LINE_NAME = "Roads"；

private const String ROUTE_NAME = "Route"；

private const String LINE_TABLE_NAME = "LineEventTab"；

private const String SPATIAL_DATA_NAME = "LineSpatialData";
private Workspace m_workspace;// Workspace:工作空间是用户的工作环境,主要完成数据的组织和管理,包括打开、关闭、创建、保存工作空间文件。
private MapControl m_mapControl;// MapControl:地图控件类。该类是用于为地图的显示提供界面的,同时为地图与数据的互操作提供途径。
private Datasource m_datasource;// Datasource:数据源类。该类管理投影信息、数据源与数据库的连接信息和对其中的数据集的相关操作,如通过已有数据集复制生成新的数据集等。
private DataGridView m_dataGridView;// DataGridView:栅格数据集类,用于描述栅格数据,例如高程数据集和土地利用图。
private Boolean flag = false;//标记是路由还是空间数据,true 为空间数据
public SampleRun(Workspace workspace, MapControl mapControl, DataGridView dataGridView)
{m_workspace = workspace;
m_mapControl = mapControl;
m_dataGridView = dataGridView;
m_mapControl. Map. Workspace = m_workspace;
m_mapControl. Map. IsAntialias = true;//设置一个布尔值指定是否反走样地图
Initialize();}
///打开数据源,注册对象选择事件
private void Initialize()
{try{DatasourceConnectionInfo connectionInfo = new DatasourceConnectionInfo(@"..\..\SampleData\LinearReferencing\LinearReferencing. udb", "LinearReferencing", "");// DatasourceConnectionInfo:数据源连接信息类。包括进行数据源连接的所有信息,如所要连接的服务器名称,数据库名称、用户名、密码等。当保存工作空间时,工作空间中的数据源的连接信息都将存储到工作空间文件中。
connectionInfo. EngineType = EngineType. UDB;// EngineType:该枚举定义了空间数据库引擎类型常量。
m_datasource = m_workspace. Datasources. Open(connectionInfo);//根据指定的连接信息打开数据源
m_mapControl. GeometrySelected + = newGeometrySelectedEventHandler(m_mapControl_GeometrySelected);}
catch(System. Exception ex)
{Trace. WriteLine(ex. Message);}}
//加载用于生成路由的线数据,并初始化 DataGridView
public void AddDataForGenerateRoutes()
{//向 m_dataGridView 中添加列,m_dataGridView 用于显示生成的路由数据集的节点信息
{m_dataGridView. AutoResizeColumns();//调整所有列的宽度以适应其所有单元格

的内容

```
m_dataGridView.AutoSizeColumnsMode = DataGridViewAutoSizeColumnsMode.Fill;
m_dataGridView.Columns.Clear();//清除集合
m_dataGridView.Columns.Add("序号","序号");
m_dataGridView.Columns.Add("坐标","坐标");
m_dataGridView.Columns.Add("坐标","坐标");
m_dataGridView.Columns.Add("刻度值","刻度值");
//获取用于生成路由的线数据集
DatasetVector line = (DatasetVector)m_datasource.Datasets[LINE_NAME] as DatasetVector;
AddToMap(line,true);
//设置 mapControl 的当前操作状态
m_mapControl.Action = SuperMap.UI.Action.Select;//选择对象
m_mapControl.Map.Refresh();}//刷新
///生成路由数据集,生成方式为单字段
public void GenerateRoutes()
{try{GenerateRoutesParameter parameter = new GenerateRoutesParameter();//
```
GenerateRoutesParameter:生成路由参数类。该类提供生成路由数据集(LinearReferencing. GenerateRoutes)的生成路由的方式、刻度字段、刻度偏移量、刻度因子及输出等参数的设置。

```
parameter.Type = GenerateType.BySingleField;//使用单字段方式生成路由
parameter.MeasureEndField = "SHAPE_LENG";
parameter.IgnoringGaps = true;
parameter.OutputDatasource = m_datasource;
parameter.OutputDatasetName = ROUTE_NAME;//设置名称
if(m_datasource.Datasets.Contains(ROUTE_NAME))//检查当前数据源中是否包
```
含指定名称的数据集

```
{//删除该数据集的动态分段关系,否则无法删除该数据集
DynamicSegmentManager.RemoveDynamicSegmentInfos((DatasetVector)m_
```
datasource.Datasets[ROUTE_NAME]);// DynamicSegmentManager:动态分段管理类。利用动态分段管理类来管理数据源的动态分段信息,包括查询删除的指定数据源的所有动态分段信息;查询删除的指定数据集的所有动态分段信息。注意:不支持跨数据源。

```
m_datasource.Datasets.Delete(ROUTE_NAME);}//根据指定的名称来删除数据
```
集

```
//调用 GenerateRoutes 方法生成路由数据集
DatasetVector referenceLine = m_datasource.Datasets[LINE_NAME] as DatasetVector;
//获取数据源包含的数据集的集合对象
DatasetVector resultRoute = LinearReferencing.GenerateRoutes(referenceLine,
parameter);// LinearReferencing:线性参考类。该类提供用于动态分段的生成路由数据集、
```

校准路由数据集、创建事件表、创建空间数据及事件表的叠加与融合等功能的实现。

```csharp
    if(resultRoute! =null)
    {flag=false;
    //加载该线数据集到地图上并设置其图层风格
    Layer layer=AddToMap(resultRoute,false);//将数据集添加到当前地图上
LayerSettingVector lineSettingVector=layer.AdditionalSetting as LayerSettingVector;//设置图层的风格设置
    GeoStyle lineStyle=new GeoStyle();
    lineStyle.LineColor=Color.FromArgb(128,128,255);//设置颜色
    lineStyle.LineWidth=0.4;//设置线宽
    lineSettingVector.Style=lineStyle;//设置风格
    m_mapControl.Map.Refresh();}}//刷新
    catch(Exception ex)
    {Trace.WriteLine(ex.Message);}}
    //加载事件表,并添加事件表的信息到 m_dataGridView 中
public void AddDataForGenerateSpatialData()
    {m_dataGridView.AutoResizeColumns();//调整所有列宽
    m_dataGridView.AutoSizeColumnsMode=DataGridViewAutoSizeColumnsMode.Fill;
    m_dataGridView.Columns.Clear();//清除集合
    m_dataGridView.Columns.Add("SMID","SMID");//将具有给定列名和列标题文本添加到集合中
    m_dataGridView.Columns.Add("路由 ID","路由");
    m_dataGridView.Columns.Add("起始刻度值字段","起始刻度值字段");
    m_dataGridView.Columns.Add("终止刻度值字段","终止刻度值字段");
    for(int i=0; i< m_dataGridView.Columns.Count; i++)
    {m_dataGridView.Columns[i].ReadOnly=true;//设置一个值,指示用户是否可以编辑列的单元格}
    //添加事件表信息到 m_dataGridView 中
    DatasetVector lineEvent = m_datasource.Datasets[LINE_TABLE_NAME] as DatasetVector;//获取数据源所包含的数据集的集合对象
    m_dataGridView.Rows.Clear();//删除集合
    Recordset recordset=lineEvent.GetRecordset(false,CursorType.Static);//根据给定的参数来返回空的记录集
    m_dataGridView.RowCount=recordset.RecordCount;//设置显示行数
    recordset.MoveFirst();//用于移动当前记录位置到第一条记录,使第一条记录成为当前记录
    for(Int32 i=0; i< m_dataGridView.RowCount; i++)
    {m_dataGridView.Rows[i].Cells[0].Value=recordset.GetFieldValue("SMID");
    //根据字段名指定字段,获得数据集的属性表中当前记录该字段的值
```

```
            m_dataGridView.Rows[i].Cells[1].Value = recordset.GetFieldValue("
ROUTEID");//根据字段名指定字段,获得数据集的属性表中当前记录该字段的值
            m_dataGridView.Rows[i].Cells[2].Value = recordset.GetFieldValue("
STARTMEASURE");//根据字段名指定字段,获得数据集的属性表中当前记录该字段的值
            m_dataGridView.Rows[i].Cells[3].Value = recordset.GetFieldValue("
ENDMEASURE");//根据字段名指定字段,获得数据集的属性表中当前记录该字段的值
            recordset.MoveNext();}
        recordset.Close();//关闭记录集
        //设置图层不可选择
        for(Int32 i=0; i < m_mapControl.Map.Layers.Count; i++)
        {m_mapControl.Map.Layers[i].IsSelectable=false;}//设置中对象是否可以选择
        m_mapControl.Map.Refresh();}//刷新
        ///由路由数据集合事件表生成空间数据
    public void GenerateSpatialData()
        {try{DatasetVector referenceLineM = m_datasource.Datasets[ROUTE_NAME] as
DatasetVector;//获取数据源所包含的数据集的集合对象
        //获取用于生成空间数据的事件表
        DatasetVector eventTable = m_datasource.Datasets[LINE_TABLE_NAME] as
DatasetVector;
        GenerateSpatialDataParameter parameter = new GenerateSpatialDataParameter();
        parameter.ReferenceLineM = referenceLineM;
        parameter.RouteIDField = "ROUTEID";
        parameter.EventTable = eventTable;
        parameter.EventRouteIDField = "ROUTEID";
        parameter.MeasureStartField = "STARTMEASURE";
        parameter.MeasureEndField = "ENDMEASURE";
        parameter.ErrorInfoField = "ErrorInfo";
        parameter.OutputDatasetName = SPATIAL_DATA_NAME;
        if(m_datasource.Datasets.Contains(SPATIAL_DATA_NAME)){//检查当前数据
源中是否包含指定名称的数据集
        DynamicSegmentManager.RemoveDynamicSegmentInfos(m_datasource.Datasets
[SPATIAL_DATA_NAME] as DatasetVector);//获取数据源所包含的数据集的集合对象
        m_datasource.Datasets.Delete(SPATIAL_DATA_NAME);}//根据指定的名称来删
除数据集
        //调用GenerateSpatialData方法生成空间数据
        DatasetVector resultSpatialData = LinearReferencing.GenerateSpatialData(parameter);
        if(resultSpatialData != null)
            {flag = true;
        //对生成的空间数据制作专题图
```

```csharp
StreamReader reader = new StreamReader(@"..\..\SampleData\LinearReferencing\LinearReferencing.xml");
String xml = reader.ReadToEnd();//从流的当前位置到末尾读取流
ThemeUnique theme = new ThemeUnique();//ThemeUnique:单值专题图类。将字段或表达式的值相同的要素采用相同的风格来显示,从而用来区分不同的类别。
theme.FromXML(xml);
m_mapControl.Map.Layers.Add(resultSpatialData, theme, true);//添加图层
m_mapControl.Map.Refresh();
for (Int32 i=0; i < m_mapControl.Map.Layers.Count; i++)
{if (m_mapControl.Map.Layers[i].Dataset.Name==SPATIAL_DATA_NAME)
{m_mapControl.Map.Layers[i].IsSelectable=true;}//设置图层对象是否可以选择
else
{m_mapControl.Map.Layers[i].IsSelectable=false;}}}
catch (Exception ex)
{Trace.WriteLine(ex.Message);}}
///选中对象后显示属性信息
    private void m_mapControl_GeometrySelected(object sender, GeometrySelectedEventArgs e)
{try{ Recordset recordset = m_mapControl.Map.Layers[0].Selection.ToRecordset();//需要保证要显示信息的对象所在图层为地图的最上面
Geometry geometry = recordset.GetGeometry();
GeoLineM lineM = geometry as GeoLineM;// GeoLineM:路由对象,是一组具有X,Y坐标与线性度量值(M值)的点组成的线性地物对象。比如高速公路上的里程碑,交通管制部门经常使用高速公路上的里程碑来标注并管理高速公路的路况、车辆的行驶限速和高速事故点等。
if (geometry is SuperMap.Data.GeoLineM)
{if (flag==true)//选中的路是空间数据
{String id = recordset.GetFieldValue("EVENT_SMID").ToString();
for (Int32 i=0; i < m_dataGridView.Rows.Count; i++)
{if (m_dataGridView.Rows[i].Cells[0].Value.ToString()==id)
{m_dataGridView.Rows[i].Selected=true;//设置一个指示行是否已被选定的值
m_dataGridView.CurrentCell=m_dataGridView.Rows[i].Cells[0];//设置当前处于活动状态的单元格
return;}}}
else
{//将该路有对象的节点信息显示到 DataGridView 中
FillDataGridView(lineM);}}
recordset.Close();//关闭记录集
recordset.Dispose();}//释放资源
```

```csharp
            catch(Exception ex)
            {Trace.WriteLine(ex.Message);}
            /// 填充 DataGridView
    private void FillDataGridView(GeoLineM lineM)
            {try{m_dataGridView.Rows.Clear();//清除集合
        m_dataGridView.RowCount=lineM[0].Count;//获取显示的行数
            for(Int32 i=0; i < m_dataGridView.RowCount; i++)
                {m_dataGridView.Rows[i].Cells[0].Value=i+1;//获取与此单元格关联的值
                m_dataGridView.Rows[i].Cells[1].Value=lineM[0][i].X;
                m_dataGridView.Rows[i].Cells[2].Value=lineM[0][i].Y;
                m_dataGridView.Rows[i].Cells[3].Value=lineM[0][i].M;}}
            catch(Exception ex)
            {Trace.WriteLine(ex.Message);}}
            ///开始修改事件
    public void Modify()
            {try{//实践中的路由器起始时刻和终止时刻字段可以修改
            for(Int32 i=2; i < m_dataGridView.Columns.Count; i++)
                {m_dataGridView.Columns[i].ReadOnly=false;}}//设置一个值,指示用户是否可以编辑列的单元格
            catch(Exception ex)
            {Trace.WriteLine(ex.Message);}}
            ///确认修改完成
    public void Complete()
            {//修改完成后,使起始时刻和终止时刻字段不能修改
            for(int i=2; i < m_dataGridView.Columns.Count; i++)
                {m_dataGridView.Columns[i].ReadOnly=true;}} //设置一个值,指示用户是否可以编辑列的单元格
            /// 根据 DataGridView 的单元格修改后的值来修改事件表中对应的记录
    public void Conform(Int32 changedRow)
            {try{DatasetVector lineEvent = m_datasource.Datasets[LINE_TABLE_NAME] as DatasetVector;//获取数据源所包含的数据集的集合对象
            Recordset recordset=lineEvent.GetRecordset(false, CursorType.Dynamic);
            recordset.MoveTo(changedRow);//用于移动当前记录到指定的位置
            recordset.Edit();//锁定并编辑记录集的当前记录
            recordset.SetFieldValue("STARTMEASURE", Double.Parse(m_dataGridView.Rows[changedRow].Cells[2].Value.ToString()));//根据要修改的字段的名称,设定记录集中相应字段的值
            recordset.SetFieldValue("ENDMEASURE", Double.Parse(m_dataGridView.Rows[changedRow].Cells[3].Value.ToString()));
```

```
        recordset.Update();//用于提交对记录集的修改
        m_mapControl.Map.Refresh();}
    catch(Exception ex)
        {Trace.WriteLine(ex.Message);}}
    ///  重置
    public void Reset()
        {try{m_dataGridView.Rows.Clear();//清除集合
        m_mapControl.Map.Layers.Clear();//删除此图层集合对象中所有的图层
        m_mapControl.Map.Refresh();}
        catch(Exception ex)
        {Trace.WriteLine(ex.Message);}}
        ///  将数据集添加到当前地图上,添加图层
    private Layer AddToMap(DatasetVector datasetVector,Boolean isClearLayers)
        {if(isClearLayers)
        {m_mapControl.Map.Layers.Clear();}//删除此图层集合对象中所有的图层
        Layer layer=m_mapControl.Map.Layers.Add(datasetVector,true);//添加图层
        m_mapControl.Map.Refresh();
return layer;}
```

7. 窗体代码

(1)添加 using 引用代码

添加 using 引用代码具体如下:

using SuperMap.Data;//添加 SuperMap.Data 命名空间,使其对应的类可以使用

using SuperMap.Analyst.SpatialAnalyst;//添加 SuperMap.Analyst.SpatialAnalyst 命名空间,使其对应的类可以使用

using SuperMap.Mapping;//添加 SuperMap.Mapping 命名空间,使其对应的类可以使用

using SuperMap.UI;//添加 SuperMap.UI 命名空间,使其对应的类可以使用

using System.Diagnostics;//添加 System.Diagnostics 命名空间,使其对应的类可以使用

(2)在 public FormMain()函数前添加代码

在 public FormMain()函数前添加以下代码:

//定义类对象

private Workspace m_workspace;// Workspace:工作空间是用户的工作环境,主要完成数据的组织和管理,包括打开、关闭、创建、保存工作空间文件。

private MapControl m_mapControl;

private SampleRun m_sampleRun;

private Int32 m_changeRow=-1;

private Boolean m_isBeginChanged=false;//是否开始修改空间数据的属性表

(3)窗体的 Load 事件

窗体的 Load 事件,在窗体的属性窗口点击 ⚡,左键双击 Load,添加如下代码:
```
private void FormMain_Load(object sender, EventArgs e)
    {m_workspace = new Workspace();
    m_mapControl = new MapControl();
    m_mapControl.Dock = DockStyle.Fill;
    m_mapControl.Map.IsAntialias = true;
    this.splitContainer1.Panel1.Controls.Add(m_mapControl);
    this.splitContainer1.Panel1.Controls.SetChildIndex(m_mapControl, 0);
    //按钮的初始化状态
    ButtonsInitialStatus();
    m_sampleRun = new SampleRun(m_workspace, m_mapControl, dataGridView);}
```

(4) 设置工具条上按钮的初始状态

设置工具条上按钮的初始状态,需添加如下代码:
```
private void ButtonsInitialStatus()
    {//设置控件是否可用
    buttonLoadLine.Enabled = true;
    buttonGenerateRoutes.Enabled = false;
    buttonLoadEvent.Enabled = false;
    buttonGenerateSpatialData.Enabled = false;
    buttonModify.Enabled = false;
    buttonOK.Enabled = false;
    buttonReset.Enabled = false;}
```

(5) 添加各个控件的代码

① 加载线数据 Button 控件的 Click 事件。左键双击漫游 Button 控件,即可添加如下该按钮的 Click 事件代码,实现加载线数据:
```
private void buttonLoadLine_Click(object sender, EventArgs e)
    {try{//加载用于生成路由的数据
    m_sampleRun.AddDataForGenerateRoutes();//通过 sampleRun.AddDataForGenerateRoutes 实现添加线数据
    buttonLoadLine.Enabled = false;
    buttonGenerateRoutes.Enabled = true;
    buttonReset.Enabled = true;}
    catch(Exception ex)
    {Trace.WriteLine(ex.Message);}}
```

② 生成路由数据集 Button 控件的 Click 事件。左键双击全屏 Button 控件,即可添加如下该按钮的 Click 事件代码,实现生成路由数据集:
```
private void buttonGenerateRoutes_Click(object sender, EventArgs e)
```

{try}//生成路由数据集

m_sampleRun.GenerateRoutes();//通过sampleRun.GenerateRoutes()实现生成路由数据集

  buttonGenerateRoutes.Enabled=false;

  buttonLoadEvent.Enabled=true;

  buttonReset.Enabled=true;}

  catch(Exception ex)

  {Trace.WriteLine(ex.Message);}}

③ 加载事件表Button控件的Click事件。左键双击分析Button控件，即可添加如下该按钮的Click事件代码，实现加载事件表：

private void buttonLoadEvent_Click(object sender,EventArgs e)

  {try}//加载用于生成空间数据的事件表

m_sampleRun.AddDataForGenerateSpatialData();//通过sampleRun.AddDataForGenerateSpatialData加载事件表

  buttonLoadEvent.Enabled=false;

  buttonGenerateSpatialData.Enabled=true;

  buttonReset.Enabled=true;}

  catch(Exception ex)

  {Trace.WriteLine(ex.Message);}}

④ 生成空间数据Button控件的Click事件。左键双击分析Button控件，即可添加如下该按钮的Click事件代码，实现生成空间数据：

private void buttonGenerateSpatialData_Click(object sender,EventArgs e)

  {try}//生成空间数据

m_sampleRun.GenerateSpatialData();//通过sampleRun.GenerateSpatialData生成空间数据

  buttonGenerateSpatialData.Enabled=false;

  buttonModify.Enabled=true;

  buttonReset.Enabled=true;}

  catch(Exception ex)

  {Trace.WriteLine(ex.Message);}}

⑤ 修改事件表Button控件的Click事件。左键双击分析Button控件，即可添加如下该按钮的Click事件代码，实现修改事件表：

private void buttonModify_Click(object sender,EventArgs e)

  {try{m_isBeginChanged=true;

m_sampleRun.Modify();//通过sampleRun.Modify 修改事件表

  buttonModify.Enabled=false;

  buttonOK.Enabled=true;

buttonReset.Enabled=true;}

catch(Exception ex)

{Trace.WriteLine(ex.Message);}}

⑥ 完成修改 Button 控件的 Click 事件。左键双击分析 Button 控件，即可添加如下该按钮的 Click 事件代码，实现完成修改：

private void buttonOK_Click(object sender, EventArgs e)

{try{m_isBeginChanged=false;

m_sampleRun.Complete();//通过 sampleRun.Complete 实现完成修改

buttonOK.Enabled=false;

buttonModify.Enabled=true;

buttonReset.Enabled=true;}

catch(Exception ex)

{Trace.WriteLine(ex.Message);}}

⑦ 重置 Button 控件的 Click 事件。左键双击分析 Button 控件，即可添加如下该按钮的 Click 事件代码，实现重置：

private void buttonReset_Click(object sender, EventArgs e)

{try{ButtonsInitialStatus();

m_sampleRun.Reset();}//通过 sampleRun.Reset 实现重置

catch(Exception ex)

{Trace.WriteLine(ex.Message);}}

⑧ 全幅显示 Button 控件的 Click 事件。左键双击分析 Button 控件，即可添加如下该按钮的 Click 事件代码，实现全幅显示：

private void buttonEntireView_Click(object sender, EventArgs e)

{try{m_mapControl.Map.ViewEntire();}//通过 mapControl.Map.ViewEntire 实现全幅显示

catch(Exception ex)

{Trace.WriteLine(ex.Message);}}

⑨ dataGridView 控件的 CellValueChanged 事件。在控件的属性窗口单击 图标，左键双击 CellValueChanged，添加如下代码，实现修改 dataGridView 单元格的值时修改对应的记录：

private void dataGridView_CellValueChanged(object sender, DataGridViewCellEventArgs e)

{try{if(m_isBeginChanged)

{m_changeRow=e.RowIndex;

m_sampleRun.Conform(m_changeRow);}}//通过 sampleRun.Conform 实现修改记录

catch(Exception ex)

{Trace.WriteLine(ex.Message);}}

(5) 窗体的 FormClose 事件

窗体的 FormClose 事件，代码同前文中的相应部分。

8. 运行结果

窗体设计完成后点击 ▶ 按钮，运行该程序，运行后弹出窗体，如图 10-4 所示。图 10-4 表示动态分段的加载线数据生成路由数据、生成空间数据、事件表与空间数据的联动、事件表的修改。

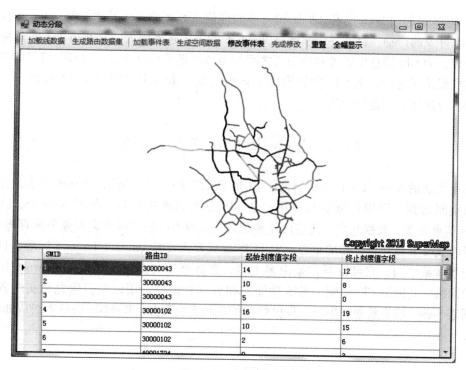

图 10-4 动态分段图

第 11 章 WebGIS 设计与开发

WebGIS 是利用 Web 技术来扩展和完善地理信息系统的一项技术，是 Internet 技术应用于 GIS 开发的产物。目前各个领域对 WebGIS 的需求越来越大，它开拓了 GIS 资源利用的新领域，为 GIS 信息提供者和使用者之间的沟通提供了有效途径，为 GIS 信息的高度社会化共享提供了可能，为 GIS 发展提供了新的机遇。本章主要围绕 WebGIS 的定义、特点、技术方法和应用进行阐述。

11.1 WebGIS 的定义及其组成

飞速发展的 Internet/Intranet 已经成为 GIS 新的系统平台。利用 Internet 技术，在 Web 上发布空间数据，供用户浏览和使用，是 GIS 发展的必然趋势。所谓 WebGIS，就是在 Internet 信息发布、数据共享、交流协作基础之上实现 GIS 的在线查询和业务处理等功能，Web 分布式交互操作是工作的重心。WebGIS 显然要求支持 Internet/Intranet 标准，具有分布式应用体系结构，它可以看作是由多主机、多数据库与多台终端通过 Internet/Intranet 组成的网络，其网络 Client 端为 GIS 功能层和数据管理层，用以获得信息和实施各种应用；网络 Server 端为数据维护层，提供数据信息和系统服务。WebGIS 的基本组成如图 11-1 所示。

图 11-1 WebGIS 组成

① Web 浏览器是用户和 WebGIS 的交互接口，用来显示地图和实现客户端的在线查询和分析功能；

② Web 服务器响应来自 Web 浏览器的请求，通过 CGI、Servlet 将请求传递给 Map 服务器，并从 Map 服务器得到请求结果发还给浏览器；

③ Map 服务器是 WebGIS 的核心，它负责将 Web 服务器转发过来的用户请求分配给相应的 GIS 服务器或空间数据库，并能够实现网络的负载平衡；

④ GIS 服务器是 WebGIS 的底层 GIS 软件，它提供了空间数据的存取、查询、分析、

处理等功能；

⑤ 空间数据库用来存储和管理空间数据；

⑥ 浏览器和服务器之间是通过超文本传输协议 HTTP 来发送请求和结果数据，数据传输的格式有基于栅格的、基于矢量的和基于 XML 的；

⑦ 以上不同的服务器可以部署在不同的计算机上。

11.2 WebGIS 的特点

WebGIS 使各类用户能够通过浏览器对空间数据进行访问，实现检索、查询、制图输出和编辑等 GIS 基本功能，具有传统 GIS 的所有特点，在以下几个方面，WebGIS 还显示出与传统 GIS 的根本区别：

1. 基于 Internet/Intranet 标准

WebGIS 采用 Internet/Intranet 标准，以标准的 HTML 浏览器为客户端，通过 TCP/IP 和 HTTP 协议，可以访问任何地方的空间数据。

2. 分布式的体系结构

空间数据本身空间上是分布的，WebGIS 采用分布式的体系结构，形成了客户端和服务器端相互分离、协同工作的多层分布结构，通过各种均衡策略有效平衡两者之间的负载，这种结构适应了空间数据分布的特征，提高了网络资源和存储资源的利用率。

3. 服务范围广

WebGIS 服务范围的广泛主要体现在一方面可以通过网络为更加广阔范围内的用户提供空间信息服务，另一方面，WebGIS 用户可以同时访问多个位于不同地方的服务器上的最新数据，而这一 Internet/Intranet 所特有的优势极大地方便了 GIS 的数据管理。

4. 平台无关

一般来讲，WebGIS 的客户端采用的是通用浏览器，因此对客户端的软硬件没有特殊要求，在服务器端无论采用什么样的操作系统和 GIS 软件，由于通过网络将请求和处理结果发给客户端，WebGIS 服务器的处理方式对客户端而言是透明的，任意一个用户都可以通过通用浏览器访问任何得到许可的 WebGIS 服务器，这种特性使得远程异构数据共享成为可能，极大地提高了软硬件平台的独立性。

当然在具体实现方面，由于各技术的特点不同，在平台无关实现程度方面也是不一样的。从后面的阐述中，我们将看到，为了提高系统的整体效率，有些技术与具体的操作平台之间并没有达成完全的无关。

5. 成本低廉、操作简单

在 WebGIS 实现中，客户端往往只需要使用 WebGIS 浏览器（有时可能安装一些插件以处理图形数据），而数据和软件的管理与维护基本上由服务器端完成，因此系统的成本比以往的全套专业的 GIS 软件平台要少得多，客户端软件的简单性所节省的维护费用也是不容忽视的。

6. 支持地理分布存储的多源数据

WebGIS 能充分利用已有的各种空间信息资源，支持地理分布存储的多重来源和格式的空间数据，不仅有利于数据的维护和更新，而且又利于平衡系统的负载，提高存储速度。

11.3 传统 WebGIS 的分类

WebGIS 是一个分布式的处理系统，它通常包括三个基本方面：客户端、服务器和空间数据库，WebGIS 中的客户端由通用的 Web 浏览器组成，必要时再加上插件共同构成，它是用户使用 WebGIS 的界面，用户通过它提出请求，获得结果。WebGIS 服务器包括 www、mail、ftp 服务器和 GIS 服务器，通过 HTTP 协议和 TCP/IP 协议为用户提供信息交换的通道和地理信息处理的功能。空间数据库则为客户的数据请求和 WebGIS 的各种处理功能提供空间数据。根据 WebGIS 服务器的组成结构和其与空间数据库关系的不同，可以把 WebGIS 分为以下两种。

1. 基于浏览器/服务器模式的 WebGIS

客户通过 HTTP 协议向 Web 服务器请求数据服务，服务器返回以 HTML 方式描述的界面。在实现形式上，这种结构的 WebGIS 分为动态和主动两类。

(1) 动态 WebGIS

WebGIS 最早出现的时候，只是简单地将地图图片链接到网页上，对于任意一个用户查询，服务器端也仅仅是把预先形成的相同的地形文件和数据返回给客户端，由于页面固定、数据量大，在多用户并发访问时，很容易形成网络堵塞，此后的改进方案就是在服务器端使用公共网关接口(CGI)技术，由 CGI 程序负责处理客户请求，将请求指令发往运行于后台的 GIS 服务器，再将服务器处理的结果返回给用户，这是一个动态的操作空间数据库并生成相应的查询结果的方式，被称为动态 WebGIS，或是被动的 WebGIS。

(2) 主动的 WebGIS

与被动的 WebGIS 不同，在主动的 WebGIS 中，由服务器向客户端发送一段能运行在客户机上的程序，由该程序处理用户的简单请求，需要矢量数据时，直接向服务器申请；对于那些复杂的客户请求(如空间分析)，则由服务器处理，处理的结果也以矢量的形式返回至客户端。

2. 基于中间件技术的 WebGIS

基于浏览器/服务器模式的 WebGIS 在早期可以满足许多应用需求，但随着应用复杂度的提高和数据量的膨胀，客户对 GIS 服务器的访问频率增加，单一的服务器和应用复杂的应用程序无法快速地处理大量的地理信息系统服务请求，与此同时，中间件技术这一软件设计和开发的新模式得到迅速的发展，WebGIS 适时地引入了这一新技术，极大地改进了传统模式 WebGIS 的体系结构和系统的运行效率。

中间件技术应用于 WebGIS 时，客户端的请求均通过中间件处理。GIS 服务器包括由中间件组成的分布式的多个进程，由于存在多个中间件，中间件和中间件之间的关系比较复杂，它们可以相互调用，一个中间件的进程可能是其他中间件的客户，同时也可能是其他中间件的进程的服务，中间件内的进程所访问的空间数据库也不再是单一的数据库，可能是分布式的异质、异构和多元数据库。

基于中间件技术的 WebGIS 是一个多浏览器和多服务器模式的复杂系统，各中间件的组织通过既定的接口实现，而用户的调用成动态特征，即只有当接收到客户请求时才动态地装载中间件并处理地理信息。因此,浏览器和服务器之间的负载是动态的,需要解决它们之间的动态负载平衡问题。目前,支持分布式的计算的主要中间件技术有 CORBA、DCOM、J2EE、.NET 等。

11.4 WebGIS 的数据模型

在万维网服务器端,利用类似于 CGI 的 Web 服务器环境,可以对某种空间数据库进行功能强大的访问;或者用独立于平台的 Java Applet 和特定平台的 ActiveX 及浏览器插件技术增强客户端的功能;或者把两者结合起来,达到万维网应用系统的最佳状态,三种客户—服务器结构如图 11-2 所示。

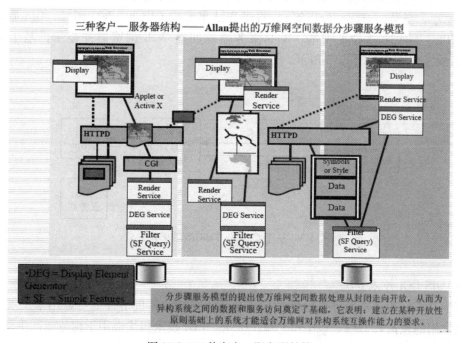

图 11-2 三种客户—服务器结构

① 客户端请求地图图像的方式。在这种结构下,作为客户端的浏览器只进行图像的显示,而把选择空间数据、生成显示元素序列和地图图像的步骤放在服务器端。浏览器通过服务器的 CGI 接口以 JPEG 或 GIF 图像格式请求地图图像。

② 客户端请求图形元素的方式。客户端由地图生成和显示两部分组成,通过 Java Applet、ActiveX 来实现,由它们向服务器请求要显示的图形元素或地图图像。随着 SVG(Scalable Vector Graphics)和 WebCGM 成为万维网协会(World Wide WebConsortium,W3C)的标准,如果用它们来编码矢量空间数据,则浏览器就可以直接显示。

③ 客户端请求空间数据的方式。服务器端只执行查询,从空间数据源中得到需要的空间数据,然后把数据发送到客户端。由浏览器上的 Java Applet、ActiveX 或浏览器插件来进行后面的工作。浏览器生成最终结果时,还会向服务器请求必要的显示符号信息。

11.5 传统 WebGIS 的实现方法

1. CGI——通用网关接口

GIS 厂商在其产品基础上发展 Internet 解决方案通常采用 CGI，即提供专用空间数据库的 Web 接口。CGI 通用网关接口是较早应用于 WebGIS 开发的方法，CGI 程序与 HTML 结合实现交互式动态通信。CGI 是连接应用软件和 Web 服务器的标准技术，HTML 的功能扩展，在超文本文件与 Web 服务器应用程序之间传递信息，将 Web 服务器和数据库服务器结合起来，实时、动态地生成 HTML 文件，基于 CGI 的 WebGIS 的工作原理是：Web 服务器用户发出 URL 及 GIS 数据操作请求到服务器上，Web 服务器接受请求后，通过 CGI 脚本，将用户的请求传递给 GIS 服务器，GIS 服务器接受请求，对 GIS 数据进行处理（如放大、缩小、漫游、查询、分析等），将操作结果形成 GIF 和 JPEG 图像，GIS 服务器将 GIF 或 JPEG 图像通过 CGI 脚本、Web 服务器返回给 Web 浏览器并显示，基于 CGI 的 WebGIS 的工作方式如图 11-3 所示。

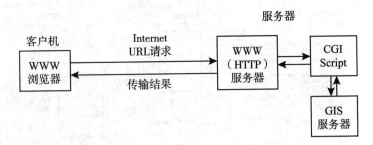

图 11-3 基于 CGI 的 WebGIS 的工作方式

(1) GIS CGI 技术方法

由于 CGI 是服务器上可执行程序，基本上所有的计算机语言都可以用来发展 CGI 程序，最常用的包括 C/C++、Perl 和 Visual Basic。CGI 技术很快被用于构造能产生动态地图的 Web 网站，根据程序特点，GIS CGI 技术方法可分为以下两类：

① CGI 启动制图软件以批处理方式运行：这种系统的长处是程序简单，运行速度快，但功能不足，而且大多数 GIS 软件不能以批处理方式运行。

② CGI 启动后端 GIS 程序：一般 GIS 软件都基于图形用户界面"事件"驱动，一旦启动，就可以一直后台运行等待触发事件。于是网络浏览器上的事件，通过 CGI 很容易传到后端 GIS 软件上，CGI 和后台 GIS 软件的信息交换是通过"进程间通讯协议"IPC 来完成的，常见的 IPC 协议有传统 UXIX 上的远程过程调用 RPC，Windows 的动态数据交换 DDE 和实体连接与嵌入 OLE 以及 MAC 机上的 Apple event。利用这种 CGI 方法，只要用户在网络浏览器按一下，信息就通过网络传回到网络服务器上，然后由 CGI 程序将此信息通过 IPC 传到后端的 GIS 软件上。例如，可以要求 GIS 软件将地图某个地区放大，然后将放大后的地图屏幕图像传回给用户。

(2) 基于 CGI 的 WebGIS 系统的优势

基于 CGI 的 WebGIS 系统的优点在于运行速度较 CGI 启动制图软件快，因为它不需要每次启动后端的 GIS 软件，同时可以利用商业化 GIS 软件产生高质量的地图。事实上，GIS 软件的所有功能都可以被利用起来。

由于所有的 GIS 操作都是由 GIS 服务器完成的，因此基于 CGI 的 WebGIS 系统具有客户端小、处理大型 GIS 操作分析的功能强、充分利用现有的 GIS 操作分析资源等优势。

由于在客户机端使用的是支持标准 HTML 的 Web 浏览器，操作结果是以静态的 GIF 或 JPEG 图像的形式表现，因而客户机端与平台无关。

(3) 基于 CGI 的 WebGIS 系统的不足

① 增加了网络传输的负担。由于用户的每一步操作，都需要将请求通过网络传给 GIS 服务器；GIS 服务器将操作结果形成新的栅格图像，再通过网络返回给用户。因而网络的传输量大大增加了。

② 服务器的负担重。所有的操作都必须由 GIS 服务器解释执行，服务器的负担很重；信息(用户的请求和 GIS 服务器返回的图像)通过 CGI 脚本在浏览器和 GIS 服务器之间传输，势必影响信息的传输速度。

③ 同步多请求问题。由于 CGI 脚本处理所有来自 Web 浏览器的输入和解释 GIS 服务器的所有输出。对于每一个客户机的请求，都要重新启动一个新的服务进程。当有多用户同时发出请求时，系统的功能将受到影响。

④ 静态图像。在浏览器上显示的是静态图像，因而用户既不能漫游、缩放，又不能通过几何图形如点、线、面来选择显示其关心的地物。

⑤ 用户界面的功能受 Web 浏览器的限制，影响 GIS 资源的有效使用。

2. GIS Plug-in 插件

基于 CGI 的系统仅提供给用户端(client)有限的 GIS 功能，传给用户的信息都是静态的，而且用户的 GIS 操作都需要由服务器来处理。解决这个问题的方法是把一部分服务器上的功能移到用户端上，这样不仅加快了用户操作的反应速度，而且也减少了互联网上的流量。标准万维网浏览器只提供了一些最基本的浏览和导航功能，而缺乏处理地理空间数据的能力。一种方法是安装额外能和网络浏览器交换信息的专门 GIS 软件。这种增加网络浏览器功能的方法就叫"插入法"(Plug-in)。为便于其他软件厂商发展插入型软件，Netscape 公司专门提供了一套应用程序接口(API)。目前这种插入软件已被普遍采用，在多媒体领域尤为明显。这种插入软件不但可以增加网络浏览器处理地理空间数据的能力，使人们更容易获取地理数据，而且可以减少网络服务器的信息流量从而使服务器更有效地为更多的用户服务，因为大多数用户的数据处理功能可以由网络浏览器插入软件来完成。与传统应用软件类似，插入软件也需要先安装再使用，因而传统软件不同版本之间的不兼容性及版本管理问题仍然存在。

GIS Plug-in 是在浏览器上扩充 Web 浏览器的可执行的 GIS 软件。GIS Plug-in 的主要作用是使 Web 浏览器支持处理无缝 GIS 数据，并为 Web 浏览器与 GIS 数据之间的通讯提供条件。GIS Plug-in 直接处理来自服务器的 GIS 矢量数据。同时，GIS Plug-in 可以生成自己的数据，以供 Web 浏览器或其他 Plug-in 显示使用。Plug-in 必须安装在客户机上，然后才能使用。其体系结构如图 11-4 所示。

(1) Plug-in 模式的工作原理

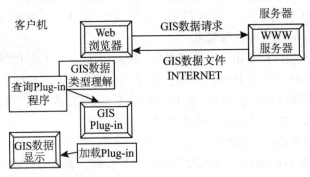

图 11-4 基于 Plug-in 的 WebGIS 的工作方式

Web 浏览器发出 GIS 数据显示操作请求；Web 浏览器接受到用户的请求，进行处理，并将用户所要的 GIS 数据传送给 Web 浏览器；客户机端接受 Web 服务器传来的 GIS 数据，并对 GIS 数据类型进行理解；在本地系统查找与 GIS 数据相关的 Plug-in（或 Helper）。如果找到相应的 GIS Plug-in 用它来显示 GIS 数据；如果没有，则需要安装相应的 GIS Plug-in，加载相应的 GIS Plug-in 来显示 GIS 数据。GIS 的操作如放大、缩小、漫游、查询、分析皆由相应的 GIS Plug-in 来完成。

（2）基于 Plug-in 的 WebGIS 系统的优势

① 无缝支持与 GIS 数据的连接。由于对每一种数据源都需要有相应的 GIS Plug-in，因而 GIS Plug-in 能无缝支持与 GIS 数据的连接。

② GIS 操作速度快。所有的 GIS 操作都是在本地由 GIS Plug-in 完成，因此运行的速度快。

③ 服务器和网络传输的负担轻。服务器仅需提供 GIS 数据服务，网络也只需将 GIS 数据一次性传输。服务器的任务很少，网络传输的负担轻。

（3）基于 Plug-in 的 WebGIS 系统的不足

① GIS Plug-in 与平台相关。对同一 GIS 数据，不同的操作系统如对 UNIX、Windows、Macintosh 而言，需要有各自不同的 GIS Plug-in 在其上使用。对于不同的 Web 浏览器，同样需要用相对应的 GIS Plug-in。

② GIS Plug-in 与 GIS 数据类型相关。对 GIS 用户而言，使用的 GIS 数据类型是多种多样的，如 ArcInfo、MapInfo、AtlasGIS 等 GIS 数据格式。对于不同的 GIS 数据类型，需要有相应的 GIS Plug-in 来支持。

③ 需要事先安装。用户如想使用，必须下载安装 GIS Plug-in 程序。如果用户准备使用多种 GIS 数据类型，必须安装多个 GIS Plug-in 程序。GIS Plug-in 程序在客户机上的数量增多，势必对管理带来压力。同时，GIS Plug-in 程序占用客户机磁盘空间。

④ 更新困难。当 GIS Plug-in 程序提供者已经将 GIS Plug-in 升级了，须通告用户进行软件升级。升级时，需要重新下载安装。

⑤ 使用已有的 GIS 操作分析资源的能力弱，处理大型的 GIS 分析能力有限。

3. Java Applet

尽管插件可以和网络浏览器一起有效处理空间数据，但这种方法仍有不少缺陷。首先,它将导致用户端负担过重,因为几乎每个软件厂商都希望它的软件能与互联网兼容。这显然不符合标准网络浏览器便宜简单的设计思想。其次,众多的插入软件的管理会成为信息技术部门的一个大问题,因为任何人只要可以连接上互联网都可以接收最新的插入软件,为解决上述各种问题,互联网程序语言应运而生。互联网程序语言的出现标志着 Web GIS 的开始,目前最普及的互联网程序语言是由 SUN 公司开发的 Java。

Java 是一种面向对象的计算机语言,它借鉴了 C、Smalltalk、Object C++ 和 Cedar/Mesa 等面向对象语言的优点,其特点是简洁动态、适应性强,运行稳定、安全,对网络而言,与计算机结构体系无关,容易移植,在一种系统下发展的应用软件可以直接在完全不同的系统下运行。事实上,Java 编译器产生的是一种独立于任何操作系统的字节码 Bitecode,这种字节码程序可以在任何一台 Java 虚拟机(Java Virtual Machine-JVM)上运行,任何系统只要支持 Java 虚拟机就可以运行 Java 程序,而与程序在何种系统下开发和编译无关,目前 Netscape 和微软公司的网络浏览器都直接支持 Java 程序。基于 GIS Jave Applet 的 WebGIS 的工作方式如图 11-5 所示。

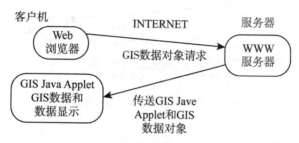

图 11-5 基于 GIS Java Applet 的 WebGIS 的工作方式

(1)基于 GIS Java Applet 的 WebGIS 系统的优势

① 体系结构中立,与平台和操作系统无关。在具有 Java 虚拟机的 Web 浏览器上运行。写一次,可到处运行。

② 动态运行,无须在用户端预先安装。由于 GIS Java Applet 是在运行时从 Web 服务器动态下载的,所以当服务器端的 GIS Java Applet 更新后,客户机端总是可以使用最新的版本。

③ GIS 操作速度快。所有的 GIS 操作都是在本地由 GIS Java Applet 完成,因此运行的速度快。

④ 服务器和网络传输的负担轻。服务器仅需提供 GIS 数据服务,网络也只需将 GIS 数据一次性传输。服务器的负担很小,网络传输的负担轻。

(2)基于 GIS Java Applet 的 WebGIS 系统的不足

① 使用已有的 GIS 操作分析资源的能力弱,处理大型的 GIS 分析能力有限。

② GIS 数据的保存、分析结果的存储和网络资源的使用能力有限。

4. ActiveX 方法

第二种互联网程序语言是由微软公司提出的 ActiveX,其实 ActiveX 是在 OLE 技术基础上发展起来的,ActiveX 仍然依赖现有 OLE 编程体系来达到增加互联网的交互性目的,方法之一是利用一个 OLE 文件实体 DocObject 作为一个通用控制容器 Container,例如,微软公司的网络浏览器 Explorer 就可以发展成为一个文件实体的控制容器，然后加入 GIS 引擎作为文件实体的服务器函数。这样，扩充后的网络浏览器就能显示和处理地理空间数据。ActiveX 也可以作为服务器，例如，微软公司的分布式 COM（Distributed Common Object Model），将使在用户和服务器两端的 Active 控件互相交换信息，从而把整个网络上的负荷分布到各个不同的子网上。与 Java 相比，ActiveX 目前还没有解决非常重要的网络安全问题。

（1）GIS ActiveX 的工作原理

Web 浏览器发出 GIS 数据显示操作请求；Web 服务器接受到用户的请求，进行处理，并将用户所要的 GIS 数据和 GIS ActiveX 控件传送给 Web 浏览器；客户机端接受到 Web 服务器传来的 GIS 数据和 GIS ActiveX 控件，启动 GIS ActiveX 控件，对 GIS 数据进行处理，完成 GIS 操作。工作方式如图 11-6 所示。

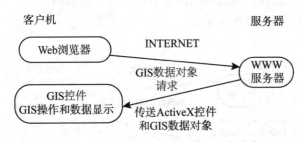

图 11-6　基于 ActiveX 的 WebGIS 的工作方式

（2）基于 GIS ActiveX 控件的 WebGIS 系统的优势

GIS ActiveX 模式具有 GIS Plug-in 模式的所有优点。同时，ActiveX 能被支持 OLE 标准的任何程序语言或应用系统所使用，比 GIS Plug-in 模式更灵活，使用方便。

（3）基于 GIS ActiveX 控件的 WebGIS 系统的不足

系统需要下载；占用客户机端机器的磁盘空间；与平台相关；对不同的平台，必须提供不同的 GIS ActiveX 控件；与浏览器相关；GIS ActiveX 控件最初只使用于 Microsoft Web 浏览器；在其他浏览器使用时，须增加特殊的 Plug-in 予以支持；使用已有的 GIS 操作分析资源的能力弱，处理大型的 GIS 分析能力有限。

5. WebGIS 系统构造模式优缺点对比

WebGIS 系统构造模式优缺点对比详见表 11-1。

表 11-1　　　　　　　　　　WebGIS 系统构造模式优缺点

类型	工作模式	实例	优点	缺点
基于 CGI 的 Internet GIS	CGI	IMS，ProServer	客户端很小；充分利用服务器的资源	JPEG 和 GIF 是客户端操作的两种形式；互联网和服务器的负担重，CGI 的应用程序一般都是可执行程序

续表

类型	工作模式	实例	优点	缺点
基于 Server API 的 Internet GIS	Server API	GeoBeans IMS	客户端很小；充分利用服务器的资源，以动态链接库的形式存在	JPEG 和 GIF 是客户端操作的两种形式；互联网和服务器的负担重
基于 Plug-in 的 Internet GIS	Plug-in	MapGuide	具有动态代码模块。比 HTML 更灵活，可直接操作 GIS 数据	与平台和操作系统相关；不同的 GIS 数据需要不同的 Plug-in 支持；必须安装在客户机的硬盘上
基于 ActiveX 的 Internet GIS	ActiveX	GeoMedia Web Map	具有动态代码模块。通过 OLE 与其他程序、模块和互联网通信，是一种通用的部件	需要下载、安装，占用硬盘空间；与平台和操作系统相关；不同的 GIS 数据需要不同的 ActiveX 控件支持
基于 Java Applet 的 Internet GIS	Java Applet	ActiveMap, GeoBeans	在支持 Java 的互联网浏览器上运行，与平台和操作系统无关；完成 GIS 数据解释和 GIS 分析功能	处理较大的 GIS 分析任务的能力有限；GIS 数据的保存、分析结果的存储和网络资源的使用能力有限

6. 模式评价

WebGIS 系统构造模式评价详见表 11-2。

表 11-2　　　　　　　　　　WebGIS 系统构造模式评价

执行能力		基于 CGI	基于 API	基于 Plug-in	Java Applet	ActiveX 控件
执行能力	客户机	很好	很好	好	好	好
	服务器	差到好	好	好	很好	很好
	网络	差	好	好	好	好
	总体	一般	好	好	好到很好	好到很好
相互作用	用户界面	差	好	好	很好	很好
	功能支持	一般	好	好	很好	很好
	本地数据支持	否	否	是	否	是
可移动性		很好	很好	差	好	一般
安全		很好	很好	一般	好	一般

传统的 WebGIS 的开发模式存在诸多弊端，采用瘦客户端开发模式及其技术，是一种纯粹的侧重服务器端的策略，在系统运行时弊端将突出表现在以下 3 个方面：

① 增加网络传输负担。由于用户的每一步操作，都需要将请求通过网络传给 GIS 服务器，GIS 服务器将操作结构形成新的栅格图像，再通过网络返回给用户，因而网络的传输量大大增加了。

② 服务器的负担重。所有的操作必须由 GIS 服务器来执行，再加上对于每个请求，服务器端都要生成一个新的系统进程来处理请求，导致服务器负载大，效率低。

③ 客户端操作限制大。由于最终的结果都是以静态图像的形式返回，一些极为频繁的操作，比如放大、缩小、漫游、点选地物等实现起来困难且响应延迟明显。

客户开发模式是纯粹的侧重客户端的策略，客户端与服务器端的数据交互是海量的空间数据，这将造成如下弊端：

① 频繁的海量数据的传输，在当前整个 Internet 体系结构下，将造成网络传输瓶颈，且在接下来的客户端的地图显示中，处理过程也会非常缓慢，严重影响系统性能。

② 空间分析能力不足，无法进行复杂的操作和分析，而这部分操作仍将放在服务器端执行。

综合这两种策略，形成一种混合型的解决办法，来更好地发挥服务器和客户端的优势与潜力，涉及繁重的数据库操作或复杂分析的任务让性能的一方来承担，一般来说是服务器，涉及用户控制的任务让客户端承担，这样，混合策略综合考虑客户机、服务器计算能力和网络通信量，适当地分布 GIS 任务，以充分使用客户机和服务器的计算功能，提高互操作性和系统性能，是一种非常行之有效的方法。但在现有的商业软件产品中，没有一种可以达到这种效果，寻求一种新方法，一种新的开发模式，一种廉价的开发方式，降低因使用商业软件而造成的昂贵的开发成本，对于 WebGIS 的发展就变得颇为重要。基于 JavaScript 技术的 WebGIS 开发模式就是在这种环境下设计的，这种模式应用的就是混合策略。

11.6 SuperMap iClient 7C for JavaScript

1. JavaScript 的优点

JavaScript 可以使多种任务仅在客户端就可以完成而不需要网络和服务器的参与，从而支持分布式的运算和处理，因此把 JavaScript 技术应用于 WebGIS，大大减轻了网络传输和服务器的负担。在这种技术下，所有的 GIS 操作都是在本地完成的，服务器仅需提供 GIS 数据服务，网络也只需将 GIS 数据一次性传输。

JavaScript 具有如下优点：

(1) 简单性

JavaScript 是一种基于 Java 基本语句和控制流之上的简单而紧凑的设计。它的变量类型是采用弱类型，并未使用严格的数据类型，不需要对程序进行预先编译而产生可执行的机器代码，使它比编译性语言更加易于编程和应用。

(2) 动态性

JavaScript 是动态的，它可以直接对用户或客户输入做出响应，无须经过 Web 服务程序。它对用户的反应响应是采用以事件驱动的方式进行的。所谓事件驱动，就是指在主页 (Home Page) 中执行了某种操作所产生的动作，就称为"事件"(Event)。比如按下鼠标、移动窗口、选择菜单等都可以视为事件。当事件发生后，可能会引起相应的事件响应。

(3)跨平台性

JavaScript 依赖于浏览器本身，与操作环境无关，只要能运行浏览器的计算机，并支持 JavaScript 的浏览器就可正确执行，从而实现了"编写一次，走遍天下"的梦想。

(4)安全性

JavaScript 是一种安全性语言，它不允许访问本地的硬盘，并不能将数据存入到服务器上，不允许对网络文档进行修改和删除，只能通过浏览器实现信息浏览或动态交互，从而有效地防止数据的丢失。

2. SuperMap iClient 7C for JavaScript 产品简介

SuperMap iClient7C for JavaScript 是一款在服务式 GIS 架构体系中，面向 HTML5 的应用开发，支持多终端、跨浏览器的客户端开发平台。SuperMap iClient 7C for JavaScript 采用 HTML+CSS+JavaScript 的开发组合，无须安装任何插件，便可在终端浏览器上实现美观的地图呈现，动态实时的要素标绘，以及与多源 GIS 服务的高效交互，快速构建内容丰富、响应迅速、体验流畅的地图应用，同时支持离线存储与访问地图功能，满足用户在离线状态下的地图应用。SuperMap iClient 7C for JavaScript 与其他产品的架构关系如图 11-7 所示。

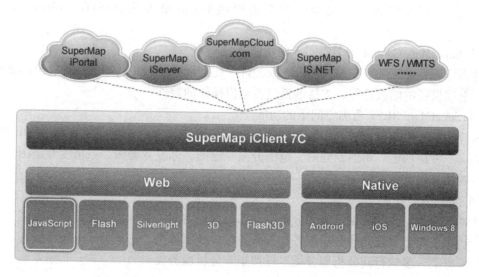

图 11-7　SuperMap iClient 7C for JavaScript 与其他产品的架构关系

3. SuperMap iClient 7C for JavaScript 的技术特点

(1)灵活的交互设计与丰富的数据呈现方式

SuperMap iClient 7C for JavaScript 产品面向 HTML 5 应用开发，可在 HTML 页面利用丰富的图形、图表、图像以及动画等实现 GIS 数据的动态呈现与灵活交互。基于 HTML 5 用于绘画的新元素 Canvas 实现了地图图片的高效、稳定呈现。SuperMap iClient 7C for JavaScript 脱离了客户端插件的限制，可灵活构建多终端、跨浏览器的服务式 GIS 应用。

(2) 支持多终端访问

SuperMap iClient 7C for JavaScript 支持多终端模式的 Web 应用开发，包括个人电脑、平板电脑、手机等多种终端上的浏览器应用，为用户的系统构架提供了丰富选择。

(3) 支持多源地图数据

SuperMap iClient 7C for JavaScript 支持 SuperMap iServer Java 7C 服务及多种标准第三方服务，并支持各种服务在客户端的无缝聚合。包括 OpenGIS 协会制定的 WMS、WFS、KML 等标准格式服务及 Google Maps、ArcGIS Map、Yahoo! Map、KaMap、MSVirtualEarth 等第三方服务。同时支持超图云服务，包括在线地图服务或通过第三方 API 开发得到的地图应用云。SuperMap iClient 7C for JavaScript 产品将在服务式 GIS 体系下为用户提供更好的 Web 应用支撑。

(4) 支持地图离线缓存

SuperMap iClient 7C for JavaScript 基于 PhoneGap 开源开发框架实现地图离线缓存的插件化，可将已有的地图应用直接打包生成支持 Android 的应用程序，满足用户在离线状态下的地图应用。

4. SuperMap iClient 7C for JavaScript 提供的功能

(1) SuperMap.Layer 命名空间下的功能

SuperMap 云服务图层类、分块缓存图层类、分块动态 REST 图层类、SuperMap iServer 7C 定义的图层类的基类、矢量要素渲染图层类（渲染方式有 SVG、VML、Canvas、Canvas2）。

(2) SuperMap.Control 命名空间下的功能

地图拖拽控件、图层选择控件、鹰眼控件、比例尺控件、平移缩放控件、绘制要素类控件、触摸设备的缩放控件、支持触摸设备触摸操作的控件。

(3) SuperMap.REST 命名空间下的功能

量算功能、距离量算、面积量算、查询功能、距离查询、几何对象查询、SQL 查询、范围查询、专题功能、单值专题图、范围分段专题图、标签专题图、点密度专题图、等级符号专题图、统计专题图。

(4) 空间分析功能

缓冲区分析、叠加分析、表面分析、动态分段、空间关系分析、网络分析功能、最近设施分析服务类、最佳路径分析服务类、服务区分析服务类、选址分区分析服务类、旅行商分析服务类、多旅行商分析服务类、耗费矩阵分析服务类。

(5) 数据功能

数据集 ID 查询服务类、数据集几何查询服务类、数据集缓冲区查询服务类、数据集 SQL 查询服务、数据集编辑服务类。

(6) 交通换乘分析功能

UGC 图层服务功能、UGC 栅格图层类、UGC 影像图层类、UGC 矢量图层类、UGC 专题图图层类。

5. SuperMap iClient 7C for JavaScript 的目录结构说明

(1) apidoc

apidoc 存放产品的类参考，点击 index.html 文件可以查看产品的控件和所有接口的列表。

（2）examples

examples 存放 SuperMap iClient 7C for JavaScript 产品页面及其相关资源，产品页面包括产品介绍、开发指南、示范程序、类参考、技术专题文档，其中：

① 产品介绍向用户介绍了产品是什么，其体系架构、功能和产品包结构等。该部分主要从整体上帮助使用者了解产品；

② 产品开发指南介绍了产品包获取方法以及产品包的使用方法和基本开发流程，方便用户快速掌握产品的基本开发方法；

③ 示范程序提供了产品所有示范代码的使用，给开发者提供案例参考；

④ 类参考提供所有控件、对象的接口列表，为开发者提供接口使用参考；

⑤ 技术专题介绍了产品的一些关键技术，主要涉及高性能矢量渲染、动态分段、离线缓存与 App 等。

（3）libs

libs 文件夹存放产品的库文件以及用于国际资源化的文件。

（4）resource

resource 为产品使用的资源。目前包括 PhoneGap、WinRTApp、AppPackages 文件夹。

（5）PhoneGap

存放 PhoneGap 框架以及基于此框架实现的离线存储范例。通过 PhoneGap 开发框架实现离线存储与访问地图功能，满足用户在离线状态下的地图应用。

（6）WinRTApp

存放基于 Windows 8 JavaScript 版 Windows 8 应用商店程序的开发的源码文件，该源码程序基于 SuperMap 云服务图层，实现基本的地图浏览、缩放及量测功能。

（7）AppPackages

存放基于 Windows 8 JavaScript 版 Windows 8 应用商店程序的打包文件，下载后可以直接在 Windows 8 系统的 PC 直接安装。

（8）theme

theme 文件夹存放产品使用的主题文件。包括 default、image 两个文件夹，其中：default 存放产品类库默认样式文件；image 存放控件的图像资源。

(9)index.html

产品首页，启动 index.html 文件可查看产品的兼容性和产品的变更信息，通过首页，可以链接到产品介绍、产品开发指南、示范程序、类参考以及技术专题等页面，如图 11-8 所示。

6. SuperMap iClient 7C for JavaScript 的主要参考类

（1）SuperMap.Map

地图类。用于实例化 map 类创建一个新地图，实现地图在客户端的交互操作，可通过给创建的 map 添加图层和控件来扩展应用，在创建地图时，如果没有添加指定的控件，则默认 Navigation、PanZoomBar 控件。

1）SuperMap.Map 的 Constants

① Z_INDEX_BASE：{Object} 不同的类的基本 z-index 值。

图 11-8　SuperMap iClient 7C for JavaScript 的目录结构说明图

② EVENT_ TYPES：支持应用事件的类型。
{Array(String)}支持应用事件的类型。为一个特殊的事件注册一个监听对象使用下面的方法：

> map. events. register(type, obj, listener);

监听对象将会作为事件对象的参考，事件的属性将取决于所发生的事情。

③ 所有的事件对象都有以下属性：

a. object{Object}　发出浏览器事件的对象。

b. element{DOMElement}浏览器接收事件的 DOM 元素。

④ 浏览器事件有以下的附加属性：

xy{SuperMap. Pixel}事件触发的像素位置(相对于地图视口)。其他的属性来源于浏览器事件。

⑤ 支持地图事件类型包括以下几种：

a. preaddlayer：图层在被添加之前被触发。事件对象包含 layer 属性来指明添加的图层。当监听函数返回"false"，表示无法添加图层。

b. addlayer：图层被添加之后被触发。事件对象包含 layer 属性来指明添加的图层。

c. preremovelayer：图层被移除之前触发。事件对象含 layer 属性来指明要被移除的图层。当监听函数返回"false"，表示无法移除该图层。

d. removelayer：图层被移除后触发。事件对象包含 layer 属性来指明要被移除的图层。

e. changelayer：当一个图层的名称(name)、顺序(order)、透明度(opcity)、参数个数(params)、可见性(visibility)(由可见比例引起)或属性(attribution)发生改变时被触发。监听者会接收到一个事件对象，该事件对象包含 layer 和 property 属性，其中 layer 属性指明发生变化的图层，property 属性是代表变化属性的关键字(name、order、opacity、params、visibility 或 attribution)。

f. movestartdrag，pan 或 zoom：操作开始时会触发该事件。

g. movedrag，pan 或 zoom：操作开始后被触发。

h. moveenddrag，pan 或 zoom：操作完成时被触发。

i. zoomendzoom：操作完成后被触发。

j. mouseover：鼠标移至地图时被触发。

k. mouseout：鼠标移出地图时被触发。

l. mousemove：鼠标在地图上移动时被触发。

m. changebaselayer：改变基础图层时触发事件。

2）SuperMap. Map 的 Properties

① events：{SuperMap. Events} 事件对象，负责触发地图上的所有事件。

② allOverlays：{Boolean} 地图所有图层都被当作叠加图层来使用，默认是 false。如果设置为 true，则图层相互叠加，最先绘制的图层可以被视为是底图（显示效果上的底图，其 isBaseLayer 为 false）。此外，如果将此属性设置为 true，所有将要添加的图层的"isBaseLayer"属性在添加的时候都会被默认修改成 false。

提示：如果将 map. allOverlays 设置为 true，则不能设置 map. setBaseLayer 或者 layer. setIsBaseLayer。当设置了 allOverlays 属性为 true 时，位于显示索引最下边的图层会被当作"base layer"，所以，如果想要更改"base layer"，使用 setLayerIndex 或者 raiseLayer 将图层的 index 设置为 0。

③ div：{DOMElement | String} 装载地图的 DOM 元素（或一个元素的 id 值），如果调用 SuperMap. Map 构造函数的时候传递两个参数，它将作为第一个参数。此外，构造函数也能够只接收选项（options）对象作为唯一的参数，这种情况下，div 属性可能会也可能不会被提供。如果 div 属性没有被提供，稍后可以使用 render 方法指定地图渲染的地方。

提示：如果在地图定义以后使用 render 来指定渲染位置，不要将 maxResolution 属性设置为 auto，而是使用 maxExtent 指定你所期望的最大范围。

④ layers：{Array(SuperMap. Layer)} 地图上图层的有序列表。

⑤ popups：{Array(SuperMap. Popup)} 地图上的弹窗列表。

⑥ baseLayer：{SuperMap. Layer} 用来确定 min/max zoom level、projection 等属性。

⑦ tileSize：{SuperMap. Size} 在地图选项中被设置，用来设置地图上默认瓦片尺寸。

⑧ projection：{String} 在地图选项中被设置，用来指定地图添加图层的默认投影。同时也可以选择设置 maxExtent、maxResolution、和 units 等属性。该项默认值是"EPSG：4326"。

⑨ units：{String} 地图的单位。默认是"degrees"。可选值为"degrees"（或者"dd"），"m"，"ft"，"km"，"mi"，"inches"。

⑩ resolutions：{Array(Float)} 降序排列的地图分辨率数组（地图上每个像素代表的尺寸），如果在构造图层时没有设置该属性，可通过比例尺相关的属性（例如：maxExtent，maxResolution，maxScale 等）计算获得。

⑪ maxResolution：{Float} 用于地图实例化的时候设置最大分辨率（设置该值可以使地图在分辨率达到某个值的时候无法再缩小）。当不想将整张地图都展示在通过 tileSize 指定大小的一张瓦片上的时候设置该值。设置为 auto 的时候地图会自适应视口大小。

⑫ minResolution：{Float} 用于地图实例化的时候设置最小分辨率（设置该值可以使地图在分辨率达到某个值的时候无法再放大）。

⑬ maxScale：{Float} 用于地图实例化的时候设置最大比例（设置该值可以使地图在比例达到某个值的时候无法再放大）。

⑭ minScale：{Float} 用于地图实例化的时候设置最小比例（设置该值可以使地图在比例达到某个值的时候无法再缩小）。

⑮ maxExtent：{SuperMap. Bounds}用于地图实例化的时候设置地图的最大范围。默认是(-180，-90，180，90)。

⑯ minExtent：{SuperMap. Bounds}用于地图实例化的时候设置地图的最小范围。默认值是 null。

⑰ restrictedExtent：{SuperMap. Bounds}限定地图缩放范围。如果设置了 restrictedExtent，平移将会被限制在指定的边界内。

⑱ numZoomLevels：{Integer}用于地图实例化的时候设置地图缩放级别的数量。默认值 16，当需要的时候可以地图选项中设置其他的值。

⑲ theme：{String}要加载的主题风格的 CSS 文件的相对路径。如果想通过直接在页面添加层叠样式表链接或样式声明来设置样式，那就应该在地图属性(options)对象中指定为 null(例如{theme：null})。

⑳ displayProjection：{SuperMap. Projection}将投影设置为除 EPSG：4326 或 EPSG：900913/EPSG：3857 以外的投影，需要 proj4js 的支持。该投影用于通过控件(controls)向用户展示数据。如果设置了该属性，它就会被设置到任何一个在添加到地图上的时候 displayProjection 属性为 null 的控件上。

㉑ fallThrough：{Boolean}用来设置页面元素是否会接收地图触发的事件，默认是 true。

㉒ eventListeners：{Object}如果在构造方法中设置此选项，事件监听对象将注册。

㉓ panMethod：{Function}作用于平移时的动画效果方法，设置为 null 时，动画平移将关闭。

3) SuperMap. Map 的 Constructor

Parameters 主要包括以下内容：

① div：{DOMElement | String}页面上装载地图的 DOM 元素，当设置了 div 选项(option)或稍后调用 render 方法时可以省略。

② Options：{Object}地图的选项设置，如果在此参数中没有对 Controls 进行设置，则默认 Navigation、PanzoomBar 控件。

例如：

```
//在 id 为 map1 的页面元素中创建一个 map 地图对象。
var map=new SuperMap. Map("map1");
var options = {maxExtent: new SuperMap. Bounds(-200000, -200000, 200000, 200000), maxResolution: 156543, units:'m', projection:"EPSG：41001"};
//在 id 为 map2 的页面元素中创建一个 map 地图对象。
var map=new SuperMap. Map("map2", options);
//为地图的属性设置 div、options 参数。
var map = new SuperMap. Map ({div:" map _ id ", maxExtent: new SuperMap. Bounds(-200000, -200000, 200000, 200000), maxResolution: 156543, units:'m', projection:"EPSG：41001"});
```

```
//为地图的属性设置 options 参数。
var map = new SuperMap.Map({
maxExtent: new SuperMap.Bounds(-200000,-200000,200000,200000),
maxResolution:156543, units:'m', projection:"EPSG:41001"});
//用 addControls 方法为地图添加控件
var map = new SuperMap.Map('map',{controls:[]});
map.addControl(new SuperMap.Control.Navigation());
//创建 map 地图对象，通过设置 options 参数添加指定控件
var map = new SuperMap.Map ('map',{controls: new SuperMap.Control.LayerSwitcher(), new SuperMap.Control.ScaleLine(), new SuperMap.Control.PanZoomBar(), new SuperMap.Control.Navigation()]});
```

4) SuperMap.Map 的 Functions

SuperMap.Map 的 Functions 详见表 11-3。

表 11-3　　　　　　　　　　**SuperMap.Map 的 Functions**

render	在指定的容器中渲染地图
destroy	销毁地图(注意，如果从 DOM 中移除 map 容器，需要在 destroy 前执行)
setOptions	设置地图的 options
getTileSize	获取地图瓦片的大小
getBy	获取一个对象数组，并通过给定的属性匹配其中的项
getLayersBy	获取与给定属性和匹配字符串匹配的图层列表
getLayersByName	获取根据名称匹配得到的图层列表
getLayersByClass	根据类名获取的图层列表
getControlsBy	根据给定的属性和匹配字符串匹配到的控件列表
getControlsByClass	根据给定类的类名匹配到的控件列表
getLayer	根据传入参数 ID 获取图层
addLayer	向地图中添加指定的单个图层
addLayers	向地图中添加指定的多个图层
removeLayer	通过删除可见元素(即 layer.div 属性)来移除地图上的图层。然后从地图的图层列表中移除该图层，同时设置图层的 map 属性为 null
getNumLayers	获取地图上的图层数量
getLayerIndex	获取图层在地图上的索引值(索引值从零开始)
setLayerIndex	移动图层到图层列表中的指定索引值(索引值从零开始)的位置改变它在地图显示时的 z-index 值。使用 map.getLayerIndex()方法查看图层当前的索引值(注意，该方法不能将底图移动到叠加图层之上)

续表

raiseLayer	通过给定的增量值(delta 参数)来改变给定图层的索引值。如果增量值为正,图层就会在图层堆栈中向上移;如果增量值为负,图层就会向下移。这个方法同样不能将底图移动到叠加图层之上
setBaseLayer	允许用户指定当前加载的某一图层为地图新的底图
addControl	为地图添加控件。可选的位置参数用来指定控件的像素位置
addControls	将控件添加到 map 上。可以将 options 的第二个数组通过像素对象控制控件的位置两个数组匹配,如果 pixel 设为 null,控件会显示在默认位置
getControl	通过 id 值获取控件对象
removeControl	移除控件
addPopup	在地图中添加弹出窗口
removePopup	移除指定的弹出窗口
getSize	获取当前地图容器大小
updateSize	通过动态调用 updateSize( )方式,来改变地图容器的大小
getCenter	获取当前地图的中心点坐标
getZoom	获取当前地图的缩放比例级别
pan	根据指定的屏幕像素值平移地图
panTo	平移地图到新的位置,如果新的位置在地图的当前范围内,地图将平滑地移动
setCenter	设置地图中心点
getProjection	该方法返回代表投影的字符串
getProjectionObject	返回 baseLayer 的投影
getMaxResolution	返回 baseLayer 最大的分辨率
getMaxExtent	获取地图的最大范围
getNumZoomLevels	获取当前底图的缩放级别总数,在底图存在的情况下与底图缩放级别(numZoomLevels)相同
getExtent	获取当前地图的范围
getResolution	获取当前地图的分辨率
getUnits	获取地图的当前单位
getScale	获取当前地图的缩放比例
getZoomForExtent	通过给定的范围获取比例级别
getResolutionForZoom	根据缩放级别获取对应分辨率
getZoomForResolution	根据分辨率获取对应缩放级别
zoomTo	缩放到指定的级别
zoomIn	在当前缩放级别的基础上放大一级

续表

zoomOut	在当前缩放级别的基础上缩小一级
zoomToExtent	缩放到指定范围，重新定位中心点
zoomToMaxExtent	缩放到最大范围，并重新定位中心点
zoomToScale	缩放到指定的比例
getViewPortPxFromLonLat	根据指定的地理位置，返回其相对于当前地图窗口左上角的像素位置
getLonLatFromPixel	根据相对于地图窗口左上角的像素位置，返回其在地图上的地理位置。依据当前 baselayer 转换成 lon/lat(经度/纬度)形式
getPixelFromLonLat	获取地图上的像素坐标。依照当前 baselayer，将指定的地理点位置坐标转换成其相对于地图窗口左上角点的像素坐标
getViewPortPxFromLayerPx	
getLayerPxFromViewPortPx	根据视图窗口像素点坐标获取图层像素点坐标
getLayerPxFromLonLat	根据传入的大地坐标获取图层坐标对象

（2）SuperMap.Layer

1）SuperMap.Layer 的 Properties

① id：{String}图层 id，唯一标识图层，默认为 null，在初始化时会动态创建唯一的值

② name：{String}图层名称，默认为 null。初始化图层时可以外部传参进行修改，可以通过图层管理器（LayerSwitcher）查看当前所有图层的名称。

③ div：{DOMElement}存放图层的界面元素 div，默认为 null。

④ alwaysInRange：{Boolean}当前地图显示的分辨率在图层的最大最小分辨率范围内，如果图层以非比例尺显示，此变量设置为 true。

⑤ 其他 SuperMap.Layer 的 Properties 见表 11-4。

表 11-4　　　　　　　　　　**SuperMap.Layer 的 Properties**

events	{SuperMap.Events}
map	{SuperMap.Map}图层所关联的地图，默认为 null。当图层添加到地图上时设置此变量
isBaseLayer	{Boolean}当前图层是否为基础层，默认为 false。需在子类中单独设置此属性
displayInLayerSwitcher	{Boolean}是否在图层管理器（LayerSwitcher）中显示图层名字，默认为 true
visibility	{Boolean}图层是否可见，默认为 true
eventListeners	{Object}监听器对象，在构造函数中设置此参数。通过 SuperMap.Events.on 注册
gutter	{Integer}瓦片间的交接间距（像素），默认为 0

续表

projection	{SuperMap.Projection} or {String} 地图投影，默认为 null。创建图层时，在图层的 options 上可以设置当前图层默认的投影字符串，如"EPSG:4326"还需要设置 maxExtent、maxResolution、units
units	{String} 地图单位，可以为 "degrees"（or "dd"）, "m", "ft", "km", "mi", "inches"，默认为 "degrees"
scales	{Array} 降序排列的比例尺数组，默认为 null。一般会通过 resolutions 来自动计算，也可以在图层初始化的时候设置
resolutions	{Array} 降序排列的地图分辨率列表。如果在创建 layer 时没有设置 resolutions，则需要计算，此时需设置 resolution 计算相关的属性（maxExtent, maxResolution, maxScale 等）
maxExtent	{SuperMap.Bounds} 在图层实例化的时候设置图层的最大范围，平移过程中边界中心点不会偏离可视窗口。不同的投影下范围不同，如 world 在 "EPSG:4326"下一般为左下：(-180.0, -90.0)，右上：(180.0, 90.0) 而在 "EPSG:3857"下为左下：(-20037508.34, -25776731.36)，右上：(20037508.34, 25776731.36)
minExtent	{SuperMap.Bounds} 在图层实例化的时候设置图层最小范围
maxResolution	{Float} 在图层实例化的时候设置图层最大的分辨率，默认最大的是 360 度/256 像素（投影为 4326），相当于缩放级别为 0 级。不同的投影下 maxResolution 会不同，内部会进行计算
minResolution	{Float} 在图层实例化的时候设置图层最小分辨率，默认最小的是 360 度/(256×16) 像素（投影为 4326），相当于缩放级别为 16 级。不同的投影下 minResolution 会不同，内部会进行计算
numZoomLevels	{Integer} 在图层实例化的时候设置缩放级别，一般为 16
minScale	{Float} 在图层实例化的时候设置最小比例尺，在不同的投影下根据 minResolution 计算
maxScale	{Float} 在图层实例化的时候设置最大比例尺，在不同的投影下根据 maxResolution 计算
displayOutsideMaxExtent	{Boolean} 判断请求地图瓦片是否完全超出了当前图层的最大范围，默认为 false
wrapDateLine	{Boolean} 当地图平移到日期变更线外边后是否仍然继续循环显示地图。当为 false 时不显示，默认为 false

2) SuperMap.Layer 的 Constants

① EVENT_TYPES：{Array(String)} 支持事件的类型，注册监听事件方法如下所示：

layer.events.register(type, obj, listener);

所有监听对象具备如下属性：

a. object：{Object} object 引用。

b. element：{DOMElement} element 引用。

支持事件类型：

a. loadstart：当图层开始加载时触发事件。

b. loadend：当图层结束加载时触发事件。

c. loadcancel：当图层取消加载时触发事件。

d. visibilitychanged：当图层可见性发生变化时触发事件。

e. move：当图层移动时触发此事件（拖拽时每次鼠标移动触发此事件）。

f. moveend：当图层移动结束时触发此事件。

g. added：图层加载到 map 上触发此事件。

h. removed：图层从 map 上移除后触发此事件。

i. tileloaded：每个瓦片下载完成所触发的事件，返回该瓦片对象。

例如：

```
//需要将 layer 的 bufferImgCount 设置为 0，并且将页面在服务端发布出来
layer = new SuperMap.Layer.TiledDynamicRESTLayer("World", DemoURL.china,
{transparent: true, cacheEnabled: true, redirect: true}, {maxResolution:"auto",
bufferImgCount: 0});
layer.events.on({tileloaded: function(evt){var ctx=evt.tile.getCanvasContext();
if(ctx){var imgd=ctx.getImageData(0, 0, evt.tile.size.w, evt.tile.size.h);
imgd=modify(imgd);
ctx.putImageData(imgd, 0, 0);
evt.tile.drawImgData(ctx.canvas.toDataURL(), evt);}}});
function modify(imgPixels){for(var y=0, h=imgPixels.height; y<h; y++){for(var
x=0, w=imgPixels.width; x<w; x++){
var i=(y * 4) * w+x * 4;
var gray = 0.299 * imgPixels.data[i] + 0.587 * imgPixels.data[i+1] + 0.114 *
imgPixels.data[i+2];
imgPixels.data[i]=gray;
imgPixels.data[i+1]=gray;
imgPixels.data[i+2]=gray;}}
return imgPixels;}
```

② Resolution_Properties：{Array} resolutions 计算使用的属性数组，这些属性包括：scales、resolutions、maxScale、minScale、maxResolution、minResolution、numZoomLevels、maxZoomLevel。

3) SuperMap.Layer 的 Functions

SuperMap.Layer 的 Functions 详见表 11-5。

表 11-5　　　　　　　　　　　**SuperMap. Layer 的 Functions**

函数	说明
setName	将新的名字赋给当前图层，可以触发地图上的 changelayer 事件
addOptions	通过新的 Options 覆盖以前的 Options 参数。Parameters：newOptions-{Object}新的 Options 参数。reinitialize-{Boolean}如果设为 true，并且当前的 baseLayer 的 resolution 发生变化，则 map 需要重新定位有效的 resolution，并且触发 changebaselayer 事件
onMapResize	此函数在子类中复写实现
redraw	重新绘制图层，对于图片图层，该方法销毁掉该图层的 div 以及地图图片，然后重新组织 div，重新请求地图图片。如果图层被重绘，返回 true，否则返回 false。在子图层控制时，修改图层信息后，调用该方法重新绘制图层显示改变后的效果。该方法不适用于覆盖物图层，例如 SuperMap. Layer. Markers
removeMap	从地图中移除图层
getImageSize	获取瓦片的大小。Parameters：bounds-{SuperMap. Bounds}瓦片的边界选项。可以被子类用来处理图层上不同瓦片的边缘范围，例如 Zoomify
setTileSize	设置瓦片的大小
getVisibility	获取当前图层可见性。Returns：{Boolean}是否可见（当前地图的 resolution 在最大和最小的 resolution 之间）
setVisibility	设置图层可见性，设置图层的隐藏、显示、重绘的相应的可见标记
display	临时隐藏或者显示图层。通过对 CSS 控制产生即时效果，重新渲染失效。一般用 setVisibility 方法来动态控制图层的显示和隐藏
calculateInRange	计算当前地图显示的分辨率是否在图层的最大最小分辨率范围内。Returns：{Boolean}图层以当前地图分辨率显示，如果'alwaysInRange'设置为 true，则此函数返回 true
setIsBaseLayer	设置当前图层性质（底图或普通图层），一旦图层性质改变，会触发 changebaselayer 事件。Parameters：isBaseLayer-{Boolean}
getResolution	获得当前图层分辨率
getExtent	获得边界范围
getZoomForExtent	获得当前的缩放级别。Parameters：extent-{SuperMap. Bounds} closest-{Boolean}查找最接近指定范围边界的缩放级别。默认为 false
getResolutionForZoom	根据指定的缩放级别返回对应的分辨率。Parameter：zoom-{Float}缩放级别，范围一般在[0，16]内
getScaleForZoom	通过指定的缩放级别返回对应的比例尺。Parameter：zoom-{Float}缩放级别，范围一般在[0，16]内
getZoomForResolution	根据指定的分辨率返回对应的缩放级别。Parameters：resolution-{Float}分辨率大小。closest-{Boolean}查找当前显示的分辨率对应的缩放级别，默认为 false

续表

getLonLatFromViewPortPx	根据指定的像素点位置返回经纬度坐标。Parameters：viewPortPx-{SuperMap.Pixel}传入的像素点
getViewPortPxFromLonLat	根据指定的纬度坐标返回像素点位置。Parameters：lonlat-{SuperMap.LonLat}经纬度
setOpacity	设置图层的不透明度，取值[0-1]之间

（3）SuperMap.Geometry

1）SuperMap.Geometry 的 Properties

SRID：{Interger}投影坐标参数。通过该参数，服务器判断 Geometry 对象的坐标参考系是否与数据集相同，如果不同，则在数据入库前进行投影变换。

2）SuperMap.Geometry 的 Constructor

SuperMap.Geometry：创建一个几何图形的对象。

3）SuperMap.Geometry 的 Functions

SuperMap.Geometry 的 Functions 详见表 11-6。

表 11-6 **SuperMap.Geometry 的 Functions**

clone	创建克隆的几何图形。克隆的几何图形不设置非标准的属性
getBounds	获得几何图形的边界。如果没有设置边界，可通过计算获得
calculateBounds	重新计算几何图形的边界（需要在子类中实现此方法）
distanceTo	计算两个几何图形间的最小距离（x-y 平面坐标系下，需要在子类中实现此方法。）Parameters：geometry-{SuperMap.Geometry}目标几何图形
getVertices	返回几何图形的所有顶点的列表（需要在子类中实现此方法）。Parameters：nodes-{Boolean}如果是 true，线则只返回线的末端点，如果 false，仅仅返回顶点，如果没有设置，则返回顶点
getCenterid	计算几何图形的质心（需要在子类中实现此方法）。Returns：{SuperMap.Geometry.Point}采集的质心

（4）SuperMap.Popup

1）SuperMap.Popup 的 Properties

① autoSize：{Boolean}根据弹窗内容自动调整弹窗大小，默认为 false。

② minSize：{SuperMap.Size}允许弹出内容的最小尺寸。

③ maxSize：{SuperMap.Size}允许弹出内容的最大尺寸。

④ panMapIfOutOfView：{Boolean}是否移动地图以确保弹窗显示在窗口内。默认为 false。

⑤ keepInMap：{Boolean}如果 panMapIfOutOfView 设为 false，keepInMap 设为 true，弹窗则将一直适应当前的地图空间显示。默认情况下，不会在基类中设置。如果在地图边缘

附近创建不允许平移的弹窗,并且有固定的相对位置,此方法设为 false 比较好,子类需要重写此设置。默认为 false。

⑥ closeOnMove:{Boolean}当地图平移时,关闭弹窗。默认为 false。

2)SuperMap.Popup 的 Constructor

① SuperMap.Popup:创建弹窗。在地图上可以打开或关闭,通常情况下点击一个 icon 打开弹窗,弹窗直接加载到 map 上,不需要创建图层,可用 SuperMap.Map.addPopup 方法在地图上添加使用。例如:

```
var popup = new SuperMap.Popup("chicken", new SuperMap.LonLat(5,40), new SuperMap.Size(200,200),"example popup", true);
popup.closeOnMove = true;
map.addPopup(popup);
```

Parameters(参数):

a. id:{String}弹窗的唯一标识,如设为 null,则将会自动生成。

b. lonlat:{SuperMap.LonLat}地图上弹窗显示的位置。

c. contentSize:{SuperMap.Size}弹窗内容的大小。

d. contentHTML:{String}弹窗中显示的一个 HTML 要素的字符串。

e. closeBox:{Boolean}在弹出窗口的里面是否显示关闭窗。

f. closeBoxCallback:{Function}关闭弹窗触发该回调函数。

3)SuperMap.Popup 的 Functions

SuperMap.Popup 的 Functions 详见表 11-7:

表 11-7 **SuperMap.Popup 的 Functions**

updateSize	自动调整弹窗大小适应其弹出内容,弹窗大小受限制于当前地图空间大小
setBackgroundColor	设置弹出框的背景颜色
setOpacity	设置弹出框的透明度
setBorder	设置弹出窗体的边框样式
getSafeContentSize	

(5)SuperMap.Handler

用于处理 Control 事件的事件处理器。事件处理器(Handler)内部封装了浏览器事件监听及其相应的处理方法,当一个事件处理器被激活时,Handler 中定义的浏览器事件监听及其相应的方法被注册到浏览器监听器,在浏览器事件被触发之后,首先会调用 Handler 中处理该浏览器事件的方法来做事件的确认和信息封装,然后才再传递给 Control 等做具体响应处理。当一个处理器被注销,这些方法在事件监听器中也会相应地被取消注册。

事件处理器通过 active 和 deactive 两个方法,实现动态的激活和注销。一个控件被激活时,其对应的事件处理器会被激活,相应的浏览器事件被触发之后,通过事件处理器(Handler)来监听、处理和响应这些事件;同样,当一个控件被注销时,其相应的处理器也会被注销,包括了 Box、Click、Drag、Feature、Hover、Keyboard、MouseWheel、Path、Pinch、Point、Polygon、RegularPolygon 等几种类型的事件处理器。

1) SuperMap. Handler 的 Properties

① control:{SuperMap. Control}控件对象。

② keyMask:{Integer}使用位运算符和一个或者多个 Handler 类型的常量来创建键盘编码符(KeyMask)。

Example:

//事件仅仅在 Shift 键按下时响应

handler. keyMask = SuperMap. Handler. MOD_ Shift;

//事件仅仅在 Shift 键或 Ctrl 键按下时响应

handler. keyMask = SuperMap. Handler. MOD_ Shift

SuperMap. Handler. MOD_ Ctrl;

2) SuperMap. Handler 的 Constructor

SuperMap. Handler:实例化事件处理器类对象

3) SuperMap. Handler 的 Functions

① activate:激活处理器,如果当前处理器对象已经激活,则返回 false

② deactivate:关闭处理器,如果当前处理器已经是关闭状态,则返回 false

4) SuperMap. Handler 的 Constants

SuperMap. Handler 的 Constants 详见表 11-8。

表 11-8　　　　　　　　　　**SuperMap. Handler 的 Constants**

SuperMap. Handler. MOD_ NONE	如果设置这个 KeyMask,按下任意键,checkModifier 方法都返回 false;默认值为 0
SuperMap. Handler. MOD_ SHIFT	如果设置这个 KeyMask,在按下 Shift 键的时候,checkModifiers 方法返回 false;默认值为 1
SuperMap. Handler. MOD_ CTRL	如果设置这个 KeyMask,在按下 Ctrl 键的时候,checkModifiers 方法返回 false;默认值为 2
SuperMap. Handler. MOD_ ALT	如果设置这个 KeyMask,在按下 Alt 键的时候,checkModifiers 方法返回 false;默认值为 4

(6) SuperMap. Projection

坐标转换类。这个类封装了与 proj4js 投影对象进行交互的几种方法。SuperMap 默认支持 EPSG:4326,CRS:84,urn:ogc:def:crs:EPSG:6.6:4326,EPSG:900913,EPSG:3857,EPSG:102113,EPSG:102100 投影间的转换。对于 SuperMap 不支持或者用户想要自定义投影类型,可通过下载 proj4js 产品包,并引入产品包中的 proj4js. js 实现自定义的投影转换。具体方法可以参见超图软件网站上的开发指南《坐标投影转换》。

目前,proj4js 支持的投影种类有:WGS84,EPSG:4326,EPSG:4269,EPSG:

3875，EPSG：3785，EPSG4139，EPSG：4181，EPSG：4272，EPSG：4302，EPSG：21781，EPSG：102113，EPSG：26591，EPSG：26912，EPSG：27200，EPSG：27563，EPSG：41001，EPSG：42304，EPSG：102067，EPSG：102757，EPSG：102758，EPSG：900913，GOOGLE

1) SuperMap. Projection 的 Constructor

Parameters(参数)主要包括：

① projCode：{String}影编码。

② options：{Object}设置图层上的的附加属性。

例如：

> var geographic = new SuerMap. Projection("EPSG：4326")；
> Returns
> {SuperMap. Projection} 投影对象。

2) SuperMap. Projection 的 Functions 和 Properties

SuperMap. Projection 的 Functions 和 Properties 见表 11-9。

表 11-9　　**SuperMap. Projection 的 Functions 和 Properties 表**

getCode	获取 SRS 代码字符串
getUnits	获取投影的单位字符串。如果 proj4js 不可用则返回 null
defaults	{Object} SuperMap 默认支持 EPSG：4326，CRS：84，urn：ogc：def：crs：EPSG：6.6：4326，EPSG：900913，EPSG：3857，EPSG：102113，EPSG：102100 投影间的转换。defaults 定义的关键字为坐标系统编码，相应的属性值为 units，maxExtent(坐标系统的有效范围)和 yx(当坐标系统有反向坐标轴时为 true)
addTransform	设置自定义投影转换方法。在 proj4js 库不可用或者自定义的投影需要处理时使用此方法
transform	点投影转换
nullTransform	空转换，有用的定义投影的别名时 proj4js 不可用；当 proj4js 不可用时，用来定义投影的别名
projCode	{String}投影编码

(7) SuperMap. Control

控件类，提供了多种控件，比如比例尺控件，鹰眼控件，缩放条控件，等等。用于处理 Control 事件的事件处理器 Handler，内部封装了一系列的浏览器事件，在控件(control)实现过程中可调用 Handler，通过 active 和 deactive 两个方法，实现动态的激活和注销。可见，控件不需要使用 activate 方法激活便可以使用。

对于非可见控件来说，带有 autoActivate 属性并且默认值为 true 的控件会在使用时自动激活。

对于不具备 autoActivate 属性的控件，由于父类 Control 类中的 autoActivate 属性默认值为 false，所以使用这些控件需要调用 activate 方法进行激活控件，也可通过设置父类的 autoActivate 属性为 true 来激活。

控件影响地图显示和地图操作，在没有指定控件的情况下，地图默认添加 Navigation、PanZoomBar 控件。也可通过参数 options 传入的 div 添加控件到一个外部的 div。

例如，下面的示例演示如何在地图上添加多种控件的方法：

```
var map=new SuperMap.Map('map',{controls:[]});
map.addControl(new SuperMap.Control.LayerSwitcher({'ascending':false}));
map.addControl(new SuperMap.Control.MousePosition());
map.addControl(new SuperMap.Control.OverviewMap());
map.addControl(new SuperMap.Control.KeyboardDefaults());
```

下面的代码片段是一个关于用户如何截获按住 Shift 键的同时移动鼠标拖出的边界框的范围事件的简单的例子，例如：

```
var control=new SuperMap.Control();
SuperMap.Util.extend(control,{draw:function(){this.box=new SuperMap.Handler.Box(control,{"done":this.notice},{keyMask:SuperMap.Handler.MOD_SHIFT});this.box.activate();},notice:function(bounds){}});
    map.addControl(control);
```

1) SuperMap.Control 的 Properties

SuperMap.Control 的 Properties 详见表 11-10。

表 11-10   **SuperMap.Control 的 Properties**

id	{String}控件的 id
div	{DOMElement}包含控件的 DOM 元素，如果不指定，则控件放置在地图内部
type	{Number}控件的类型
title	{string}此属性用来在控件上显示一个提示框
autoActivate	{Boolean}当控件添加到地图上时激活此控件。默认为 false
active	{Boolean}控件是激活状态(只读)，用 activate、deactivate 可以改变控件的状态
eventListeners	{Object}如果设置为构造函数的一个选项，eventListeners 将被 SuperMap.Events.on 注册，对象结构必须是一个监听器对象，具体例子详细参见 events.on 方法
events	{SuperMap.Events}注册控件特定事件的监听器实例。

2) SuperMap.Control 的 Constants

EVENT_ TYPES 表示 {Array(String)} 支持的事件类型。注册特定事件的监听器，例如：

```
control.events.register(type, obj, listener);
```

event 对象具备以下属性：object-{Object} 发出浏览器事件的对象。element-{DOMElement} 浏览器接收事件的 DOM 元素。

支持的地图事件类型：activate-控件激活时触发此事件。deactivate-控件失效时触发此事件。

3) SuperMap.Control 的 Functions

① activate：激活控件及其相关的处理事件(handler)，控件失效调用 deactivate 方法。

② deactivate：使控件及其相关的处理事件(handler)失效。

4) SuperMap.Control 的 Constructor

SuperMap.Control 表示创建控件，options 作为参数传递直接扩展控件。例如：

```
var control=new SuperMap.Control({div: myDiv});
```

重写默认属性值为 null 的 div。

(8) SuperMap.Marker

标记覆盖物，对地图上的点进行标注，可以自定义选择标注的图标，需添加到 Markers 图层上显示。

1) SuperMap.Marker 的 Properties

events：{SuperMap.Events} the event handler.

支持事件类型有：

a. click：当鼠标单击 maker 时触发此事件。

b. dblclick：当鼠标双击 maker 时触发此事件。

c. mousedown：当鼠标在 maker 上按下时触发此事件。

d. mouseup：当鼠标在 maker 上按下并放开时触发此事件。

e. mousemove：当鼠标移过 maker 时触发此事件。

f. mouseout：当鼠标移出 maker 时触发此事件。

g. mouseover：当鼠标移进 maker 时触发此事件。

h. rightclick：当鼠标右键单击 maker 时触发此事件。

例如，点击 marker 弹出 popup：

marker.events.on({"click": openInfoWin,"scope": marker});

function openInfoWin(){ var marker=this;

var lonlat=marker.getLonLat();

var contentHTML="<div style=\'font-size:.8em;opacity:0.8;overflow-y:hidden;\'>";

contentHTML+="<div>"+marker.sm_capital+"</div></div>";

var popup = new SuperMap.Popup.FramedCloud("popwin", new SuperMap.LonLat(lonlat.lon, lonlat.lat), null, contentHTML, null, true);

map. addPopup(popup);}

2) SuperMap. Marker 的 Constructor

① SuperMap. Marker：创建标记。通常通过调用 SuperMap. Layer. Markers 将标记添加到指定的标记图层。例如：

var markers = new SuperMap. Layer. Markers("Markers");
map. addLayer(markers);
var size = new SuperMap. Size(21, 25);
var offset = new SuperMap. Pixel(-(size. w/2), -size. h);
var icon = new SuperMap. Icon('.. img/marker. png', size, offset);
markers. addMarker(new SuperMap. Marker(new SuperMap. LonLat(0, 0), icon));

② Parameters(参数)如下：

a. lonlat：{SuperMap. LonLat}当前标记的位置。

b. icon：{SuperMap. Icon} 当前标记的图标。

3) SuperMap. Marker 的 Functions

SuperMap. Marker 的 Functions 详见表 11-11。

表 11-11　　　　　　　　　　　　SuperMap. Marker 的 Functions

getLonLat	获取 marker 的当前坐标
destroy	清除标记，需要首先移除图层上添加的标记，在标记内不能执行此操作，因为不知道标记连接到哪个图层
isDrawn	获取标记是否绘制。Returns：{Boolean}标记是否被绘制

(9) SuperMap. StyleMap

图层样式组类，定义 vector 图层不同状态下的样式信息。属性 styles 中存储由 key('state')，value('Style')组成的状态-样式键值对信息。

1) SuperMap. StyleMap 的 Properties

styles：Hash of {SuperMap. Style}属性 styles 中存储由 key('state')，value('Style')组成的状态-样式键值对构成。默认支持四种样式"default"，"temporary"，"select"，"delete"。

2) SuperMap. StyleMap 的 Constructor

① SuperMap. StyleMap：构造 StyleMap 新实例，可通过三种方式构造 StyleMap 新实例。

② Parameters(参数)如下：

style：{Object}传入的 style 参数。

样式参数分多种情况：

a. 设置单一样式对象 Style，结果会将四种状态样式都设置为此，即相当于没有了状态区别；

b. 样式组信息，由状态标签-样式组成的键值对 key：Style(或者 Object)，读取提供的样式，覆盖默认值，如果后面的值不是 Style 对象，内部对其封装成 Style 后使用；

c. 有样式具体信息构成的 Object 对象，直接将其封装成 Style 样式，然后同①；

d. 当默认支持的四种状态不能满足用户需求时，用户也可自定义状态及其样式。

以上几种情况示例代码如下所示：

a. 直接设置单一 style，实现代码如下，依照下面代码定义样式之后，四种状态都有下面的 style 样式：

var myStyles=new SuperMap. StyleMap(

new SuperMap. Style ({fillColor:"#ffcc33", strokeColor:"#ccff99", strokeWidth: 2, graphicZIndex: 1}));

b. 直接设置状态-样式键值对，代码如下所示，需要注意的是，代码中只定义了 default 和 select 两个状态的样式，所以仅仅这两个状态有样式，另外两种状态 temporary、delete 的样式为 null：

var myStyles = new SuperMap. StyleMap({"default": new SuperMap. Style({fillColor:"#ffcc33", strokeColor:"#ccff99", strokeWidth: 2, graphicZIndex: 1}),

"select": {fillColor:"33eeff", strokeColor:"3366aa", graphicZIndex: 2}});

c. 直接传入样式信息，实现代码如下：

var myStyles = new SuperMap. StyleMap({fillColor:"#ffcc33", strokeColor:"#ccff99", strokeWidth: 2, graphicZIndex: 1});

d. 自定义状态及其样式，代码如下：

var myStyles = new SuperMap. StyleMap({"default": new SuperMap. Style({fillColor:"#ffcc33", strokeColor:"#ccff99", strokeWidth: 2, graphicZIndex: 1}),

"click": {fillColor:"33eeff", strokeColor:"3366aa", graphicZIndex: 2}});

options: {Object} 此类开出来的属性。

（10）SuperMap. REST

SuperMap. REST 的 Constants 有：

① SuperMap. REST. GeometryType: {Object} 几何对象枚举类。该类定义了一系列几何对象类型。

② SuperMap. REST. QueryOption: {Object} 查询结果类型枚举类。该类描述查询结果返回类型，包括只返回属性、只返回几何实体以及返回属性和几何实体。

③ SuperMap. REST. JoinType: {Object} 关联查询时的关联类型常量。该类定义了两个表之间的连接类型常量，决定了对两个表之间进行连接查询时，查询结果中得到的记录的情况。

④ SuperMap. REST. SpatialQueryMode: {Object} 空间查询模式枚举类。该类定义了空间查询操作模式常量。

⑤ SuperMap. REST. SpatialRelationType: {Object} 数据集对象间的空间关系枚举类。该类定义了数据集对象间的空间关系类型常量。

⑥ SuperMap. REST. MeasureMode: {Object} 量算模式枚举类。该类定义了两种测量模式：距离测量和面积测量。

⑦ SuperMap. REST. Unit: {Object} 距离单位枚举类。该类定义了一系列距离单位类型。

⑧ SuperMap. REST. EngineType：{Object}数据源引擎类型枚举类。
⑨ SuperMap. REST. ThemeGraphTextFormat：{Object}统计专题图文本显示格式枚举类。
⑩ SuperMap. REST. ThemeGraphType：{Object}统计专题图类型枚举类。
⑪ SuperMap. REST. GraduatedMode：{Object}专题图分级模式枚举类。
⑫ SuperMap. REST. RangeMode：{Object}范围分段专题图分段方式枚举类。
⑬ SuperMap. REST. ThemeType：{Object}专题图类型枚举类。
⑭ SuperMap. REST. ColorGradientType：{Object}渐变颜色枚举类。
⑮ SuperMap. REST. TextAlignment：{Object}文本对齐枚举类。
⑯ SuperMap. REST. FillGradientMode：{Object}渐变填充风格的渐变类型枚举类。
⑰ SuperMap. REST. AlongLineDirection：{Object}标签沿线标注方向枚举类。
⑱ SuperMap. REST. LabelBackShape：{Object}标签专题图中标签背景的形状枚举类。
⑲ SuperMap. REST. LabelOverLengthMode：{Object}标签专题图中超长标签的处理模式枚举类。
⑳ SuperMap. REST. DirectionType：{Object}网络分析中方向枚举类。在行驶引导子项中使用。
㉑ SuperMap. REST. SideType：{Object}行驶位置枚举类。表示在行驶在路的左边、右边或者路上的枚举，该类用在行驶导引子项类中。
㉒ SuperMap. REST. SupplyCenterType：{Object}资源供给中心类型枚举类。该枚举类定义了网络分析中资源中心点的类型，主要用于资源分配和选址分区。资源供给中心点的类型包括非中心、固定中心和可选中心。固定中心用于资源分配分析；固定中心和可选中心用于选址分析；非中心在两种网络分析时都不予考虑。
㉓ SuperMap. REST. TurnType：{Object}转弯方向枚举类。用在行驶引导子项类中，表示转弯的方向。
㉔ SuperMap. REST. BufferEndType：{Object}缓冲区分析 BufferEnd 类型。
㉕ SuperMap. REST. OverlayOperationType：{Object}叠加分析类型枚举。
㉖ SuperMap. REST. SmoothMethod：{Object}光滑方法枚举类。用于从 Grid 或 DEM 数据生成等值线或等值面时对等值线或者等值面的边界线进行平滑处理的方法。
㉗ SuperMap. REST. SurfaceAnalystMethod：{Object}表面分析方法枚举类。通过对数据进行表面分析，能够挖掘原始数据所包含的信息，使某些细节明显化，易于分析。
㉘ SuperMap. REST. DataReturnMode：{Object}数据返回模式枚举类。该枚举类用于指定空间分析返回结果模式，包含返回数据集标识和记录集、只返回数据集标识（数据集名称@数据源名称）及只返回记录集三种模式。
㉙ SuperMap. REST. EditType：{Object}要素集更新模式枚举类。该枚举类用于指定数据服务中要素集更新模式，包含添加要素集、更新要素集和删除要素集。
㉚ SuperMap. REST. TransferTactic：{Object}公交换乘策略枚举类。该枚举类用于指定公交服务中要素集更新模式，包含添加要素集、更新要素集和删除要素集。
㉛ SuperMap. REST. TransferPreference：{Object}公交换乘策略枚举类。该枚举类用于指定交通换乘服务中设置地铁优先、公交优先、不乘地铁、无偏好等偏好设置。

㉜ SuperMap. REST. GridType：{Object}地图背景格网类型枚举类。
㉝ SuperMap. REST. ColorSpaceType：{Object}色彩空间枚举。由于成色原理的不同，决定了显示器、投影仪这类靠色光直接合成颜色的颜色设备和打印机、印刷机这类靠使用颜料的印刷设备在生成颜色方式上的区别。针对上述不同成色方式，SuperMap 提供两种色彩空间，分别为 RGB 和 CMYK。RGB 主要用于显示系统中，CMYK 主要用于印刷系统中。
㉞ SuperMap. REST. LayerType：{Object}图层类型。
㉟ SuperMap. REST. StatisticMode：{Object}字段统计方法类型。
㊱ SuperMap. REST. PixelFormat：{Object}栅格与影像数据存储的像素格式枚举类。
㊲ SuperMap. REST. SearchMode：{Object}内插时使用的样本点的查找方式枚举。
㊳ SuperMap. REST. InterpolationAlgorithmType：{Object}插值分析的算法的类型。
㊴ SuperMap. REST. VariogramMode：{Object}克里金(Kriging)插值时的半变函数类型枚举。
㊵ SuperMap. REST. Exponent：{Object}定义了泛克里金(UniversalKriging)插值时样点数据中趋势面方程的阶数。
㊶ "DOUBLE"，每个像元用 8 个字节来表示，只提供给栅格数据集使用。

11.7 矿用对象采销信息管理系统的设计与实现

1. 系统建设的背景

随着计算机技术的发展，目前国内各个企业纷纷建立了自己的管理信息系统，这些系统主要侧重各个组织部门的各种资料的管理，系统涉及的内容模块比较全面，包括人力资源管理、资料管理、物流管理等，但是主要针对物资采购以及销售的系统比较少，专门针对矿用对象采销信息管理的系统也比较少见，基于 WebGIS 的采购和销售系统更是鲜见。职能部门对物资采购和销售系统中销售商和购买商的所有信息进行统一管理，GIS 技术以其强大的图形数据和属性数据管理能力可以发挥重要的作用。WebGIS 具有传统 GIS 的优点，同时也具有数据共享的特色，因此利用 WebGIS 技术进行采购和销售系统的开发具有广阔的前景。矿山企业属于国家控制的大型产业和国家经济的重要支柱产业，每个企业都会面临物资的采购以及产品的销售问题，如果处理得当，选择合适的供应商以及购买商，就会降低成本，增加利润，如果不能很好地协调，将会导致物资积压，周转不灵，严重的还会使企业停工停产。基于 WebGIS 技术建立矿山企业的矿用对象采购销售信息管理系统，实现对矿用对象采销信息的可视化，使用户可以查询所有的供应商的位置信息以及属性信息，并对查询结果进行分析之后，再进行最后决策，最终选择出更合理的供应商和物资，对所有的产品购买商进行可视化管理，进而有效地为矿山企业减少采购成本，提高经济效益。

2. 系统总体框架设计

本系统的服务对象主要是大、中、小矿山企业以及矿业集团公司，功能模块主要包括销售、采购、库存管理和系统管理。具体的用户是矿山企业的采购人员、销售人员和库存管理人员。采购人员需要提交相关申请，由管理员审核通过后才能执行相关采购操作。销售人员可以通过查找销售信息进行销售操作,库存管理人员可以通过查询、入库、出库等操作对产品和货物进行管理。

系统目标是实现一个基于 WebGIS 的矿山企业矿用对象进行的采销信息管理系统,综合运用 WebGIS 技术、数据库技术、ASP.NET 技术。基于 SuperMap iClient 7C for JavaScript 客户端产品进行二次开发,SuperMap iServer 7C(简称 iServer)提供地图服务,SuperMap iClient 7C for JavaScript 使用 Ajax 技术(即异步 JavaScript 和 XML 技术)进行地图瓦片的局部更新,同时能够与 iServer 结合,实现空间分析或处理等功能,以 HTML5 和 OpenLayers 的库为基础,在 GIS 功能上实现与 iServer 服务器产品的无缝对接。其原理是浏览器发送 Ajax 或 HTTP 请求给 Web 服务器,中间层 Web 服务器接收到请求后转交给应用服务器处理,服务器端主要通过对应的业务逻辑层的业务跳转控制,调用空间数据引擎访问接口进行数据库访问,实现数据的提取与处理,再返回给浏览器相应的结果。iServer 作为一个地图服务器,它集成了上述 web 服务器和应用服务器功能,用户只需要发送请求,内部逻辑由 iServer 自动完成并返回给用户处理后的数据。

采用 B(Browser)/S(Server)三层结构模式,即 Web 浏览器、Web 服务器和数据库服务器,在 B/S 体系结构下,用户界面完全通过浏览器实现,不需要在客户端安装专用的应用程序,简单事务逻辑在前端实现,复杂事务逻辑在服务器端实现,SuperMap iServer 7C 服务器负责存储和管理数据并完成对数据的处理,操作简单。系统体系结构如图 11-9 所示。

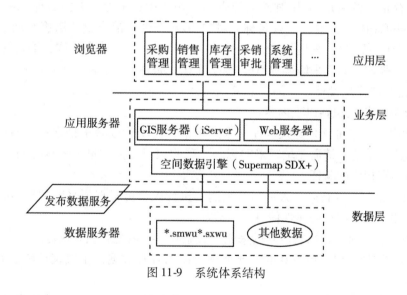

图 11-9 系统体系结构

3. 系统总体设计

系统的总体设计包括系统功能模块设计和空间数据结构设计,其中空间数据结构设计包括图形数据图层设计和属性数据表设计。

(1)空间数据结构设计

① 图形数据图层设计:空间数据类型包括地图中的点、线、面等空间实体的图形表示。本系统采用的测试空间数据是全国地图,底图是从谷歌遥感影像截取的卫星图像,利

用 SuperMap iDesktop 矢量化和符号化后得到图形数据，见表 11-12。

表 11-12　　　　　　　　　　　空间数据图层管理

类型	所含图层
点数据集	Park(公园层)、School(学校层)、BusStop(公交站点层)、Hotel(宾馆酒店)、ResidentialPoint(街区层)、kyobjects(各个矿山企业)、scllcr(销售商)、buyer(购买商)
线数据集	RoadLine(道路层)、BusLine(公交线路层)
面数据集	Vegetable(绿地层)、WaterPoly(水系层)、Bengjing(背景层)、ResidentialArea(街区层)
文本标注数据集	主要道路、各个地市区域划分

② 属性数据表设计：属性数据表主要是针对矿用对象、供应商、购买商以及矿山企业等实体的管理设计属性数据，主要包括系统用户表 TuserInfo(用户编号、主键、用户登录名、用户密码、用户真实姓名、用户电话、邮箱地址)、权限表 Tauthority(权限编号、主键、权限类型，主键、权限名称、备注)、用户权限关系表 TUser_ Anthority(编号，主键、用户编号、权限编号、备注)、矿用对象表 TKYObject(ID、主键、对象名称、对象类型、对象库存量、备注)、矿用对象类型表 KYType(ID、主键、类型名称、备注)、销售记录表 TsaleInfo(ID、主键、销售名称、对象 ID、外键、销售量、销售时间、销售客户 ID、操作员 ID、备注)、采购记录表 Tpurchase(ID、主键、采购名称、采购对象 ID，外键、采购量、采购时间、采购客户 ID、操作员 ID、备注)。

(2) 系统功能模块设计

① 采购管理模块包括以下内容：

a. 采购申请、审批：采购人员提交采购申请，由主管领导审批通过，通过后由执行人员执行采购。采购申请需要填写采购标题、采购项目、采购量、采购时间、客户名称。提交申请时会有相应的验证。

b. 执行采购：由执行采购人员执行采购，执行人员可根据要采购的项目在地图上查找供应点，并下载供应点的属性信息。

c. 采购验收：对已经执行成功的采购生成验收单，由主管领导执行。

d. 查询与统计：以简单直观的方式查询所需的各种数据，生成统计图表和报表，增强对数据的综合利用。

e. 采购信息查询：采购人员可以根据需要采购的项目在地图上查找符合条件的商家，查找出来后在地图上高亮显示查找结果。用户可以查看商家的详细信息，包括联系电话、地址、邮箱地址等。

② 销售管理模块包括以下内容：

a. 销售开单：每销售一种矿用对象都要将销售记录登记入库，由销售人员执行，填写记录时需要填写销售标题、销售对象、销售量、销售时间和客户名称等信息，每一项在提交时都会有验证，验证不通过无法入库。

b. 销售退货：接收用户的退货申请，用户需要填写相关的表格。销售人员处理客户的退货申请。

　　c. 查询与统计：以简单、直观的方式，查询所需的各种信息，生成各类查询表格和各种常见的统计分析报表，增强对数据的综合利用。

　　d. 客户信息管理：对客户信息进行整合分类显示，方便查询，可通过多种方式查询客户信息。例如，根据客户所在的位置在地图上查找，还可以根据条件来查询客户信息。

　　e. 销售信息查找：销售人员可以根据自己的销售项目在地图上查找符合条件的商家的详细信息，并且在地图上通过气泡的形式进行高亮显示。

　③ 库存管理模块包括以下内容：

　　a. 查看库存信息：管理员可对各种对象的库存进行管理，包括增、删、改、查。

　　b. 对象报损上报：对于已经损坏的矿用对象要进行上报操作，需要填写相关的表格。

　　c. 查询与统计：根据对象的库存信息生成统计图，直观地显示各种库存信息。

　④ 采销审批管理模块：销售人员在线提交销售采购申请，提交的信息包括采购名称、采购项目、申请提交时间、采购量和客户名称，并带有相应的验证，可以进行报表打印，由主管领导审核。

　⑤ 系统管理模块：本系统操作人员分为几种，即采购人员、销售人员、库存管理人员和管理员，每种角色对应自己的权限，管理员具有最高权限，可以管理客户信息、添加客户、删除客户等。

　4. 系统主要功能的实现

　(1) 采购信息查找功能

　　在实际采购过程中，采购人员需要根据公司需要确定采购的项目以及选择合适的供应商，在查找合适的供应商的过程中，本系统基于 GIS 功能，提供可视化的查询结果。如图 11-10 所示，用户填写自己需要采购的项目，单击"查找"按钮，中间的列表框中就会显示出所有的供应商名称，用户在列表中对自己感兴趣的供应商名字上进行单击，地图就会高亮显示出此供应商，用户单击此供应商之后就会弹出对话框来显示商家的详细信息，方便用户掌握供应商的地理位置及详细信息，这个过程是根据属性查询功能从属性数据库中查询出满足要求的图形。

　　本功能主要是通过 JavaScript 语言实现的，所有的操作地图的操作都是异步的，通过 JavaScript 客户端的语言操作空间数据库的一个好处就是加载速度快，空间分析和空间查询速度快。地图是通过切片处理的，这样加快地图加载的速度并且减少了系统对服务器的压力。先是通过属性查找出相应的图形在地图上的显示，然后通过气泡的形式显示商家的详细信息。

　(2) 采购申请功能

　　用户在执行采购之前需要提交采购申请，提交申请之后管理员根据库存信息以及公司需求确定是否对此申请进行审批，只有管理员的审核通过才能执行采购，通过之后又有专门的人员来执行采购，填写采购申请时需要填写采购标题、采购项目、采购量(吨)、采购时间和客户名称等项。填写每一项都带有验证，以验证用户输入的内容是否合法。

　　如图 11-11 所示，用户在提交采购申请页面填写采购标题、采购量以及采购时间，通过下拉框选择系统预设好的采购项目以及客户名称，输入或选择完成后单击"提交申请"

第 11 章　WebGIS 设计与开发

图 11-10　采购信息查找

按钮，系统就会将刚才的申请信息保存下来，使这条记录成为待审核记录，等待系统管理员进行审核。

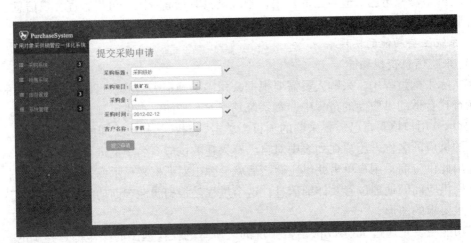

图 11-11　采购申请

本系统都是通过 Ajax 验证的，整个页面局部刷新。本功能所用的控件都是客户端控件，通过 jQuery 获取表单数据，向服务器端发送请求，然后通过 jQuery 获取服务器的响应，再将获取的数据放在 html 页，这样就实现了局部刷新。局部刷新技术能给用户好的体验，不用等整个页面刷新完毕就可以执行其他的操作。采购量验证是通过 JavaScript 正则表达式实现的，用正则表达式和文本框的内容进行匹配，验证是否符合数字规则。

(3) 采购统计功能

传统的查看数据的方式是以报表和表格等形式，这样面对大量繁杂的数据时，无疑增加了用户的工作量和工作难度，并且难以直观地看出数据的统计信息和预测某些趋势，统计图表的功能是 GIS 的一个对数据的展现形式，它形象直观并且有可预见性。本系统可以实现对采购信息进行统计，统计结果以柱状图的形式进行显示，使用户根据统计图直观地看出采购内容的趋势，并根据这些内容和趋势做出相应决策。

系统会列出所有提交的采购申请，包括管理员审核通过的和审核未通过的，列表中可以对已经提交的申请进行删除，并且系统以柱状图的形式以每一个季度为单位对每一种采购项目进行统计。列表中列出的项目包括：采购名称、采购项目、采购量、采购人、采购时间、采购客户和状态。如图 11-12 所示，界面上部分以表格的形式列出了采购的记录表，包括采购名称、采购对象等，用户可以在操作栏中对记录进行删除的操作，用户只需单击操作栏的删除按钮，就可以将此条采购记录进行删除。

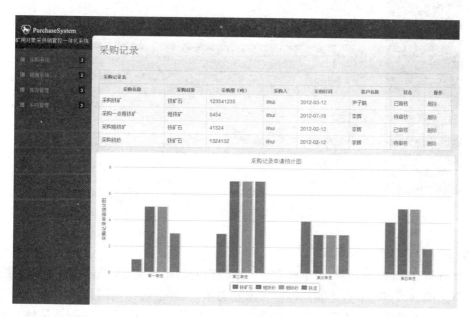

图 11-12　采购记录

本功能的列表是通过 Ajax 加载，html 页中是一个空白的<table></table>标签，通过 JavaScript 向服务器发送请求提取数据，服务器向客户端发送 json 串，客户端将 json 串解析成数组，然后再通过 JavaScript 动态生成 html 追加待<table></table>内，这样就动态加载了列表。列表下面的统计图是一个柱状图，是一个 jQuery 插件，通过 Ajax 获取数据作为统计图的数据源，统计图加载时具有动态的效果，用户体验性良好。统计图是通过季度来分组的，统计四个季度的数据进行显示。

(4) 缓冲区分析功能

在查找原料供应商和产品经销商的过程中，采购和销售部门人员可能要考虑距离因素，需要查找距离本公司或者公司某一库存点的一定范围内的供销商，本系统的缓冲区分析即可实现此功能，它是根据的点的缓冲区建立的，系统可以对选中的某一库存点对象或公司对象进行缓冲，缓冲半径由用户按照实际需求进行手动输入，从而系统可以查找出距离感性对象某一范围内的供销商，并在地图上进行显示，用户可以很直观地在地图上进行查看，极大地提高了供销商选择过程的效率。

如图 11-13 所示，用户只需要点击"请拾取缓冲对象"按钮，并在缓冲区距离文本框中根据需要输入缓冲距离，然后在地图中鼠标单击选择需要缓冲的对象，然后点击"搜索"按钮，就可以实现对选中对象按选定距离进行缓冲，落在缓冲区范围内的经销商点就是满足用户搜索要求的供销商点，用户可以从地图中直观地看出来，并可以在地图中通过鼠标双击来查看其属性信息。

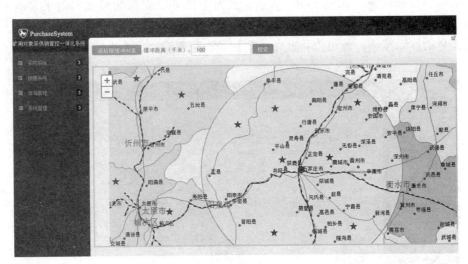

图 11-13　缓冲区分析

(5) 最短路径功能

网络分析是空间分析的一个重要方面，是依据网络拓扑关系（线性实体之间、线性实体与节点之间、节点与节点之间的连接、连通关系），并通过考察网络元素的空间、属性数据，对网络的性能特征进行多方面的分析计算。网络分析中最基本的问题是最短路径问题，它作为许多领域中选择最优问题的基础，在交通网络分析系统中占有重要地位。从网络模型的角度看，最短路径分析就是在指定网络中两节点间找一条阻碍强度最小的路径。根据阻碍强度的不同定义，最短路径不仅仅指一般地理意义上的距离最短，还可以引申到其他的度量，如时间、费用、线路容量等。相应地，最短路径问题就成为最快路径问题、最低费用问题等。

在实际采购和销售过程中，选定供应商或经销商后，需要将选定的矿用对象从供应商处输送到公司库存点或者将产品输送到经销商处，这个过程中要耗费大量的运输成本，因此通过选择公司到供销商之间的最短路径，可以减少运输成本。本系统开发的最短路径分析功能即可以实现这一要求。

如图 11-14 所示，用户首先点击"请拾取起点"按钮，然后在地图中用鼠标选择起点，点击"请拾取终点"按钮，并在地图中鼠标选择终点，再点击"最短路径"按钮，地图中就会自动高亮显示出从起点到终点最短的路径。

(6) 销售信息查找功能

在公司实际生产过程中，公司生产的产品要及时进行输出，输出到不同的商家，在这个过程中要对各个商家进行选择和查找，以确定最优的销售商，本系统的销售信息查找功

11.7 矿用对象采销信息管理系统的设计与实现

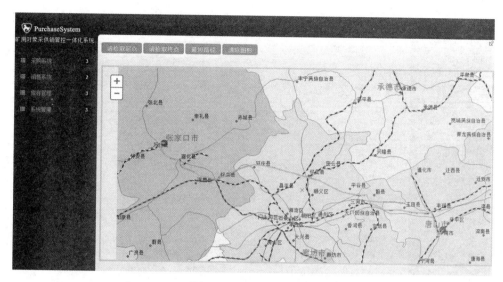

图 11-14　最短路径功能

能是基于 GIS 的查询功能实现的，能实现可视化的查询结果。

销售人员根据自己需要销售的项目在地图上查找满足要求的商家，并且在地图上进行显示。用户单击地图上的查找出来的商家图标可以查看商家的详细信息，商家的详细信息包括商家名称、商家地址、商家联系电话等信息。

如图 11-15 所示，用户首先在输入框中输入需要销售的对象名称，并单击后面的"查找"按钮，系统就会高亮显示出所有的经销商点，用户可以在地图上直观地看到，并且在感兴趣的经销商点上进行单击就能弹出消息框，通过气泡进行高亮显示，主要是用来显示此经销商的详细信息，包括商家的名称、地址和联系电话，方便用户做出选择决策。

（7）销售记录登记

在实际生产过程中，对产品进行销售输出后并不是就结束了，还需要对每项销售过程进行记录，方便以后的查询工作，销售记录被登记以后该条记录就会保存在后台数据库中，并且在销售记录列表中出现。

在本系统中，销售管理人员需要将已经执行的销售进行登记入库，在填写记录的时候需要填写销售标题、销售对象、销售量、销售时间和客户名称等信息，其中销售标题、销售量和销售时间需要用户手动输入，而销售对象和客户名称则是通过下拉列表框的形式让用户在下拉列表中进行选择，这些下拉列表中的内容是用户预先设置好的可选择内容，这些内容分别与矿用对象表和客户表中的矿用对象名和客户名相对应。而且每一项在提交时都会有验证，验证不通过无法入库。验证信息包括多种，比如销售标题不能为空、销售量必须是数字、销售时间栏必须填写正确的日期格式等。

本系统验证都是通过 Ajax 验证，整个页面局部刷新。本功能所用的控件都是客户端控件，通过 jQuery 获取表单数据，向服务器端发送请求，通过 jQuery 获取服务器的响应，然后将获取的数据放在 html 页，这样就实现了局部刷新。局部刷新技术能给用户好的体验，不用等整个页面刷新完毕就可以执行其他的操作。销售量验证是通过 JavaScript 正则

图 11-15　销售信息查找

表达式实现的,将正则表达式和文本框的内容进行匹配,验证是否符合数字规则。如图 11-16 所示。

图 11-16　销售记录登记

(8) 销售记录列表

传统的查看数据的方式是以报表和表格等形式,这样在面对大量繁杂的数据时,无疑增加了用户的工作量和工作难度,并且难以直观地看出数据的统计信息和预测某些趋势,统计图表的功能是 GIS 的一个对数据的展现形式,它形象直观并且有可预见性。本系统可以实现对销售信息进行统计,统计结果以柱状图的形式进行显示,使用户根据统计图直观地看出产品的销售情况的趋势,并根据这些内容和趋势做出相应决策。

系统将已经执行的销售进行列表显示，并且系统根据每个季度的销售记录进行统计显示，以柱状图的形式显示统计信息。如图 11-17 所示，界面上部分以表格的形式列出了销售情况记录表，包括销售标题、销售项目、销售量、销售执行人、销售时间和客户名称，用户可以在操作栏进行记录删除的操作，用户只需单击操作栏的删除按钮，就可以将此条销售记录删除，该条记录会及时在界面的表格中删除，同时后台数据库中的这条记录也会被删除掉。

本功能的列表是通过 Ajax 加载，html 页中是一个空白的<table></table>标签，通过 JavaScript 向服务器发送请求，提取数据，服务器向客户端发送 json 串，客户端将 json 串解析成数组，然后再通过 JavaScript 动态生成 html 追加待<table></table>内，这样就动态加载了列表。列表下面的统计图是柱状图，是一个 jQuery 插件通过 Ajax 获取数据作为统计图的数据源，统计图加载时具有动态的效果，用户体验性良好。统计图是通过季度来分组的，统计四个季度的数据并进行显示。

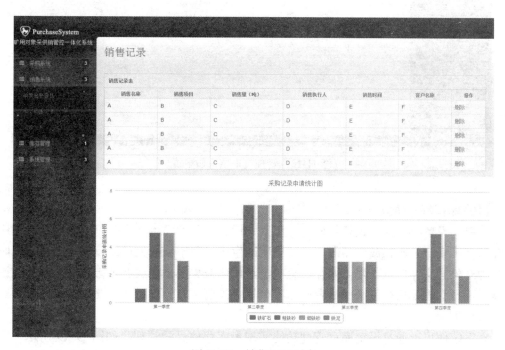

图 11-17　销售记录列表

(9)待审核采购申请

在实际生产过程中，采购部门在采购前要提交采购申请，只有在审核部门根据公司实际库存情况进行判断后，才决定是否同意该申请。因此本系统设计了待审核采购申请功能，使实际公司生产过程中的一些申请、审核等操作在计算机上得以实现，实现了管理过程的工作流式管理，减少了实际人工操作过程中由于各部门人员流动办公而造成的时间和成本浪费，从而大大提高了工作效率。本系统的待审核采购申请就是列出所有的未通过的采购申请。

采购人员填写采购申请，提交到数据库，由管理员来审核，本功能就是列出采购人员提交的申请，由管理员查看是否通过，如图 11-18 所示，界面中列出了采购名称、采购量、采购人、采购时间、客户名称、状态以及管理员可以进行的操作，其中有两种操作，通过或不通过，管理员根据实际情况通过选择"通过"或者"不通过"按钮，来对某一项采购申请进行处理。

本功能的列表也是同 Ajax 技术加载，方便快捷。"通过"和"不通过"按钮的事件也是通过客户端语言 JavaScript 向服务器端发送请求，服务器端给客户端发送一个 json 串，客户端进行解析动态加载到 html 页中。

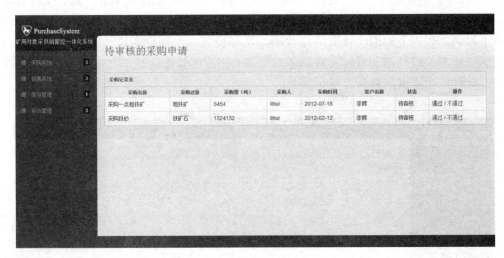

图 11-18　管理员审核采购申请

（10）库存管理功能

原料、设备、备件、燃料等矿用对象采购后，或者产品生产出来以后，都会有相应的库存信息，来记录原料或产品的存量，在实际生产过程中采购人员以及销售人员要及时掌握各种对象的库存信息，根据库存信息确定对矿用对象的采购或者销售决策。只有在原料库存量不足，需要进行采购，或者产品存量过大需要及时售出时才需要进行相应的采购或者销售行为。

本系统便可以通过设定库存管理功能实现对库存信息的管理。如图 11-19 所示，库存管理功能以表格的形式实现，表格中列出了所有对象的库存信息，包括对象名称、库存量以及操作。操作即管理员可以对相应的库存对象进行删除的操作，当库存中没有某种对象了，或者某种对象不需要进行库存的管理了，就可以对该对象进行删除操作，具体方法就是在对应对象的操作栏点击"删除"按钮，这个对象的库存信息就会在系统中被删除，同时属性表中的信息也会被删除。

本功能的列表是通过 Ajax 加载，html 页中是一个空白的 <table></table> 标签，通过 JavaScript 向服务器发送请求提取数据，服务器向客户端发送 json 串，客户端将 json 串解析成数组，然后再通过 JavaScript 动态生成 html 追加待 <table></table> 内，这样就动态加载了列表，列表中的信息是从后台的 SQL Server 数据库表中提取的。

（11）添加系统用户功能

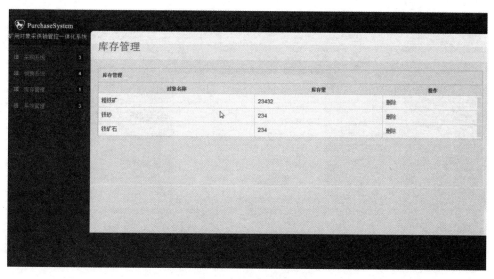

图 11-19　库存管理功能

在实际生产过程中，可能由于公司人员的增加，使用本系统的人员可能会有变动，会不时地有新用户来使用此系统，因此系统要有必要的添加用户功能。

系统管理员可以给系统添加用户，以使不同的人能对系统进行操作，添加的用户类型包括采购人员、销售人员、库存管理员和系统管理员，每个角色对应各种不同的权限，系统管理员具有最高的权限，这些用户类型信息和角色以及权限信息对应于 SQL Server 数据表中用户表和权限表中的信息，系统添加用户并和表中的数据相联系来确定不同用户的权限以及所能做的操作。

如图 11-20 所示，在界面中输入用户名、密码、真实姓名、用户类型、电话以及邮箱，其中用户名、密码、真实姓名等信息需用户手动输入，用户类型通过下拉列表方式进行选择。然后点击"确认添加"按钮，用户信息即添加成功，输入的信息会被保存到后台数据库，此用户以后便可以通过正确的登录信息来登录和使用本系统。

添加用户界面是通过弹出一个 div 来进行添加的，首先在页面上隐藏一个 div 标签，标签带有相应的表单标签。给页面上的添加按钮一个 JavaScript 事件弹出隐藏的 div 来添加用户，添加用户操作也是通过 JavaScript 向服务器端发送一个 json 串，json 串的内容包括添加用户所需要的数据，服务器端执行完毕之后再由 JavaScript 接受响应，在客户端表现出来，所以所有的操作都是异步的。

（12）管理用户功能

系统需要对添加的用户进行管理，例如，当有用户不再使用此系统时，管理员要对此用户进行删除，系统便可以实现对用户的管理。如图 11-21 所示，界面中列出了所有的用户信息，当某个用户不再使用本系统，管理员就需要在此用户管理页面把该用户的信息删除。

（13）添加客户功能

在公司实际生产经营的过程中，公司的客户不是固定不变的，当有新的客户与公司建

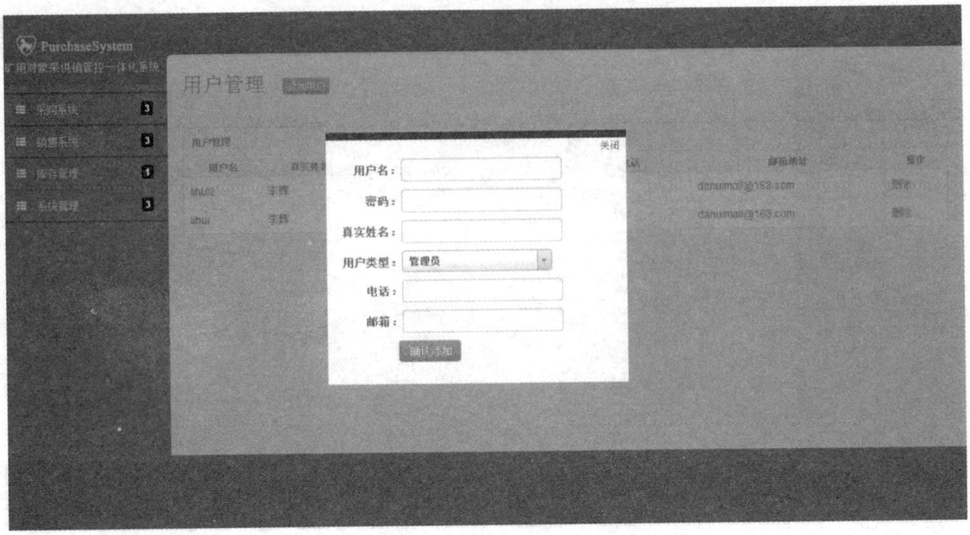

图 11-20　添加用户

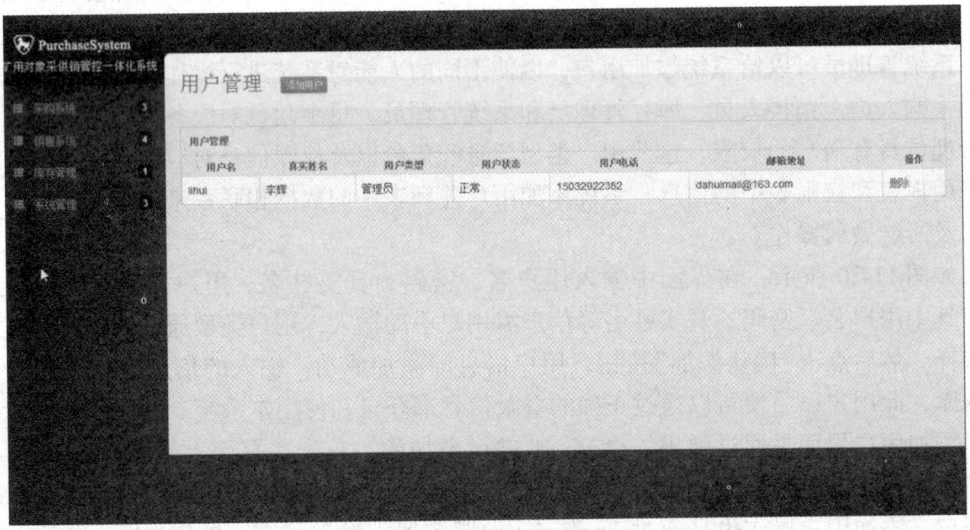

图 11-21　用户管理

立合作关系时，系统需要添加相应的客户信息，方便以后对客户的管理，本功能就可以实现这个实际需要。如图 11-22 所示，在需要添加客户的时候，只需要在界面中输入客户名称、业务名称、客户状态以及客户地址，其中客户状态通过下拉列表的形式进行选择"是/否"启用，输入或选择完毕后点击"确认添加"按钮，客户即添加成功。以后其他功能的操作中就可以使用此用户。本系统的添加用户功能是由系统管理员来进行的，添加客户操作也是通过弹出一个小窗口来执行的。这个小窗口是一个 jQuery 插件，窗口内的元素都是通过 JavaScript 动态生成插在 html 中的。这个小窗口可以拖动。添加用户操作也是通过 Ajax 实现的，页面局部刷新。

图 11-22　添加客户

(14) 管理客户功能

系统需要对公司的客户进行及时的管理，例如，当公司不再跟某个客户合作时，管理员要对此客户进行删除。本系统便可以实现对客户的管理。

如图 11-23 所示，界面中列出了所有的客户信息，当公司不再与某个客户合作时，管理员就需要在此页面把该客户的信息删除，具体操作只需要在该客户的操作栏点击"删除"按钮，该条客户记录就会被删除。

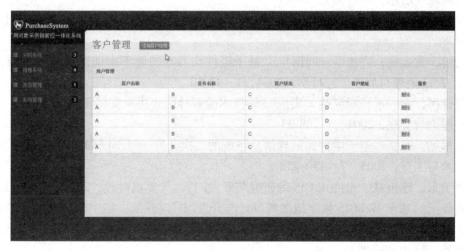

图 11-23　客户管理

参考文献

[1] 吴信才. 地理信息系统软件设计与开发[M]. 北京：电子工业出版社，2009.
[2] 李满春，陈刚，陈振杰，等. GIS设计与实现[M]. 北京：科学出版社，2011.
[3] 张丰，杜震洪，刘仁义. GIS程序设计教程：基于ArcGIS Engine的C#开发实例[M]. 杭州：浙江大学出版社：2012.
[4] 黄杏元，马劲松. 地理信息系统概论(第3版)[M]. 北京：高等教育出版社，2008.
[5] http://support.supermap.com.cn/ProductCenter/ResourceCenter/ProductNotebook.aspx.
[6] 王兴举. C#环境下的SuperMap Objects组件式开发[M]. 北京：中国铁道出版社，2013.
[7] 张正栋，胡华科，钟广锐，等. SuperMap GIS应用与开发教程[M]. 武汉：武汉大学出版社：2006.
[8] 刘亚静，马赛，李辉. 基于WebGIS的矿用对象采销信息管理系统的设计与实现[J]. 煤炭工程，2014(7)：57-61.
[9] 孟令奎，史文中，张鹏林，等. 网络地理信息系统原理与技术(第2版)[M]. 北京：科学出版社，2010.
[10] 杨姗姗，王明军，杜清运，等. JavaApplet与JavaScript交互方法的探讨[J]. 测绘与空间地理信息，2005(8)：26-29.
[11] 房体盈. 基于JavaScript技术的WebGIS设计与实现[D]. 大连：大连理工大学，2008.
[12] 杨旭，黄家柱，许建军，闾国年. 基于组件式GIS的地下水动态管理系统设计与开发[J]. 地理与地理信息科学，2004，1：47-50.
[13] 万鲁河，李一军，臧淑英. 集成"3S"技术的森林防火决策支持系统研究[J]. 系统工程理论与实践，2004，7：88-93.
[14] 王军见，张弘. 基于组件式地理信息系统的二次开发——SuperMap Objects[J]. 科学技术与工程，2005，7：450-453.
[15] 罗予东，陈伟君. 组件式GIS的研究与开发[J]. 计算机时代，2004，2：11-13.
[16] 杨舒奎. 基于WebGIS数字地图的设计与开发[D]. 武汉：华中科技大学，2005.
[17] 李永红，邓红艳，赵敬东，等. 组件式GIS开发的实践[J]. 计算机工程与设计，2005，4：1090-1092.
[18] 王彬. 基于组件式GIS的区域污染源管理信息系统开发[D]. 成都：四川大学，2006.
[19] 杨旭，黄家柱，陈锁忠，等. 组件式GIS在地下水资源管理系统开发中的应用[J]. 水文，2003，1：10-14.

[20] 杨海荣．基于组件式GIS的地籍管理信息系统研究［D］．长沙：长沙理工大学，2004．

[21] 孙治贵，黎贞发，李杰，等．基于组件式GIS技术的水稻生产管理信息系统开发研究［J］．农业工程学报，2004，3：137-140．

[22] 胡亚，李永树．基于组件式GIS-SuperMap Objects的二次开发［J］．四川测绘，2004，1：3-5．

[23] 侯顺艳．基于组件式GIS的精准农业变量施肥处方系统应用研究［D］．保定：河北农业大学，2003．

[24] 李兵．基于GIS的土地利用规划管理信息系统建设研究［D］．重庆：西南农业大学，2003．

[25] 杨旭，黄家柱，袁丁，等．基于GIS的水文地质基础数据管理系统的开发［J］．计算机应用与软件，2005，3：19-21+36．

[26] 陈国良，吴立新，李秋，等．组件式GIS及其在房地产信息查询系统开发中的应用［J］．北京电子科技学院学报，2003，1：27-32．

[27] 陶丽娜，唐胜传，陈谦应．基于组件式GIS技术的边坡支护方案优化设计系统［J］．岩石力学与工程学报，2004，16：2824-2829．

[28] 王志恒，杨国东，吴琼，等．基于ArcEngine的虚拟校园信息管理系统的设计与实现［J］．地理信息世界，2008，3：80-84．

[29] 何瑞珍，张颖，张敬冬，等．基于组件式GIS的森林资源管理信息系统的设计与开发——以郑州市森林资源管理信息系统设计与开发为例［J］．林业资源管理，2005，5：77-80．

[30] 赖剑菲，江舟．基于ArcObjects的组件式GIS的开发与研究［J］．国土资源信息化，2005，3：29-32．

[31] 李琼．基于组件式GIS的土地利用规划实施管理信息系统研究［D］．乌鲁木齐：新疆农业大学，2007．

[32] 江俊福．基于组件式GIS的城市道路管理系统设计研究［J］．微计算机信息，2006，6：190-192．

[33] 刘宇晶．基于GIS的区域地质灾害信息管理系统的研发［D］．北京：中国地质大学，2006．

[34] 丛方杰，王国利，肖传成，等．基于GIS组件技术的地下水资源管理信息系统的研究与开发［J］．水文，2006，4：60-63．

[35] 朱怀松．基于组件式GIS的城镇土地定级估价信息系统研究与实现［D］．乌鲁木齐：新疆大学，2004．

[36] 杨斌．组件式GIS技术在流域水资源管理信息系统中的应用研究［D］．乌鲁木齐：新疆农业大学，2006．

[37] 唐攀科，杨兵，袁继明，等．利用组件式GIS开发地质矿产数据库［J］．地质与勘探，2003，1：62-65．

[38] 杨福运．基于组件式GIS的农业经济信息服务系统的研究与开发［D］．北京：中国农业科学院，2008．

[39] 周启超. 基于组件式 GIS 的吉林省水利信息系统设计与实现研究[D]. 长春：东北师范大学, 2007.

[40] 王海军, 詹长根, 张玉梅. 组件式 GIS 技术在河道信息系统中的应用[J]. 武汉大学学报(工学版), 2004, 1: 59-62.

[41] 王杰, 童新华. 组件式 GIS 在土地利用信息系统中的应用[J]. 广西师范学院学报(自然科学版), 2005, 1: 78-81.

[42] 刘爽. 基于 MapObjects 组件的 GIS 二次开发[J]. 大连民族学院学报, 2007, 5: 87-91.

[43] 安小刚. 基于 GIS 的组件式电网规划和理论线损的研究[D]. 保定：华北电力大学, 2007.

[44] 叶张煌, 杨志, 刘林清. GIS 在龙虎山旅游管理和开发中的应用[J]. 地理空间信息, 2006, 6: 16-18.

[45] 王枝军. 基于组件 GIS 的土地利用建库系统研究与设计[D]. 上海：同济大学, 2007.

[46] 刁海亭. 组件式 GIS 支持下的城镇土地定级信息系统[D]. 泰安：山东农业大学, 2003.

[47] 戚文云, 刘惠德, 郭向坤. 基于 GIS 的矿山水文地质信息管理系统的研究[J]. 矿业快报, 2007, 3: 45-47.

[48] 刘铁英, 宋雨, 边小凡, 等. 基于组件式 GIS 技术的数字校园模型研究[J]. 河北大学学报(自然科学版), 2006, 2: 213-218.

[49] 贺振, 贺俊平, 张卫星. 基于 SuperMap Objects 组件式 GIS 的开发与研究[J]. 商丘师范学院学报, 2008, 9: 102-104.

[50] 杨小雄, 何志明, 苏夏, 等. 基于组件式 GIS 的地价评估与管理信息系统的设计与开发[J]. 资源开发与市场, 2006, 4: 321-323+330.

[51] 张晓磊, 方堃. 基于 GIS 的桂林旅游地理信息系统开发研究[J]. 黑龙江对外经贸, 2007, 7: 90-91+94.

[52] 庞敏, 周荣福, 卢红星, 等. 基于组件式 GIS 的瓦斯地质图管理系统设计与开发[J]. 煤矿安全, 2008, 10: 21-24.

[53] 杨康年. 基于 ComGIS 的区域污染源管理信息系统的研究[D]. 成都：四川大学, 2005.

[54] 秦建新, 谭子芳. 基于 ComGIS 的城镇宗地估价系统设计与实现[J]. 计算机工程与应用, 2008, 24: 85-89.

[55] 陈朝镇, 王彬, 罗文锋. 基于组件式 GIS 的区域污染源管理信息系统开发[J]. 安徽农业科学, 2008, 21: 9223-9227.

[56] 段滔, 刘家彬, 韦燕飞. 基于组件式 GIS 的城镇基准地价动态更新系统的开发与应用[J]. 广西师范学院学报(自然科学版), 2003, S1: 184-186.

[57] 侯顺艳, 李志远, 王秀. 基于 ComGIS 的精准农业变量施肥处方推荐系统研究[J].

河北农业大学学报，2005，6：101-103.

[58] 程丽丽. 基于WebGIS的地籍管理信息系统开发研究[D]. 长春：东北师范大学，2007.

[59] 倪东英. 基于WebGIS的森林资源管理信息系统的研建[D]. 北京：北京林业大学，2010.

[60] 贺炳彦. 基于WebGIS的校园地理信息发布系统的设计与研究[D]. 西安：长安大学，2003.

[61] 赵昂. 基于WebGIS的信息系统的分析、设计与实现[D]. 武汉：武汉理工大学，2002.

[62] 金鑫. 基于WebGIS的县级山洪灾害预警系统的设计与实现[D]. 上海：东华理工大学，2012.

[63] 庞丽峰. 基于WebGIS省级林业信息共享平台的设计与开发[D]. 北京：中国林业科学研究院，2004.

[64] 余镜周. 基于WebGIS的物流配送车辆监控系统研究与实现[D]. 北京：北京邮电大学，2010.

[65] 刘宝锋. 基于WebGIS数字矿山软件体系架构[D]. 西安：西安科技大学，2010.

[66] 田洪阵，刘沁萍，刘军伟. WebGIS的现状及其发展趋势[J]. 许昌学院学报，2004，2：48-52.

[67] 段五星. 基于Java和WebGIS的水质预测系统的研发[D]. 广州：广东工业大学，2013.

[68] 赵杨. 基于J2EE-WebGIS应急调度系统的研究与实现[D]. 武汉：武汉理工大学，2012.

[69] 李富兵. 基于ArcIMS克什克腾国家地质公园旅游信息系统构建与实现[D]. 北京：中国地质大学，2005.

[70] 吴运超，王汶，牛铮，等. Ajax在WebGIS中的应用[J]. 地理与地理信息科学，2007，2：43-46.

[71] 陈春平. 基于Java的WebGIS的研究与实现[D]. 合肥：合肥工业大学，2010.

[72] 杜栋. 基于SuperMap IS Java的特种设备安全监察系统设计与实现[D]. 杭州：浙江大学，2007.

[73] 孙恒. 基于ASP.NET和WebGIS的校园信息管理系统开发及应用[D]. 上海：华东师范大学，2010.